CELL DIFFERENTIATION RESEARCH DEVELOPMENTS

CELL DIFFERENTIATION RESEARCH DEVELOPMENTS

LAURA B. IVANOVA
EDITOR

Nova Biomedical Books
New York

Errata for *Cell Differentiation Research Developments*
The ISBN of the book should be ISBN 13: 978-1-60021-938-2; ISBN 10: 1-60021-938-1

For permission to use material from this book please contact us:
Telephone 631-231-7269; Fax 631-231-8175
Web Site: http://www.novapublishers.com

LIBRARY OF CONGRESS CATALOGING-IN-PUBLICATION DATA

Cell differentiation research developments / Laura B. Ivanova (editor).
p. ; cm.
Includes bibliographical references and index.
ISBN-13: 978-1-60021-939-9 (hardcover)
ISBN-10: 1-60021-939-X (hardcover)
1. Cell differentiation. I. Ivanova, Laura B.
[DNLM: 1. Cell Differentiation. 2. Cell Transformation, Neoplastic. QU 375 C39267 2008]
QH607.C464 2008
571.8'35--dc22 2007030895

Published by Nova Science Publishers, Inc. ✦ New York

Contents

PREFACE

Cellular differentiation is a concept from developmental biology describing the process by which cells acquire a "type". The morphology of a cell may change dramatically during differentiation, but the genetic material remains the same, with few exceptions.

A cell that is able to differentiate into many cell types is known as pluripotent. These cells are called stem cells in animals and meristematic cells in higher plants. A cell that is able to differentiate into all cell types is known as totipotent. In mammals, only the zygote and early embryonic cells are totipotent, while in plants, many differentiated cells can become totipotent with simple laboratory techniques.

This new book presents lead-edge research from around the world in this field.

Chapter I - Development of the oral and craniofacial region is a complex and fascinating set of processes which require a sequential integration of numerous biological steps. For medical and dental doctors, interest is particularly high in this region, because it is composed of three blastoderms - ectoderm, mesoderm, and endoderm - as well as neural crest cells. There are many different types of neoplasms in this region. In general, proliferation, development and cytological differentiation of the neoplastic cells reflect the normal physiological development of the outbreak mother cells and/or tissues. Collected human neoplasm cases, such as osteosarcoma appearing in the oral and craniofacial region, are examined regarding the immunohistochemical expression of some morphogenesis regulation factors. Furthermore, examination of Notch signaling is also conducted for some odontogenic neoplasms. This chapter mainly describes the examination results of some morphogenesis regulation factors, such as Notch signaling, in the neoplastic cells originating in the oral and craniofacial region, especially in the odontogenic neoplasms, in both well-differentiated and poorly-differentiated neoplasms of tooth germ enamel organ-derived neoplasm. In general, these morphogenesis regulation factors are responsible for cytological regulation of cell fate, morphogenesis and/or development. The results suggest that these factors play some role in cytological differentiation or acquisition of tissue specific characteristics in neoplastic cells. Furthermore, there would appear to be a relationship between the cytological differentiation in the oral and craniofacial neoplastic cells and the physiological development and differentiation of their originating mother cells and tissues of the oral and craniofacial region.

Chapter II - Cellular therapies are promising approaches in the treatment of several neurological diseases such as Parkinson's disease [Isacson et al., 2001] or Huntington's disease [Dunnett et al., 2000], but also for spinal cord injury [Hall et al., 2001]. One main problem concerns the origin and nature of the cells to be used for such procedures. In this context, recent studies suggest that somatic stem cells (stem cells from foetal or adult tissues) might be able to exhibit more plasticity than previously thought as they seem able to differentiate into many cell types, including cell types which are not encountered in their tissue origin. This last property, named phenotypic plasticity of somatic stem cells, is thus the capacity for a stem cell to develop in several phenotypes depending on their environment. Several recent reports suggest that bone marrow mesenchymal stem cells (MSC) could be a source of somatic stem cells suitable for cell replacement strategies in the treatment of central nervous system (CNS) disorders. MSC can differentiate into many types of mesenchymal cells, i.e. osteocytes, chondrocytes and adipocytes, but can also differentiate into non-mesenchymal cell, i.e. neural cells in appropriate *in vivo* and *in vitro* experimental conditions [Kopen et al., 1999; Brazelton et al, 2000; Mezey et al, 2000; Wislet-Gendebien et al., 2003, 2005]. Some works have attributed the neural phenotypic plasticity to "transdifferentiation" [Krause et al., 2001; Orlic et al., 2001; Priller et al., 2001; Wislet-Gendebien et al., 2005], while some other works suggested that this neural plasticity could be explained by cell fusion [Terada et al., 2002; Ying et al., 2002; Vassilopoulos et al., 2003; Alvarez-Dolado et al., 2003]. These observations could suggest that mesenchymal cells are heterogeneous and there are two cell populations able to adopt a neural phenotype: one which is able to fuse with already-present neurons and a second one which is really able to differentiate in neurons. In the first part of this chapter, the authors will review the studies realized on the potential neural phenotypic plasticity of the mesenchymal stem cells. The second part of this chapter will focus on recent studies demonstrating that stem cells isolated from adipose, skin and umbilical cord cells have the ability to differentiate into neural cells [Nagase et al., 2007; McKenzie et al., 2006; Fallahi-Sichani et al., 2006]. This ability could be attributed to the presence of neural crest stem cells in those tissues [Fernandes et al., 2007; Crane and Trainor, 2006]. Consequently, the authors will address the question of the potential presence of neural crest stem cells in bone marrow.

Chapter III - Spermatozoa are formed in the mammalian testes by a complex differentiation process collectively referred to as spermatogenesis. The process involves multiple molecular events during mitotic cell division, meiosis and spermiogenesis. The last event is the final phase of the complex process when a non-dividing ordinary looking round spermatid is transformed into a uniquely shaped spermatozoon containing a well-developed sperm head (with a nucleus and a new organelle, the acrosome) and a fully formed flagellum. It is important to emphasize that the constituent proteins/glycoproteins, present within the acrosome of a testicular spermatozoon, are synthesized in testicular germ cells during spermatogenesis. However, our knowledge on the origin and molecular processes that regulate the expression and processing of acrosomal enzymes in the germ cells is very limited. In this article, the authors have described multiple biochemical and immunohistochemical approaches to examine the synthesis, processing and localization of two acid glycohydrolases in spermatogenic cells from rodent testes. Mixed germ cells, prepared from the rat or mouse testis by enzymatic digestion, were separated by unit gravity sedimentation using a linear gradient of 2-4% bovine serum albumin. Fractions rich in spermatocytes, round spermatids, and condensed/elongated spermatids (>95% pure cell populations) were pooled separately and the enriched cells were assayed for several acid glycohydrolase activities. All these cell populations as well as the cauda epididymal spermatozoa were found to contain several acid glycohydrolase activities. The presence of two enzymes, namely β-D-galactosidase and β-D-glucuronidase in the germ cells was further confirmed using immunocytochemical and biochemical approaches. Immunocytochemical approaches at the electron microscopic level revealed that the two enzymes were present in the Golgi membranes, Golgi-associated vesicles, and lysosome-like structures in the late spermatocytes and round spermatids. Indirect immunofluorescence studies at the light microscopic level revealed intense immunopositive reaction in the forming/formed acrosome of the round spermatids, elongated spermatids and the cauda epididymal spermatozoa. A combination of radiolabeling of spermatocytes and round spermatids in cell culture, immunoprecipitation of the [^{35}S]methionine-labeled (newly synthesized) β-D-galactosidase, SDS-PAGE and autoradiography procedures were used to examine the synthesis and processing of the enzyme in rat spermatogenic cells. Data from these approaches demonstrate that the β-D-galactosidase is synthesized in the spermatocytes and round spermatids as high molecular weight precursor forms of 90/88-kD that undergo processing to the lower molecular weight mature forms in a cell-specific manner. The net result is the formation of predominantly 64-kD and 62-kD forms in spermatocytes and round spermatids, respectively. Taken together, our studies demonstrate that the two acid glycohydrolases present in the acrosome of the epididymal spermatozoa are first expressed in spermatocytes. The authors' final intention is to briefly discuss the significance of acrosomal glycohydrolases in sperm function and fertilization. The authors hope that various studies reported in this article will contribute to a better understanding of the sperm acrosome and its potential role in fertilization.

Chapter IV - Spermatogenesis is a very complex terminal cell differentiation process, essential for all the species with sexual reproduction.

Despite of its importance for sexual reproduction and as a main source of variability, mammalian spermatogenesis is still poorly understood at the molecular level, mainly due to some difficulties it presents for its study. One of the main drawbacks is that the testis is a very complex tissue, with the spermatogenic cells of different stages - somatic spermatogonia, meiotic cells and haploid spermatids in different phases of elongation - coexisting with somatic Sertoli cells in the seminiferous tubules of adult animals. Besides, spermatogenic cells do not maintain the differentiation process *in vitro*, thus reducing the possibilities of the use of germ cells culture for experimentation. Therefore, a prerequisite for the analysis of differential gene expression during spermatogenesis is the availability of methods that allow rapid obtainment of highly enriched populations of a certain germ cell type.

In spite of these drawbacks, a few experimental approaches have been recently developed in some laboratories for the analysis of gene expression in the germ line of the male. The studies are revealing testis as a very interesting system for the analysis of differential gene expression because of the astonishingly large number of genes that are differentially expressed and the peculiar patterns of regulation that govern gene expression in spermatogenic cells. Moreover, candidate genes for roles in the regulation of fertility and possible contraceptive targets are starting to be identified.

Here, different approaches for the analysis of gene expression along spermatogenesis employed in various laboratories as well as in the authors own will be reviewed, and the main results will be discussed.

Chapter V - So far there are only few works concerning changes in male reproductive system after extremal hypoxic and ischemic influences. In the current study the authors have investigated the state of spermatogenesis of white male rats after modeling of clinical death and acute hypobaric hypoxia. It has been demonstrated that the observed impairment and death of the considerable part of germ cells as well as Sertoli cells and Leydig cells was due to increased production of free radicals. The level of free-radical oxidation was higher in the early reperfusion period than in the early post-hypoxic period. It has been found that lactate intratesticular level decreased during first hours after clinical death modeling whereas it elevated after acute hypobaric hypoxia modeling; it provided the more expressed preservation of germ cells in the second case. By the 21st day of the both experiments the tendency towards the increase in number of germ epithelial cells as well as Leydig and Sertoli cells has been revealed yet by the 60th day of the observation the cellular picture of the testicles has not been fully recuperated (compared to the intact animals). The low intensity of the processes of cell division and maturation in the late post-reanimation and post-hypoxic periods is obliged to the small number of both Sertoli and Leydig cells, as these cells provide metabolic and hormonal support of gametogenesis. . The found disorders of spermatogenesis lead to qualitative and quantitative changes of ejaculate indices.

Chapter VI - Activated human T cells produce receptor activator NF-κB ligand (RANKL), interleukin-17 (IL-17), and interferon-γ (IFN-γ). All regulate human osteoclastogenesis; RANKL and IL-17 potently induce osteoclastogenesis, while IFN-γ directly inhibits it *in vitro*. However, the authors have demonstrated that IFN-γ-producing T cells induce human osteoclastogenesis from monocytes through the expression of RANKL. Strongly supporting the authors findings, it has been reported that peripheral blood T cells from patients with early rheumatoid arthritis (RA) promote osteoclastogenesis in autologus monocytes through the expression of RANKL, although the T cells express IFN-γ. In addition, it has been reported that IFN-γ has indirect pro-osteoclastogenic properties *in vivo*; IFN-γ induces bone resorption in three mouse models of osteoporosis under conditions of estrogen deficiency, infection, and inflammation. On the other hand, the expression or function of some cytokines shows differences between humans and mice in bone cell biology. In the current article, the authors review recent findings on the role of T cells in human osteoclastogenesis and bone destruction of rheumatoid arthritis (RA), including the effects of dexamethasone and tacrolimus on the function of T cells.

Chapter VII - As a sexual reproductive mode, a zygote after cell fusion between sperm and egg, as a destiny, certainly divides into embryo. If it is the dicots, the embryo passes the stages of globular, heart, torpedo and complete one with two cotyledons and hypocotyl that finally becomes a seed. Here, the authors describe the process of somatic embryogenesis from asexual cells derived from protoplast culture, protoplast fusion, interspecific hybrid embryo culture, and asexual tissue cultures. When the immature globular-stage embryos of tomato and its wild species were taken out and cultured on suitable medium, on which somatic embryos were formed with globular, heart and torpedo stages. When the cultures of cotyledon-derived protoplasts and protoplast fusion of tomato were carried out, somatic embryogenesis was observed, according to the stages from the globular and heart to torpedo. When leaflet tissues from guineagrass and bahiagrass were cultured on suitable medium, embryogenic calli were formed and they were observed by scanning electron microscopy, indicating somatic embryo structure with a scutellum and a coleoptile. The characterizations of the somatic embryo were similar to those previously reported. Multiple shoot and plant regeneration were obtained from somatic embryos. Those somatic embryos above were analyzed with RAPD, RFLP, chromosome, gene cloning and transgenic plants, and the results were also discussed in the text.

Chapter VIII - Monocytes represent about 5-10% of blood leukocytes in mice or men and are established circulating precursors for tissue macrophages and dendritic cells (DCs). Monocyte-derived macrophages and DCs fulfill critical roles in innate and adaptive immunity during inflammation and inflammatory disorders, and it is believed that monocytes also maintain these populations in some peripheral tissues during homeostasis. Recent identification of mouse monocyte subsets that closely resemble human monocyte subsets has inspired a variety of techniques wherein monocytes can be readily traced *in vivo* to address these critical questions. There are two major monocyte subsets that vary in chemokine receptor (CCR) and adhesion molecule expression, migratory and differentiation properties. In humans, 'classical' $CD14^{+}$ $CD16^{-}$ monocytes express CCR2, CD64, CD62L, whereas 'non-classical' $CD14^{low}$ $CD16^{+}$ monocytes lack CCR2. Their counterparts in mice are $CCR2^{+}$ $Gr\text{-}1^{hi}$ and $CCR2^{-}$ $Gr\text{-}1^{low}$ monocytes, respectively. $Gr\text{-}1^{hi}$ ($Ly6C^{hi}$) monocytes are recruited to inflammatory sites, e.g. inflamed skin or acutely inflamed peritoneum. $Gr\text{-}1^{hi}$ ($Ly6C^{hi}$) monocytes are precursors for the epidermal DCs, Langerhans cells, after skin inflammation. $Gr\text{-}1^{low}$ monocytes have been proposed as precursors for steady state DCs, but experimental evidence is as of yet limited. In atherosclerosis, a chronic inflammatory disease of the blood vessels, both subsets differ in their migratory capacity, chemokine receptor pathways and differentiation towards macrophages or DCs. The rapid progress in recent years led to a better understanding of monocyte biology in the steady state and inflammation.

Chapter IX - Mouse embryonic stem (ES) cells are derived from the inner cell mass of the 3.5-day-old blastocyst. These cells retain the differentiation ability after many passages in the presence of leukemia inhibitory factor. Removal of leukemia inhibitory factor results in mouse ES cells differentiation into embryoid bodies. Mouse ES cells have been regarded as a versatile biological system which has made significant advance in cell and developmental biology. They possess developmental potentials to differentiate into hematopoietic cells. Using the mouse ES/EB system, a precursor that responds to vascular endothelial growth factor and generates colonies consisting of undifferentiated blast cells was identified. These VEGF-responsive blast cell colonies were shown to contain endothelial, primitive erythroid, and various definitive hematopoietic precursors. These cells are referred to as blast colony-forming cells and are thought to be cells represent the hemangioblast precursor of blood and endothelial lineages. Other hematopoietic cells, such as progenitor B cells, mature B cells, dendritic cells, mast cells, neutrophils, and T cells can also be generated from mouse ES cell differentiation. This ES cell differentiation toward to the generation of hematopoietic cells might be a useful system to study gene function and regulation in hematopoietic development. Using this system, the authors recently identified protein tyrosine phosphatase Shp-2 is essential for mouse ES cell-derived hematopoietic differentiation and Ape1 regulate hematopoietic differentiation of mouse ES cells through its redox function domain. In this chapter, the authors review the induction of mesoderm cells, particularly hematopoietic cells from mouse ES cell differentiation. The authors also summarize and discuss the recent advance in the hematopoietic cell development using mouse ES cell differentiation system.

Chapter X - Follicle stimulating hormone (FSH) and testosterone have been described as necessary factors for proliferation of Sertoli and germ cells, as well as having a role in apoptosis. However, their role in germ cell differentiation has not been studied in detail. FSH is a proteic hormone produced in the anterior hypophysis and has receptors located in Sertoli cells, which in turn function as nurse cells for germ cells in the seminiferous epithelium. The expression of FSH receptors changes during the cycle of seminiferous epithelium. Germ cells in consecutive steps of differentiation are located in each progressive stage of the seminiferous epithelium, indicating a role for FSH signaling in the differentiation of germ cells. CREB (CRE-protein binding) is a transcription factor involved in differentiation of germ cells and it is modulated by FSH. Some CREM-regulated products have been shown to participate in germ cell differentiation. Recent studies have focused on the intracellular signaling after FSH joins with its receptor, and how it influences germ cell differentiation. Testosterone-stimulated germ cell development occurs via paracrine communication with Sertoli cells. The androgen receptor is expressed in a stage-specific manner during the seminiferous epithelium cycle. Recent studies on gene expression show a number of genes modulated by testosterone, many of which do not follow the traditional pathway of androgen/receptor/gene expression, but are activated by another signalization pathway. This chapter begins with the background necessary to understand the general role of FSH and testosterone in seminiferous epithelium and then addresses the molecular signaling initiated by these chemical messengers.

Chapter XI - Retinoic acid induced redifferentiation and apoptosis in pancreatic adenocarcinoma cell lines. Redifferentiation included early reversion into aerobic metabolism reflected by an increase of mitochondrial activity and -mass with normal membrane potential and terminal ductal cell differentiation. Cells in such state would attempt correcting their DNA abnormalities or otherwise commit suicide by apoptosis. In some cell systems such as the present case, the stem cell potential of ductal pancreatic cells gave them an alternative option, i.e., to transdifferentiate into functional endocrine normal cell type. Again due to impossibility of correcting highly corrupted genome, cells eventually succumbed apoptosis. Mitochondrial changes appear to be forcing factor for this process. Such a process, transformation - normalizing-redifferentiation - apoptosis sequence, has been shown by several studies utilizing various cell types, apoptotic inducers, biomarkers and time-frames. Although some studies have shown concomitant apoptosis and redifferentiation, others have reported apoptosis without prior redifferentiation. However, utilizing appropriate time-frame and markers of earlier mitochondrial changes, one would detect a scenario similar to the present retinoid model. This situation can be achieved by delaying apoptosis or reducing the inducer concentration in such systems. The final fate of normal differentiated cells is apoptosis. Therefore, it was suggested that a degree of normalizing redifferentiation of transformed cells might be expect prior to apoptosis, simulating the physiological fate.

Chapter XII – *Background*: Angiotensin converting enzyme inhibitors (ACEI) have pharmacological effects in all body tissues containing rennin – angiotensin system (RAS) and the related kallikrein – kinin system (KKS). In the male reproductive tract which also contains these paracrine systems, results of effects have been conflicting. Encouraged by a few successes in humans and confirmation of improved sperm count and motility in rats; the authors set out to study at tissue level, the effects of ACEI on spermatogenesis. These experiments have potential for human application.

Methods: Four groups of adult male Wistar rats received, 0.2 ml per oral of the following doses weekly: A – 5mg/ml, B – 2.5 mg/ml, C – 0.25 mg/ml, D – 0.05 mg/ml. A fifth group, serving as controls, did not receive any drug. After 2 weeks when one spermatogenetic cycle was complete, animals were sacrificed in each group weekly and testes excised for histology.

Results: The control group showed normal evidence of spermatogenesis. The group receiving 0.05 mg/ml showed a strange arrangement of seminiferous tubules; but evidence of normal spermatogenesis in the normal tubules. By 0.25 mg/ml, seminiferous tubules were of low population with incomplete spermatogenesis in many foci. At the 2.5 mg/ml dose, all stages of spermatogenesis were evident in good amounts; with abundance of glycogen deposits. The group receiving the highest dose of 5 mg/ml showed plenty sperm heads in the seminiferous tubules. Spermatogenetic response was robust. Effect at the end of the study was largely similar to the week two specimen except for the control group.

Discussion: The existence of tissue RAS and KKS in the male reproductive tract implies potential effect of ACEI drugs. Preliminary, albeit anecdotal reports in humans of improved semen quality, supported by a dose dependent improvement of spermatogenesis in rats has been borne out by this tissue based work. There appears to be a dose dependent entry of the ACEI drugs into these tissue paracrine systems resulting in stimulation of spermatogenetic process; by creating a favourable ambient for multiplication and viability of spermatozoa. These include improvement in Leydig and Sertoli cell function made possible by local accumulation of bradykinin and nitric oxide; as well as reduction in levels of Angiotensin II brought about by inhibition of the ACE system. The potential of these results for human application in treatment of male infertility should be explored in greater detail.

In: Cell Differentiation Research Developments
Editor: L. B. Ivanova, pp. 1-30
ISBN: 978-1-60021-939-9

Chapter I

CELL DIFFERENTIATION OF NEOPLASTIC CELLS ORIGINATING IN THE ORAL AND CRANIOFACIAL REGIONS

***Toshiyuki Kawakami*[1,2], *Hitoshi Nagatsuka*[1,4], *Keisuke Nakano*[1,2], *Takako Shimizu*[1], *Hidetsugu Tsujigiwa*[5], *Hiromasa Hasegawa*[2,3] and *Noriyuki Nagai*[4]**

[1]Hard Tissue Pathology Unit, Matsumoto Dental University Institute for Oral Science, Shiojiri, 399-0781 Japan;
[2]Hard Tissue Pathology Unit, Matsumoto Dental University Graduate School of Oral Medicine, Shiojiri, 399-0781 Japan;
[3]Department of Oral Pathology, Matsumoto Dental University School of Dentistry, Shiojiri, 399-0781 Japan;
[4]Department of Oral Pathology and Medicine, Okayama University Graduate School of Medicine, Dentistry and Pharmaceutical Sciences, Okayama, 700-8525 Japan;
[5]Department of Virology, Okayama University Graduate School of Medicine, Dentistry and Pharmaceutical Sciences, Okayama, 700-8525 Japan.

ABSTRACT

Development of the oral and craniofacial region is a complex and fascinating set of processes which require a sequential integration of numerous biological steps. For medical and dental doctors, interest is particularly high in this region, because it is composed of three blastoderms - ectoderm, mesoderm, and endoderm - as well as neural crest cells. There are many different types of neoplasms in this region. In general, proliferation, development and cytological differentiation of the neoplastic cells reflect the normal physiological development of the outbreak mother cells and/or tissues. Collected human neoplasm cases, such as osteosarcoma appearing in the oral and craniofacial region, are examined regarding the immunohistochemical expression of some morphogenesis regulation factors. Furthermore, examination of Notch signaling is also conducted for some odontogenic neoplasms. This chapter mainly describes the examination results of some morphogenesis regulation factors, such as Notch signaling, in the neoplastic cells originating in the oral and craniofacial region, especially in the odontogenic neoplasms, in both well-differentiated and poorly-differentiated neoplasms of tooth germ enamel organ-derived neoplasm. In general, these morphogenesis regulation factors are responsible for cytological regulation of cell fate, morphogenesis and/or development. The results suggest that these factors play some role in cytological differentiation or acquisition of tissue specific characteristics in neoplastic cells. Furthermore, there would appear to be a relationship between the cytological differentiation in the oral and craniofacial neoplastic cells and the physiological development and differentiation of their originating mother cells and tissues of the oral and craniofacial region.

INTRODUCTION

A neoplasm is an abnormal growth mass of cells and/or tissue of which exceeds and is uncoordinated with the physiological cells and/or tissues. Thus, neoplasms are said to be transformed because they continue to replicate, apparently oblivious to the regulatory influences that control physiological original cell and/or tissue growth. In neoplasm, the words "differentiation and anaplasia" are applied to the originating parenchymal cells that constitute the transformed neoplastic components of neoplasms. The word "differentiation" of parenchymal cells refer to the extent to which they resemble their physiological features, both morphological and functional in findings.

BIOLOGY OF NEOPLASMS

In benign neoplasms, the parenchymal cell components consist of well-differentiated cells that quietly resemble their originating mother cell and/or tissue. In benign neoplasms, mitoses are extremely rare in number and are of normal physiological features. In malignant neoplasms, they are characterized by a wide variety of parenchymal cell differentiation, from completely well- to un-differentiated. Malignant neoplasms are composed of poorly- to un-differentiated cells, when said to be anaplastic. Anaplasia, a lack of differentiation, is thought to be a standard of malignancy. Anaplasia means literally to "form backward". In fact, it is well accepted that a malignant neoplasm originates from stem cells in tissues. Therefore, a failure of differentiation, rather than de-differentiation of physiologically specialized cells, accounts for un-differentiated neoplasms. In general, the more malignant and the more un-differentiated (anaplastic) a neoplasm, the less likely it is to have specialized functional activity. The cells in benign neoplasms are almost always well-differentiated and resemble the physiological normal cells of the originating mother cells and tissue. The cells in malignant neoplasms are more or less differentiated, but some loss of differentiation is always present. The differentiation in these neoplastic lesions is controlled by, in part, the normal physiological regulation system. The regulation mechanism of specialization growth of a physiological organization functions on a neoplasm, and the specialized growth of neoplasms is also regulated. Therefore, it is important to examine the normal physiological cell differentiation mechanisms for understanding the regulation system of neoplastic conditions.

DEVELOPMENTAL BIOLOGY OF TEETH

Teeth develop physiologically, as deciduous teeth and permanent teeth, from oral ectodermal epithelium and neural crest derived neuroectodermal mesenchymal cells [20]. The enamel is derived from ectoderm of the oral cavity, and all other tissues differentiate from the mesenchyme derived from mesoderm; neural crest cells are imprinted with morphogenetic information before or shortly after they migrate from the neural crest. Tooth development is initiated by the inductive influence of the neural crest mesenchyme on the covering ectoderm. Tooth development is a continuous step. However, it is usually divided into a bud stage, cap stage, and bell stage.

Odontogenesis or tooth development is a complex and highly-regulated process characterized by sequential epithelial-mesenchymal interactions leading to tooth initiation, morphogenesis and cell-differentiation with eventual formation of enamel, dentin and cementum matrices [17,18]. Osteogenesis or bone formation is also a tightly-coordinated process involving many different tissues that interact with each other via a matrix-mediated inductive mechanism, and ending in the formation of a specialized tissue, bone. Both of these processes, though distinct, are closely related in that they share common signaling pathways in terms of morphological differentiation of their cells and functional differentiation of their matrix proteins. The ameloblasts and odontoblats, which are exclusive enamel matrix-producing and dentin matrix-producing cells respectively, share several molecular characteristics with the osteoblast which is the bone matrix-forming cell. Runx2, a transcription factor, is essential for osteoblast differentiation (Figure 1). In the course of odontogensis, the Runx2-knockout mice experiments results suggest as follows: Runx2 is associated with morphogenesis of teeth and matrix protein gene expression [42]. Next, Next, compared to the incisor tooth germ, the molar tooth germ is more strongly subjected to control by Runx2, suggesting the presence of factors involved in odontogenesis of the incisor tooth germ which are different from those present in osteoblasts. Furthermore, in Runx2 knockout mice differences in expression of osteopontin and osteocarcin, matrix proteins common for teeth and bone, suggest different mechanisms of cellular differentiation or transcription regulation pathways in incisor odontoblast and bone forming cells, or osteoblasts [43].

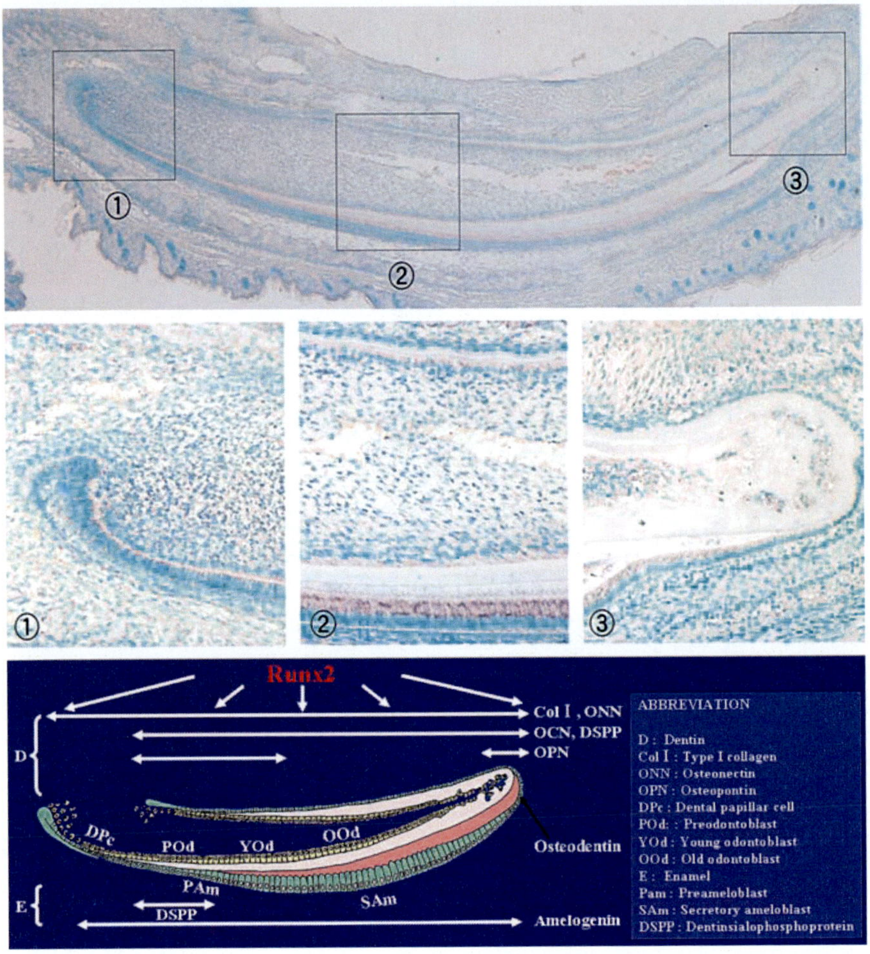

Figure 1. Runx2 regulatory control of the signaling pathway of odontogenesis. The gene expression is visible in various stages of ameloblasts (PAm, SAm, and maturation stage).

The tooth germ basal membrane interposed between the odontogenic epithelium and mesenchyme mediates the sequential and reciprocal epithelial-mesenchymal interactions essential for morphogenesis and cell-differentiation for tooth formation (Figure 2). It composes some isoforms of type IV collagen, laminin, nidogen/entacin, heparan sulfate, proteoglycan, fibronectin, and other components molecules [39,46,80]. The molecules of type IV collagen, a major framework-forming peptide of basal membrane, are heterotrimers composed of three α chains that exist in six genetically distinct forms (α1 to α6) and with at least three molecular forms [48,54,55]. The expression patterns of type IV collagen molecular forms in tooth germ organogenesis and the marked stage-specific changes in the type IV collagen distribution during the odontogenesis are limited [19]. According to the examination results using mouse developing molar tooth germ at the dental placode and bud stage in the course of odontogenesis, the basal membrane of the oral cavity epithelium expresses α1, α2, α5 and α6 chains while the gubernaculums dentis, in addition to above 4 chains, also expresses α4 chain. An asymmetrical distribution of α4, α5 and α6 chains has been observed at the bud stage in the odontogenesis. At the early bell stage, the basal membrane associated with the inner enamel epithelium of molar germ expresses α1, α2 and α4 chains while the basal membrane of the outer enamel epithelium only expresses α1 and α2 chains. With the onset of dentin formation, the collagen α chain profile of the basal membrane of inner enamel epithelium gradually disappeared. From the bell stage, however, the gubernaculums dentis consistently expressed α1, α2, α5 and α6 chains, this distribution pattern resembles the one of the fetal oral cavity epithelium. These features suggest that the odontogenic stage- and the position-specific type IV collagen α subunit distribution is according to the tooth germ odontogenesis, and its changes are essential for the morphogenesis and cell-differentiation for the tooth development [8,39,74].

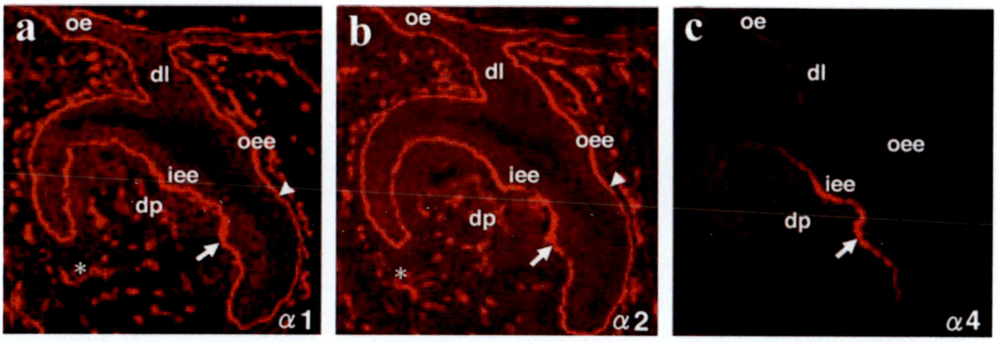

Figure 2. Distribution of α chains of type IV collagen in early bell stage of E15 mice. The basal membrane of inner enamel epithelium (iee, arrow) stains for α1 (a), α2 (b), and α4 (c) chains. oee: outer enamel epithelium, dp: dental pulp, oe: oral epithelium, dl: dental lamina, *: vascular basal membrane

Tenascin is an extra-cellular matrix glycoprotein which appears to regulate cell morphology [40]. It is more restricted to tissue distribution than fibronectin and is able to interface with the cell binding function of fibronecion. Tenascin is most typically expressed in epithelial-mesnchymal interactions in during physiological development and in the stromal tissue of malignant neoplasms. Extra cellular matrix protein is shown to play important roles in cellular growth and differentiation, in complex cell matrix interactions, in physiological organ development and neoplastic transformation course [40].

At first in dental lamina of the bell stage of human tooth germ, tenascin is present only on the submucosal connective tissue side, not on the dental follicle tissue side [6]. At this stage there are no morphological differentiations in the odontogenic epithelium on either side. Concerning fibronectin, a weak or negative localization is seen in the condensed mesenchyme surrounding the dental lamina. In the cap stage, different patterns of the distribution between tenascin and fibronectin is evident in the human tooth germ. Strong tenascin accumulation is present in the dental papilla under the basal membrane, preodontogenic layer and osteogenic tissue of alveolar bone. However, tenascin is immunohistochemically negative in the dental follicle, the fibroblastic layer developing to the periodontium [40]. Intense fibronectin is evident in the alveolar bone. The epithelial components of the tooth germ are imunohistochemically negative for both tenascin and fibronectin.

Developmental Oral Craniofacial Biology

Development of the oral and craniofacial region in human is a complex and fascinating set of processes which require a sequential integration of numerous biological progresses. The mechanism has been a source of fascination and an object of intensive examination since early scientific medical studies. Regarding the regulation in the development, biological cellular regulation during the processes contain the entire scale of values ranging from ion to molecular interactions.

The oral and craniofacial regions of a 4-week-old human embryo somewhat resemble these regions of a fish embryo at a comparable stage of development. This explains the former use of the adjective "branchial", which is derived from the Greek word "branchia". The pharyngeal (branchial) apparatus consists of the pharyngeal arch, pouch, groove, and membrane. These embryonic structures contribute greatly to the formation of the oral and craniofacial region. The oral and facial primordial begin to appear early in the 4th week around the big *stomodeum*. Facial development depends on the inductive influence of the prosencephalic and rhombencephalic organizing centers. Five facial primordial, the single frontonasal, the pair of maxillary, and the pair of mandibular prominences appear. The pair of facial prominences is derived from the first pair of pharyngeal arches. The prominences are formed predominantly by the neural crest cell proliferation that migrate from the regions of the neural fold into the arches among the 4th week of the embryonic stage. These cells are the major source of mesenchymal tissue components, which include cartilage, bone, and ligament tissues in the oral and craniofacial regions. Therefore in summary, during the 4th- and 5th-week of the embryonic stage, the primitive pharynx is bounded laterally by pharyngeal arches. The arch consists of a core of mesenchyme covered externally by ectoderm, and internally by endoderm. The original mesenchyme of the arch is derived from mesoderm, and later, neural crest cells migrate into the arches and almost all of their mesenchymal components.

Mandibular Bone and Cartilage Development

In the oral and craniofacial region, cartilage characteristics are slightly different from other cartilages of general portions, especially joint cartilage. This includes mandibular chondylar, angular and coronoid cartilages. Mandible is composed of mandibular bone and cartilage [52]. This cartilage is classified as a secondary cartilage together with condyle, coronoid and angle. Formation studies of bone and cartilage in the oral and craniofacial region have been conducted by many researchers [12,57-60]. With regard to mandibular angle and coronoid, however, very few reports have been published [28,61]. Their bone formation patters attracts researchers, suggesting large possibilities for both clinical and histological findings. Mandibular condylar cartilage has bone characteristics which are more significant than its cartilaginous characteristics [62,64].

Regarding the development of the mouse mandibular condylar cartilage, at embryonic day 14 there are no development features, although there is some osteoblastic cell proliferation and a small number of mandibular body bone matrices (Figure 3). At the distal upper portion of the developmental mandibular bone, mesenchymal cell proliferation and condensation with no metacholomasia reaction to toluidine blue (TB) are seen. At embryonic day 15, mandibular condylar cartilage is clearly evident, as a metacholomasia reaction to TB, which is firstly expressed at a middle zone of the proliferating mass. At embryonic day 16, the volume of condylar cartilage grows both in length and width. In this stage, articulation occurs between the mandibular bone and condylar cartilage. At the late embryonic stage, the mandibular condylar cartilage further grows both in length and width, especially at the hypertrophy layer. At the connection area of the mandibular trabecular bone and the hypertrophy layer of the condylar cartilage, endochondral ossification occurs. Furthermore, perichondral ossification occurs at the sheath of condyle. That is, direct bone formation occurs at the sheath of the condylar cartilage. At just before birth, endochondral ossification progresses further and the mandiblar condyle volume grows.

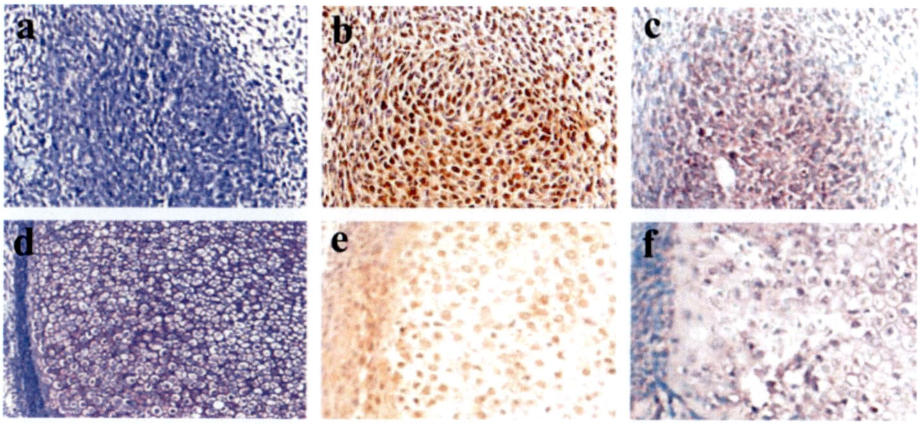

Figure 3. Developmental features of mouse mandibular chondylar cartilage. Mesemchymal cell coagulation of E14 (toluidine blue stain; a) and positive reaction of Runx2 peptide (b) and mRNA (c) are visible. At E18, direct bone formation occurs in the sheath of the cartilage (toluidine blue; d), and Runx2 positive reactions (e) are strongly detected at the sheath with gene expression (f).

The proliferating chondrocytes show immunohistochemically positive reactions to type II collagen, type I collagen and osteopontin. These data suggest that the characteristics of mandibular chondylar cartilage, as secondary cartilage, are slightly different from those of normal physiological articular cartilage. The cartilage takes on the character of bone [37]. In general, Runx2 is a transcription factor necessarily for osteoblast differentiation [21] and bone formation [33].

Mandibular condylar cartilage is recognized as an important growth site and is developed by an endochondral bone formation mode. It is recognized as secondary cartilage, which differs from primary cartilage in morphological and biochemical organization [57]. There are some different components of the extracellular matrix between the primary and secondary cartilage. Immunohistochemical studies for the expression pattern of type I and II collagens [21] have shown that, both types of collagen are simultaneously produced in chondrocytes at this area. Moreover, simultaneous expression of type I and II collagen genes have been confirmed. These findings enable the accumulation of different characteristics and aspect of this cartilage from general endochondral ossification ones. In the reported literature [64], immunohistochemically-positive reactions to osteopontin are detected in almost all layers of the cytoplasm of the mandibular chondylar chondrocytes. The findings from toluidine blue stained specimens of the early developmental stage of mandibular condylar cartilage for this study indicate that, at the distal upper portion of the developmental mandibular bone, mesenchymal cell proliferation and condensation with no metacholomasia reaction are present. Next, mandibular condylar cartilage is clearly evident as a metacholomasia reaction to TB.

Regarding the mandibular angular cartilage, the development starts nearly the same fatal of mandibular chondylar cartilage development. In other words, coagulation of mesenchymal cells have been observed at embryonic day 14, and differentiated to chondrocytes, which show a metachlomasia reaction for TB the next day. After embrionic day 17, endochondral ossification occurs with the invasion of capillaries, and perichondral ossification occurs in the periphery of the cartilage mass. In immunohistochemical examinations, the proliferating chondrocytes of the mandibular angular cartilage show positive reactions to type I collagen and osteopontin, as well as to type II collagen. Therefore, the results show that the characteristics of proliferating mandibular angular cartilage are nearly the same as mandibular condylar cartilage, and differ slightly from normal physiological articular cartilage.

Recently, various studies have shown that mandibular condylar cartilage formation is related to morphogenesis regulation factors and their signaling, such as a fibroblast growth factor receptor, a platelet-derived growth factor receptor [12]. Generally, Notch1 and Math1 are important regulation factors of morphogenesis [64]. There are no reports on mandibular condylar cartilage, although there is a report on the distribution of articular cartilage. Investigations carried out on the two essential factors of expression, Notch1 and Math1 in the mandibular condylar cartilage, reported that, these expression patterns are different from the one in the articular cartilage, and the reactions for Notch1 are present but only localized in the hypertrophic cells [33]. Math1 was distributed mainly in the hypertrophy layer and partially in the proliferate layer. Therefore, the expression patterns of Notch1 and Math1 are slightly different from those of articular cartilage. These results suggest that regulation factors of morphogenesis-Notch1 and Math1-may play some essential role in mandibular condylar cartilage [33]. Since it is inconsonant with the distribution in articular cartilage, it is presumed that the generation of the cartilage and morphogenesis mechanism does not correspond.

In general, Runx2 is a transcription factor necessarily for osteoblast differentiation and bone formation [33]. Furthermore it has been reported that Runx2 regulates chondrocyte hypertrophy during chondrogenesis in long bones. Runx2 is responsible for signaling chondrocyte maturation and endochondral ossification during mandibular condyle advancement. Because matrix protein that characterizes bone, such as type I collagen and osteopontin, has often been expressed, mandibular condylar cartilage has intense bone characteristics. As a result, at embryonic day 14, Runx2 expression is detected by means of immunohistochemical (IHC) and *in situ* hybridization (ISH) examinations, which indicates that the expression leads to the secondary cartilage differentiation. From the data, Runx2 expression is detected at embryonic day 14 as is type II collagen expression. At next embryonic day, type II collagen peptide is expressed. That explains that differentiation has started from embryonic day 14, and that area has turned into cartilaginous tissue the next day. This is in accord with the findings [59], that Runx2 is essential for the onset of formation of the mandibular condylar cartilage, as well as for normal development of Meckler's cartilage, and that muscle tissues influence mandible morphology. This finding supports the above-mentioned research result that Runx2 controls differentiation for mandibular condylar cartilage and induces differentiation [63].

Immunohistochemically, at embryonic day 14, Runx2 peptide is expressed in the nucleus and in the cytoplasm of coagulating mesenchymal cells. Next, the proliferating cells have positive products of Runx2 in cytoplasm and nucleus of almost all coagulating cells. The next day, strongly positive Runx2 reactions are detected in cells of the fibrous and proliferate layers, and weakly labeled in cells of all other layers. Furthermore, Runx2 peptide appears in cells at the sheath of the condylar cartilage. After that, Runx2 factor appear in the cells of the condylar cartilage sheath, and is also distinct in cytoplasm and nucleus. Just before birth, Runx2 positive products are observed in almost all cells of layers, and they are mostly distinct in the sheath of the condyle. Just after birth, Runx2 express in a portion of the hypertrophy cells, especially in their cytoplasm and nucleus.

Proliferating chondrocytes show positive reactions to osteopontin, through the examination periods, particularly in the cytoplasm of the proliferating chondrocytes [62]. At the early stage of the developmental day, weak labels for type II collagen are observed in a portion of coagulating mesemchymal cells. Furthermore, positive products of type II collagen exist in the cytoplasm. Just after onset of development and up to the birth, weak labels for OPN and type II collagen are presented in the deeper layer of the condylar cytoplasm and extracellular matrix [64].

ISH examination of the gene expression revealed that just at onset of development, expressions of Runx2 mRNA appear in cytoplasms of proliferating chondrocytes [63]. After that, Runx2 mRNA is detected throughout almost all cytoplasm of all layers. At just after birth, Runx2 gene expression is observed throughout almost all layers. Furthermore, the signals weakly appear in the upper layers, fibrous and proliferative. The signals are mostly distinct in the cytoplasm. Osteopontin mRNA is detected in the cytoplasm of almost all cells of all layers from onset of development to just before birth. It is distinct in cytoplasm and extracellular matrices. At just after birth, Osteopontin gene signals appear restricted in cytoplasm of maturative and proliferative layers [63].

At the embryonic day of late stage, Runx2 gene expression strongly appears in hypertrophy cartilage, probably due to the differentiation to the osteoblast. This agrees wit a report which explains that the Runx2 expression of IHC and ISH has been identified in the hypertrophy layer, and also takes part in the endochondral ossification mode. Examination results have clearly demonstrated the distribution of Runx2 expression, both of the peptide and its gene, at the cartilage in the sheath of mandibular condyle where direct bone formation is observed. The findings provide evidence for Runx2 control over and/or regulation of perichondral ossification. This is the first time anyone has stated these findings. Because of the differentiation of cartilage, Runx2 expression is generated at embryonic day 14 and the next day, and the displacement of bone from the hypertrophy cartilage induces the expression at just before birth.

In summary, for participation of Runx2 in mandibular condylar cartilage development, there are no development features of mandibular condyle. At the distal upper portion of developmental mandibular bone, mesenchymal cell proliferation and condensation with no metacholomasia reaction to toluidine blue are seen at embryonic day 14. At embryonic day 15, mandibular condylar cartilage is clearly evident as a metachlomasia reaction to TB. Immunohistochemically, at embryonic day 14, expression of Runx2 peptide is observed in the nucleus and the cytoplasm of coagulating mesenchymal cells. After the late stage of the embryonic days, Runx2 factors appear in the cells of the condylar cartilage sheath, and they are also distinct in the cytoplasm and nucleus. In gene expression at embryonic day 14 and 15, expressions of Runx2 mRNA appear in the cytoplasm of proliferating chondrocytes. Days just before birth, the mRNA is detected throughout almost all cytoplasm of all layers. These results suggest that Runx2 plays an essential role for mandibular condylar cartilage development, and that Runx2 is essential for the onset of secondary cartilage differentiation [63].

Notch Signaling in Cell Differentiation and Development

In general, Notch signaling plays an important role (Figure 4) in the regulation of cell fate, morphogenesis and/or development [5,14]. Regarding tooth development, there are some published data on how the expression of Notch1, 2, and 3 is regulated by epithelial-mesenchymal interactions in the developing mouse tooth and associated with determination of ameloblast cell fate. Jagged1 is also expressed as a ligand of Notch in the developing tooth [35]. Notch signaling is an evolutionarily-conserved cell-to-cell transmembrane interaction mechanism. Furthermore, asymmetric distribution of Notch has been observed in immature cells prior to cell division, suggesting a role in the regulation of daughter cell fate, including whether the cells remain stem cells or give rise to differentiated progeny [36]. Regarding odontogenesis, Notch1 is expressed in stellate reticulum cells, and Jagged1 is expressed in differentiated ameloblasts in the course of tooth development. During tooth development, Notch expression has been associated with the differentiation of odontogenic epithelial and mesenchymal tissues. However, Notch expression is absent in epithelial cells in close contact with mesenchyme, a feature which may be important for ameloblast cell fate. These data suggest that mesenchymal tissue negatively regulates Notch expression in epithelium. In other words, Notch expression is down regulated in odontogenic epithelium juxtaposed to mesenchyme, indicating that odontogenic epithelium needs a mesenchyme-derived signal to maintain the down regulation of Notch [35].

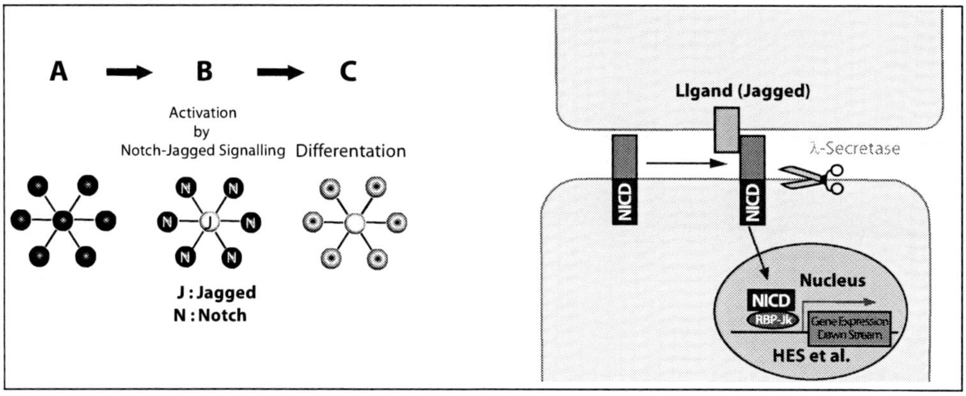

Figure 4. Notch signaling plays in the regulatoion of daughter cell fate.

Cell Differentiation in Pathological Conditions

In the oral and craniofacial region, there are many types of neoplasms and pathological conditions, such as: odontogenic neoplasms, bone and cartilage neoplasms, pathological bone and cartilage formation and/or proliferation. Especially the odonotogenic neoplasms consist of plural number blastoderms. The neoplastic cell differentiation process is complicated, and it is thought that the cell differentiation and growth pattern are copied from a physiological system of the odontogenesis.

WHO histological classification of odontogenic tumors

MALIGNANT TUMOURS	
Odontogenic carcinomas	
Metastasizing (malignant) ameloblastoma[1]	9310/3
Ameloblastic caricinoma - primary type	9270/3
Ameloblastic caricinoma - secondary type (dedifferentiated), intraosseous	9270/3
Ameloblastic caricinoma - secondary type (dedifferentiated), peripheral	9270/3
Primary intrasseous sqaumous cell carcinoma - solid type	9270/3
Primary intrasseous sqaumous cell carcinoma derived from keratocystic odontogenic tumour	9270/3
Primary intrasseous sqaumous cell carcinoma derived from odontogenic cysts	9270/3
Clear cell odontogenic caricinoma	9341/3
Ghost cell odontogenic caricinoma	9302/3
Odontogenic sarcomas	
Ameloblastic fibrosarcoma	9330/3
Ameloblastic fribrodentin-and fibro-odontosarcoma	9290/3
BENIGN TUMOURS	
Odontogenic epithelium with mature, fibrous stroma without odontogenic ectromesenchyme	
Ameloblastoma, solid/multicystic type	9310/0
Ameloblastoma, extraosseous/peripheral type	9310/0
Ameloblastoma, desmoplastic type	9310/0
Ameloblastoma, unicystic type	9310/0
Squamous odontogenic tumour	9312/0
Calcifying epithelial odontogenic tumour	9340/0
Adenomatoid odontogenic tumour	9300/0
Keratocystic odontogenic tumour	9270/0
Odontogenic epithelium with odontogenic ectomesenchyme, with or without hard tissue formation	
Ameloblastic fibroma	9330/0
Ameloblastic fibrodentinoma	9271/0
Ameloblastic fibro-odontoma	9290/0
Odontoma	9280/0
Odontoma, complex type	9282/0
Odontoma, compound type	9281/0
Odontoameloblastoma	9311/0
Calcyfing cystic odontogenic tumour	9301/0
Dentinogenic ghost cell tumour	9302/0
Mesenchyme and/or odontogenic ectonesenchyme with or without odontogenic epithelium	
Odontogenic fibroma	9321/0
Odontogenic myxoma/myxofibroma	9320/0
Cementoblastoma	9273/0
Bone-related lesions	
Ossifying fibroma	9262/0
Fibrous dysplasia	
Osseous dysplasias	
Central giant cell lesion (granuloma)	
Cherubism	
Aneurysmal bone cyst	
Simpe bone cyst	
OTHER TUMOURS	
Melanotic neuroectodermal tumour of infancy see Chapter 1, pp. 70-73	9363/0

Figure 5. WHO histopathological classification of odontogenic tumours described in the textbook [11].

Odontogenic Neoplasms (Figure 5)

Ameloblastoma is classified as a benign, locally-infiltrative odontogenic neoplasm, which is composed of proliferating odontogenic epithelial nests within a fibrous stromal tissue. Some variants have been sub-classified as follows: solid/multicystic, extraosseous/peripheral, desmoplastic, and unicystic [11]. Furthermore, other variants have been reported in the literature, and these include acanthomatous, ghost cell, and vacuolated or clear cell types [47,51,65-67,76]. Odontogenesis is a complex biological process, and this process is directly reflected in the development of odontogenic neoplasms, especially ameloblastomas [15,16]. It is thought that the above-mentioned variants are due to the developmental complex system [53].

Notch Signaling in Ameloblastomas and Ameloblastic Carcinomas

Regarding Notch signaling, the focus of our attention is on examining Notch1 and Jagged1 peptide expression, as well as their genes, in ameloblastomas and ameloblastic carcinomas (Figure 6). The speculation on their possible roles is in cytological differentiation and proliferation of ameloblastomas. In one examined case of ameloblastoma, histopathologically, the main specimens showed follicular nests consisting of islands of odontogenic epithelium within a fibrous stroma. Cells of the peripheral layer of these islands were columnar, with hyper chromatic nuclei, and lined up in a palisade fashion whereas the central cells were stellate reticulum-like. Their cytoplasms were generally vacuolated. Some nests showed central cyst formation. In small parts, the odontogenic epithelium exhibited focal basal palisading. Furthermore, occasionally a large number of nests underwent squamous metaplasia with keratinizing pearl formation. In general, degeneration of the parenchymal cells and cyst formation occurred in these ameloblastoma nests. According to the immunohistochemistry examination results, NICD-positive products were detected in most proliferating odontogenic epithelial nests of ameloblastomas by IHC. The positive reactions existed in the cytoplasm and/or nucleus. Strong reactions were seen in the pre-ameloblast-like cells or some localized cells within the nests. In some ameloblastoma nests, there were no positive reactions to NICD. Jagged1 positive reactions were also observed in the cytoplasms of same cell types in the ameloblastoma nests. Strong reactions existed at the peripheral layers. The pattern of distribution and the intensity of expression of Jagged1 were closely similar to the pattern and intensity seen in NICD. Notch1 gene signaling was localized in the cytoplasm of IHC-positive neoplastic cells. These mRNA positive signals showed variable labeling intensity. Jagged1 mRNA signals were also detected in the cytoplasm of ameloblastoma cells, and the strength pattern was nearly the same as that of Notch1. These mRNA signal expressions were not consistent with those of the transcription factor peptides. Histopathologically, the follicular type of ameloblastoma is the most common, consisting of proliferating odontogenic epithelial islands and nests in the fibrous stromal tissues. Cellular modifications, such as squamous metaplasia, keratin pearl formation, parenchyma cell degeneration and cystic changes, may also occur. Morphogenesis is a complex biological process, and this process directly reflects the development and proliferation of neoplasms. Regarding the proliferation of ameloblastomas, some morphogenesis factors are overly-expressed in ameloblastoma tissues in comparison with tooth germs. According to the analysis of gene expression in ameloblastomas and human fetal tooth germs using a cDNA microarray, there are some results are published. That analysis also included tumor-necrosis-factor-receptor-1 (TNFRSF-1), sonic hedgehog (SHH), Cadherins 12 and 13 (CDH 12 and 13), and transforming growth-factor-1 (TGF-β1), the gene expression profile identified candidate genes that might be involved in the origination of ameloblastoma, as well as several genes previously unidentified in relation to human tooth development. The expression of SHH signaling in ameloblastomas, in comparison with human tooth germs is also detected. The literature concluded that the SHH signaling might play a role in epithelial-mesenchymal interactions and cell proliferation in the growth of ameloblastomas [34].

On the morphogenesis factors Notch1 and Jagged1 in ameloblastoma, the results demonstrate that Notch1 (NICD) and Jagged1 are both detected by IHC, and their expression patterns are very similar [81]. This phenomenon means that Notch signaling is activated in the neoplastic epithelium of ameloblastoma. It is likely that the signaling plays the role of daughter cell fate regulation. Positive Notch1 reactions suggest that proliferation and cytological differentiation are probably occurring in these neoplastic cells. This explains the variation in the strength of these signals and their distribution patterns in the ameloblastoma cell nests. Furthermore, the mRNA of Notch1 and Jagged are also expressed in the ameloblatoma cells, as determined by ISH. These mRNA signal expressions are consistent with those of the transcription factor peptides. The examination of larger case series of ameloblastoma and other odontogenic epithelial neoplasms, including epithelial, mesenchymal, benign and malignant entities such as calcifying epithelial odontogenic tumor, adenomatoid odontogenic tumor, keratocystic odontogenic tumor, ameloblastic fibroma, odontoma, and odontogenic carcinoma, would help to elucidate further the role of these genes in odontogenic tumorgenesis. The results suggest that Notch signaling plays a role in cytological differentiation or acquisition of tissue-specific characteristics in these neoplastic cells of ameloblastomas [66].

Histopathologically, in ameloblastic carcinoma, proliferating polyhedral neoplastic cells show strong cellular atypia, such as mitosis and pleomorphism, especially in peripheral layers of the nests. NICD positive products are observed in most proliferating nests of benign ameloblastoma by IHC, and strong reactions are seen in the cells at the peripheral layer of the nests. In case of ameloblastic carcinomas, positive product are also been detected, and strong reactions uniformly observed. The positive reactions one are comparatively weaker in benign than in malignant tumors. In both benign and malignant cases, the gene (mRNA) expressions have been detected in the cytoplasms of IHC positive cells by ISH. In general, Notch signaling is responsible for cytological regulation of cell fate, morphogenesis and/or development. In examinations conducted by the present authors, IHC and ISH examination results have suggested that Notch signaling plays some role in cytological differentiation or acquisition of tissue specific characteristics in neoplastic cells of tooth enamel organ-derived neoplasms, including benign and malignant neoplasms, ameloblastoma and ameloblastic carcinoma (Figure 6).

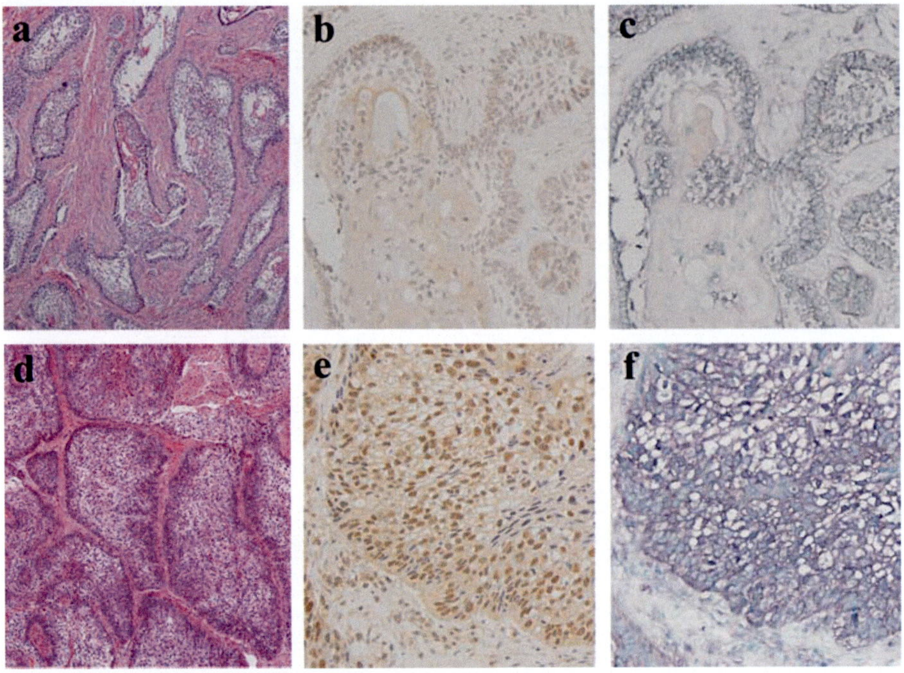

Figure 6. Proliferating follicular nests of ameloblastoma are visible (a). Immunohistochemically Notch peptide is observed in the cells at the peripheral layer (b) and the gene signals are detected in the cytoplasms (c). Polyhedral neoplastic cells are visible in proliferating ameloblastic carcinoma cell nests (d). The Notch positive products are observed uniformly in these cells (e) and the gene expression are also detected in the cytoplasms (f).

Histopathologically, follicular nests proliferate in the fibrous connective tissue in benign neoplasms. In some nests, parenchymal cyst formations or squamous metaplasia are evident. The histopathological features of the malignant neoplasms are as follows: Proliferating polyhedral neoplastic cells show strong cellular atypia, such as mitosis and pleomorphism, especially in peripheral layers of the nests. Notch intra-cellar domain positive products are observed in most proliferating nests of benign neoplasms by immunohistochemistry. Strong reactions are seen in the cells at the peripheral layer of the nests. In malignant ones, positive products are also detected, and strong reactions uniformly observed. The positive reactions in benign neoplasms are comparatively weaker than in malignant ones. In both the benign and malignant cases, the gene expressions are detected in the cytoplasm of immunohistochemistry positive cells by ISH. Our examination results suggest that Notch signaling plays some role in cytological differentiation or acquisition of tissue specific characteristics in neoplastic cells. Furthermore, there would appear to be a relationship between the cytological differentiation in the oral and craniofacial neoplastic cells and the physiological development and differentiation of their originating mother cells and tissues of the oral and craniofacial region.

CELL DIFFERENTIATION IN ODONTOGENIC NEOPLASMS

Amelogenin is a typical enamel matrix protein. The expression pattern of amelogenin genes (AMGX, AMGY) has not yet been identified in ameloblastomas. In surgical materials, amelogenin gene is expressed in all ameloblastoma cells. The mRNA of AMGY expression increases, although that of AMGX is not. This is an interesting feature in physiological normal male tooth development, in which the expression of AMGY is considerably lower than that of AMGX. This finding suggests that epigenetic change of sex chromosomes may have some correlation with tumorigenesis of ameloblastoma [77].

Regarding the collagen subunits of basal membrane components of oral neoplasms, there are some published data [2,69-71]. Regarding ameloblastomas, co expression of type IV collagen α1 and α2 chains appear as thin lines with limited areas of discontinuity along the basal membrane of neoplastic cell nests [44,45]. The expression staining is strong and in a linear continuous manner, in the periphery of the nests of the desmoplastic types. In the neoplasms, α5 and α6 chains are co-localized as continuous linear patterns demonstrating the tumoral nests from the surrounding connective tissue stroma. These collagen subunits also appear as random intracellular staining of the neoplastic cell nests [9,10]. There is no remarkable differentiation of the distribution pattern among the tumor growth patterns and various cellular subtypes within ameloblastomas (Figure 7). Distribution of α subunit of collagen in the basal membrane of ameloblastic fibromas is uniformly demonstrated in its pattern. Subunits of αl/a2, α4 and α5/α6 are distributed as liner continuous patterns that compartmentalize the neoplastic epithelial cell nests, islands and strands in manner, from the surrounding dental papilla-like ectomesenchymal cell proliferation. These α subunits of collagen are randomly expressed in the periphery preameloblast-like and central stellate reticulum-like cells.

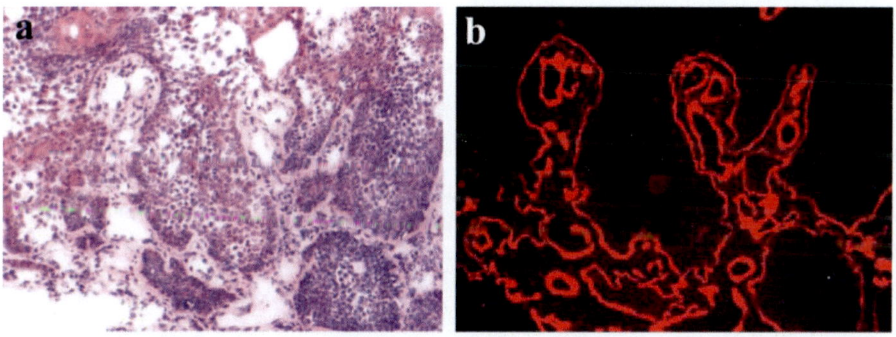

Figure 7. Histopathological feature of examined plexiform ameloblastoma (a) and immunofluoresence localization of type IV collagen α1 chain in the basal membrane zone (b).

In adenomatoid odontogenic tumors, subunits of α1/α2 and α5/a6 are strongly expressed at the area of interface between tumor cells and stromal tissues, especially in the cribriform regions. Faintly to non-positive expression of these collagen molecules is detected in the basal regions of conglomerated masses of solid epithelial whorls/rosettes/nests and duct-like structures. There are intensely positive reactions to the amorphous deposits; however, there is little or no reaction to the mineralized bodies. In malignant neoplasms, at first in ameloblastic fibro-odontosarcomas, α1/α2 and α4 chains demonstrate moderate intensity along with the periphery of the epithelial components, while α5/α6 chains are strongly co-distributed as continuous linear patterns demarcating the benign neoplastic cell nests from the surrounding sarcoma tissues. In the inductive dental hard tissue regions, no reactivity is found. In the malignant neoplasm type of ameloblastic carcinomas, collagen IV α chains demonstrate an irregular and disrupted expression pattern with specific loss of α1/α2 chains. In those regions containing poorly differentiated neoplasm cell nests, there is complete disappearance of α chain subunits. In α5/α6 subunits, there is a discontinuous and fragmented pattern. In primary intraosseous carcinomas, the expression pattern is similar that of ameloblastic carcinomas. From the above mentioned findings for cell differentiation in various benign and malignant odontogenic neoplasms, the basal membrane pattern of neoplastic epithelial cell nests yields three features: (1) The basal membrane of benign and malignant odontogenic neoplasms has distinct a chain subunits of collagen type IV; (2) modifications in the relative abundance of collagen type IV α chains in basal membrane of odontogenic neoplasms probably represent a host protective response, (3) early specific loss of α1/α2 chains proceeds the loss affecting α5/α6 chains during odontogenic neoplasm progression. Therefore, these results suggest that modification and remodeling of basal membrane collagen type IV α chains are dynamic processes crucial for odontogenic neoplastic cell growth and progression [7,45].

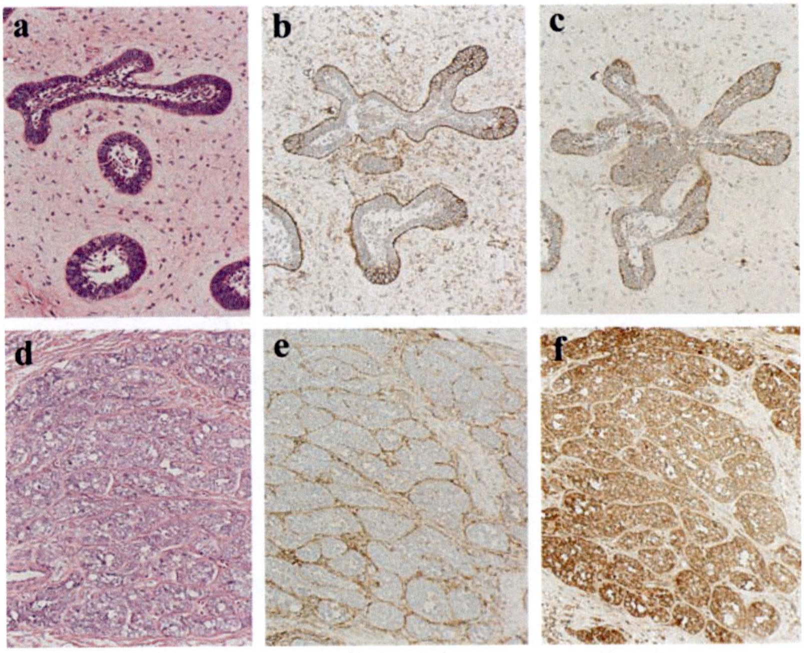

Figure 8. Ameloblastic epithelial islands within scattered dental papilla-like ectomesenchymal tissue (a). HS and heparanase immunohistochemical reactivity accentuates the cellular processes of dental papilla-like cells (b and c). Note the clear confinement and continuity of basal membrane despite the prominent localization of heparanase (b). Histopathologically cancer cell nests of ameloblastic carcinoma with increased nuclear-cytoplasmic ratio and prominent nuclei (d). HS is not detected in the cancer cells but is strongly localized in the stromal tissues adjacent to cancer nests (e). Intense and diffuse heparanase expression is observed in all cancer cells (f) in contrast to the strong staining limited in the basal cells of benign ameloblastoma (c).

Heparan sulphate (HS) and heparanase appearing in the odontogenic neoplasms are interesting molecules for these neoplastic transdifferentiation [13,50]. HS proteoglycans (HSPG) constitute a group of ubiquitous extracellular matrix macromolecules and are composed of a core protein and covalently linked HS sugar chains (Figure 8).

Although, HSPG plays critical functions in cell-to-cell and cell-to-matrix interactions through core proteins, their HS chains confer most of biological functions [3]. The negatively charged HS chains can bind and sequester numerous heparin/HS binding molecules, including growth factors, cytokines and cell adhesion molecules. HS chains also take part in the important cellar events conferred by these tethered molecules and have an influence on various developmental and pathological processes, such as neoplastic transformation, its local invasiveness and transdifferentiation [3,4]

In the examination results on ameloblastomas, HS is clearly evident on the cell surface of peripheral basal cells and also in the intercellular region of some parabasal cells, while it is not present in the central stellate cells of ameloblastoma nests. Heparanase is expressed in peripheral epithelial cells of ameloblastoma nests. The strong expression is localized at the budding region of the strands mainly in the invasive fronts in histopathological specimens. In adenomatoid odontogenic tumors, the strongly localized and limited expression is present on the surfaces of dark cuboidal cells surrounding the whorls and solid tumor cell nests. HS is also evident in luminar surfaces of some duct-like structures. In the lumen and/or duct-like structures, eosinophilic materials are HS positive. The periphery of immature calcified materials is positive to both HS and heparanase; however, completely calcified materials are negative.

In ameloblastic fibromas, as one of typical epithelial and mesenchymal mixed odontogenic neoplasms, HS exists in nearly the same location as in ameloblastomas. HS is evident in the basal membrane, but is more defined. HS is also present in the ectomesenchymal cells, unlike the stromal cells in ameloblastomas. HS and heparanase are diffusely present both in epithelial and mesenchymal tissue of the neoplasms. In some parts, heparanase exists focally in nuclei of mesenchymal cells. In ameloblastic carcinomas as malignant odontogenic neoplasms, neoplastic cells are absent to HS in contrary of benign neoplasms, such as ameloblastomas. Instead, HS clearly and strongly is present in the stromal tissues, especially in the intercellular matrices within the vicinity of neoplastic cell nests. Regarding heparanase activity, positive reactions are intense and diffuse, and occur in intracellular spaces. The above-mentioned findings are compared with the data of the physiological tooth development, in both experimental animals and human materials. Furthermore, when the results are examined using various types of oral squamous cell carcinoma, there are some differences between the types of varies histological and clinical malignancy grades. In summary, the general localization of HS and the heparanase activity in odontogenic neoplasms are temporally regulated in relation to cellular growth and function. Furthermore, heparanase over-expression is reported to promote hair follicle morphogenesis and its growth [82]. Both hair follicular morphogenesis and odontogenesis are governed by similar growth factors and signaling pathways. Heparanase may also have physiological function in tooth development through local modulation and release of HS-bound growth factors. Taken together, the facts suggest that heparanase may have physiological function in tooth development, and the increase in heparanase expression maybe an important initiating factor for odontogenic neoplastic transformation. The stromal HS sugar molecule localization and heparanase over-expression may represent the malignant progression of ameloblastoma to ameloblastic carcinoma [41].

In odontogenic neoplasms, both benign and malignant, the immunohistochemical distribution of tenascin and fibronectin is compared with that in human tooth germs [72,73]. In ameloblastomas, the extracellular matrix components of the stromal tissue of ameloblastomas exhibit considerable variety: dense and loose connective tissues, hyalinization regions, and stromal cystic spaces. In the hyalinised stroma, tenascin and fibronectin exhibit both positive and negative reactions. In cystic spaces, positive reactions of tenascin and fibronectin are seen. In follicular type of ameloblastomas, the basal membrane region reacts irregularly positive to tenascin. The fibronectin reactivity exhibits uniformly and weakly positive in the dense connective tissue of the stromal region in the follicular ameloblastomas. A partial accumulation of tenascin is found in the basal membrane. The tenascin-positive basal membrane shows fuzzy fibrillar materials, whereas the loose or myxomatous tissues of stromal region of follicular type of ameloblatoma exhibit no reaction to fibronectin. Regarding malignant odontogenic neoplasms, the data on ameloblastic carcinomas is as follows: The stromal tissue and basal membrane of ameloblastic carcinomas show an irregular and strong immunohistochemical-positive reaction to tenascin. In the epithelial cell islands of ameloblastic carcinomas, a scattered or granular positive reaction is evident. The connective stromal tissue of ameloblatic carcinomas shows an irregular and strong reaction to fibronectin. According to the localization pattern of tenascin and fibronectin in the varied types of odontogenic neoplasms, such as benign and malignant, the dental follicle of the tooth germ lacks tenascin but has fibronectin. The osteogenic tissues generally contain both tenascin and fibronectin. The ameloblastic fibromas show positive or negative distributions in the stromal tissues, which suggest a differentiation to the papilla of the tooth germ. It is also suggests that the stromal tissue cells of ameloblastic fibroma differentiate to dental follicular tissues. The relative distribution of menisci and fibronectin can be a marker in histological diagnosis of periodontal and osteogenic fibrous tissues. Furthermore, the findings also suggest that fibronectin and tenascin may be used as markers in cell-differentiation of epithelial-mesenchymal interactions during tooth development and in odontogenic neoplasm for trans-differentiation [40].

Cytokeratins with intermediate filaments characteristic of epithelial cells are very stable and range in weight from 40 to 67kD [42]. Regarding the distribution patterns of cytokeratins in some types of ameloblastomas, such as follicular ameloblastomas, the peripheral columnar cells resembling preameloblasts react positively for NSE-K (52.5kD) and 19-K (40kD) in a linear pattern along the basal membrane, but not for the markers of squamous cells, SE-K (56, 56.5, 58 and 68kD). The reaction pattern of NSE-K and 19K shows a frame-like structure in cytoplasm in peripheral columnar cells, while the central stellate reticulum-like cells show an immunohistochemical reaction with all markers of the cytketain SE-K, NSE-K and 19-K throughout the cytoplasm. In plexiform ameloblastomas, both the central spindle and peripheral cuboidal cells demonstrate a positive reaction with SE-K and 19-K, but not with NSE-K. Regarding the oral mucosa and on the developing tooth germ, there are differences for immunohistochemical reactivities of cytokeratins between the fetal oral mucosa and adult gingiva. The cells of both the fetal oral mucosa and adult gingiva demonstrate positive reactions to SE-K, and negative reactions to NSE-K. However, positive reactivity to 19-K is noted only in cells of the fetal mucosa and Merkel cells in adult gingiva. The dental lamina connecing with the basal cells of the oral mucosa have all these cytokeratins, especially 19-K. In the enamel organ, however, the immunoreactivity for SE-K, a maker of squamous differentiation, is different in each cell layer: positive in the outer enamel epithelium; slight positive to negative in the stellate reticulum and stratum intermedium; and negative in the columnar inner enamel epithelium. NSE-K and 19-K are evident in all types of cells that compose the enamel organ proper. The inner columnar enamel epithelium especially expresses a diffuse positive reaction for NSE-K and 19-K throughout the cytoplasms. These features suggest that the characteristics of the cells of the plexiform ameloblastoma are similar to the fetal oral epithelium not odontogenic epithelium. This later feature suggests that the different expression patterns of cytokeratin in ameloblastoma depend on the follicular or plexiform types. As a coincidence of cytokeratin and functional pattern is not noted among the ameloblastomas and tooth germs, these data suggest that the columnar cells of the follicular ameloblastoma have little resemblance to ameloblast-like cells in cytokeratin structure or in cellular functions [42].

Cell Differentiation in Neoplasms of Bone and Cartilage

Osteogenesis is a complex biological process, which includes recruitment of stem cells, proliferation of progenitor cells, differentiation of osteoblasts and production and assembly of bone matrix. The different steps of this process are controlled and regulated by multiple local and systemic factors, such as morphogenesis regulators. Therefore, knowledge of the complex interaction between these factors and their contribution to the development of neoplastic osteogenesis is necessary for understanding the characteristics of the neoplasms [29,30].

Histopathologically, in serious cases of osteosarcomas, spindle-shaped sarcomatous cells proliferating mesenchymal tissue directly produce neoplastic osteoid and/or coarse immature bone tissues, while variable histopathological patterns are seen in specimens of some cases. There are mainly osteoblastic and osteoid and/or immature bone matrices, as well as some spindle-shaped fibroblastic neoplastic cells. Osteoblastic neoplastic cells, located around the numerous small osteoid tissues, are comparatively monotonous, varying in size and in shape, and showing hyperchromatic nuclei and mitosis [27].

Immunohistochemically, osteopontin (OPN) peptide, as control, is expressed in almost all cells of the examined osteosarcoma. The strong expression area of OPN is in the comparatively well-differentiated regions of the osteosarcoma, the osteoblastic area containing osteoid tissues. In contrast, weak reaction products are detected in the monotonous spindle-shaped cell proliferation area. Runx2 peptide expression appears in the cytoplasm of almost all neoplastic cells (Figure 9). The expression pattern showed uniformly in the proliferating cells of almost all cases. At the bone and/or well-differentiated osteoid forming region, the positive reactions of Runx2 are slightly strong compared with other poorly-differentiated regions. Regarding the expression of NICD peptide, the peptide is detected in the cytoplasm of neoplastic cells of the comparatively well-differentiated areas of osteosarcomas, which are osteoblastic and chondroblastic containing osteoid and/or chondroid tissues, and this area is the same as the immunohistochemically strongly-stained area by OPN [49]. No expression of NICD peptide is detected in the fibroblastic and poorly-differentiated area. Delta peptide appearance is nearly the same that of NICD peptide. The positive products of Delta appear in the cytoplasms of osteoblastic and chondroblastic cells, but there are no positive reactions in the poorly-differentiated fibroblastic cell proliferation regions. On the other hand, there is no positive reaction immunohistochemically detected in negative control slides.

In general, it is important to examine the expression or localization of morphogenesis regulators to the neoplastic proliferating conditions, such as benign and malignant tumors. The expression of NICD in an Indonesian male case of osteosarcoma of the maxilla [30] has been reported. As mentioned above, the expression situation of regulation factors of morphogenesis is carried out. Regarding the relationship between these regulation factors and bone tissue, there are some reports in the literature. First, NICD is one of the important regulation factors of morphogenesis. NICD has been reported as a unique and interesting regulator for treatment of osteoporosis. Furthermore, some papers have considered NICD and bone tissue, especially the differentiation of bone forming cells [56,78]. Functional involvement of NICD in osteoblastic cell differentiation has been also reported. However, it is unclear whether Notch1 ligand Delta also induces an identical cellular response in these differentiations. Critical regulation of osteoblastic cell differentiation by Delta-activated Notch1 signaling has also been reported.

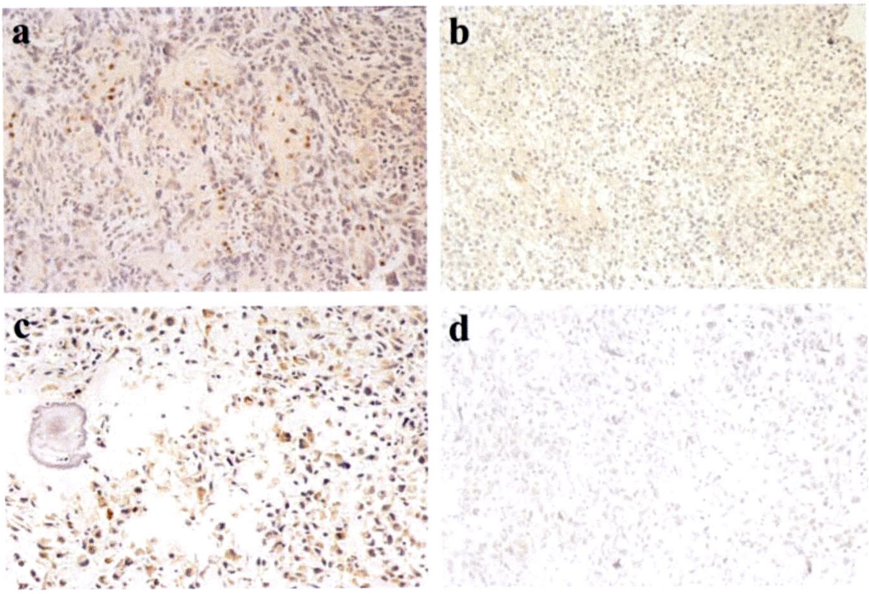

Figure 9. Runx2 positive reaction is visible in the well-differentiated area of osteosarcoma tissue (a) and weakly positive products are detected in the poorly-differentiated area of osteosarcoma tissue (b). NICD is strongly detected in the well-differentiated area of osteosarcoma tissue (c) and no-immunohistochemical reaction of NICD in the poorly-differentiated area of osteosarcoma tissue (d).

The molecular basis for inverse relationship between differentiation and oncogenesis is unknown. However, regarding Runx2, a master regulator of osteoblast differentiation belonging to that runt family of tumor suppressor genes, is consistently disrupted in osteosarcomas. Reports in the literature [75] have described that physiological coupling of osteoblast differentiation to cell cycle withdrawal is mediated through Runx2, and the process are disrupted in osteosarcoma. Furthermore, it has been reported that Runx2 is expressed constitutively in all pathology specimens of human osteosarcoma [1], and expression of Runx2 appeared in the cytoplasm of almost all neoplastic cells of examined cases, and the expression pattern showed uniformly in the proliferating cells of almost all cases. At the bone and/or osteoid forming region, as well at the differentiated area, the positive reactions of Runx2 are slightly strong in comparison with other poorly-differentiated regions. These immunohistochemical results are consistent with the above mentioned discussion. In the present investigation, the NICD peptide is expressed in the area of comparatively well-differentiated areas of osteosarcoma, as well as osteoblastic and chondroblastic areas containing osteoid and/or chondroid tissues. The results are also similar to those of a previously published Indonesian case [30]. With OPN as control peptide in this examination, expression is also detected in almost all cells; the strength pattern of OPN expression is similar to that of NICD. Therefore, Notch peptide maybe closely related to cytological differentiation or acquisition of tissue specific characteristics in neoplastic cells in osteosarcomas.

In summary, the expression of Runx2, NICD, Delta and OPN are examined in neoplastic cells in cases of osteosarcoma. The immunohistocemical expression of Runx2 appears in the cytoplasm of almost all neoplastic cells of examined cases. However, NICD appears in the localized comparatively well-differentiated areas. No expression of NICD peptide is detected in the poorly differentiated area. Delta is shown nearly the same as NICD. Expression of OPN as control appears in almost all cells and the strength of expression is shown in the area of comparatively well-differentiated tissues. Therefore, these results suggest that Runx2, Notch1, and Delta are closely related to cytological differentiation or acquisition of tissue specific characteristics in these neoplastic cells of osteosarcomas.

Cell Differentiation in BMP-Induced Heterotopic Osteogenesis

In general, it has been stated that BMPs, when implanted in heterotopic sites, induces undifferentiated mesenchymal cells to become chondrocytes in the first stage [31,32]. These cells are replaced by bone in a manner similar to that of physiological endochondral (indirect) ossification mode. However, it has been suggested that BMP-induced bone occurs through endochondral-like ossification patterns which differ from those in the physiological normal endochondral ossification process. On the other hand, intramembranous (direct) ossification has also been observed in some cases. There are two types of ossification modes in BMP-induced heterotopic osteogenesis: intramembranous (direct) and endochondral (indirect). However, the nature of osteogenesis has not been clearly detailed [22,23].

Although there is no direct evidence, we believe that the cells inviolved in "chondroid bone", experimentally induced by BMP in mice, temporarily express cartilage phenotypes. They then change directly into bone-forming cells which survive in the "chondroid bone" until the tissue is resorbed and remodeled by true bone tissue. Regarding the expression of TGF-β, in the physiological endochondroid ossification mode, the TGF-β peptide appears in the final-differentiated hypertrophic chondrocytes, and at this stage resorption and replacement by bone occur. We found that the peptide was expressed only in some chondrocites of the earl phase of BMP-induced heteotopic "transchondroid bone formation" in mice. Furthermore, the mRNA was expressed in the same chondrocytes in the early phase of the osteogenesis [23].

Therefore, it is believed that a third ossification mode "transchondroid bone formation" is displayed in BMP-induced heterotopic osteogenesis [24,25]. Chondroid bone, a tissue that has characteristics of bone and cartilage, is formed mainly in BMP-induced heterotopic osteogenesis.

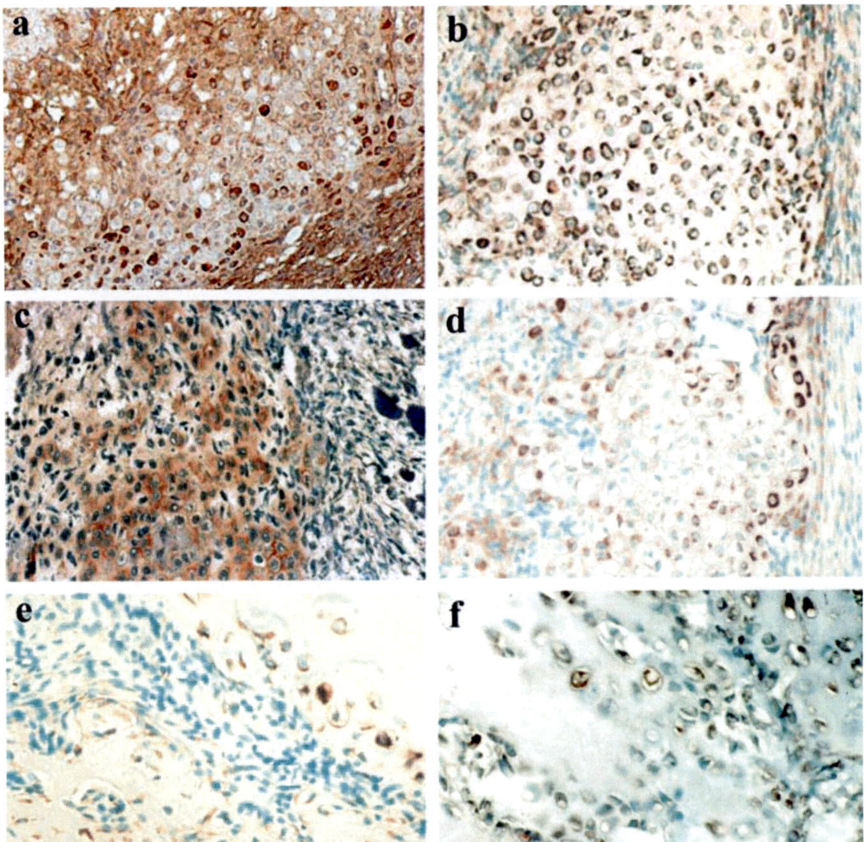

Figure 10. Immunohistochemical localization of type I collagen is observed in the 7-day proliferating cells (a) and the mRNA expression is visible in the cytoplasms (b). Positive immunostainings of type II collagen are distributed in the cytoplasms and matrices (c) and the gene expression is detected the cytoplasms (d). Osteocalcin mRNA signals are visible in the cytoplasms of proliferating cells (e). TGF-β mRNA expression is apparent in some proliferating cells (f).

The experimental results of a BMP-induced model using ddY mice are as follows (Figure 10): Histopathologically in 3-day specimens, spindle-shaped mesenchymal cells proliferated site, and the matrices are stained slightly by HE in 5- and 7-day specimens. Evidence of the proliferation of undifferentiated cells, having cytoplasm and resembling poorly-differentiated chondrocytes, can be seen at the periphery. Within 10 days, perichondral ossification has occurred, and the peripheral matrix of the cluster of cartilage has changed to chondroid bone, connected to perichondral ossification sites. Histochemically, in Mallory's azan-stained specimens, perichondral ossification sites are stained deep blue, although areas stained pale blue are visible in chondral tissue in 10-day specimens. In 10-day specimens stained with toluidine blue, ortho-metachromasia reaction occurs in the matrix, but the matrix of perichondral ossification sites does not occur [26]. In 2- and 3-week specimens, some chondroid tissues displaying ortho-metachromasia are evident in trabecular bone. Immunohistochemically, in 5-day specimens, chondro-osseous formation occurs and type-II collagen is positive in and around chondrocyte-like cells in the chondro-osseous matrix. In 7-day specimens, type II collagen is positive in the proliferating spindle cells. Type I collagen is a typical bone matrix protein, and the products of the staining reaction are seen in chondral cells. Type II collagen is also evident in the same cells. These specimens are immunostained for osteocalcin (OCN). Positive staining is observed in 7-day specimens, in particular, and OCN-positive proliferating cartilage cells are clearly visible. In 2- and 3-week specimens of lamellar bone with bone marrow, collagen type II protein disappears from the fibrillar bone matirix. OCN-positive reactions are detected in chondroid cells in 2-week specimens. Regarding these gene expressions in 3-day specimens, type II collagen mRNA are noted in the implanted sites, while OCN mRNA and type II collagen mRNA are undetectable. In 5-day specimens, both type II collagen mRNA and OCN mRNA are detected in some chondrocyte-like cells. Furthermore, in 7-day specimens, type I and II collagen mRNA and OCN mRNA are located in osteoblasts and young osteocytes in osteoid tissues. OCN mRNA is also detected in chondrocyte-like cells, osteoblasts and young osteocytes in osteoid tissues. In 2- and 3-week specimens of lamellar bone with bone marrow formed in the implanted sites, both OCN mRNA and type II collagen mRNA are detected within the newly formed bone.

Recently, a third ossification mode has been proposed: "transchondroid bone formation". Chondroid bone is formed directly by chondrocyte-like cells in the ossification site [24,25]. This form of bone has attracted limited attention, since it was the first reported several years ago. Some hypertrophic chondrocytes are believed to undergo further differentiation into osteoblast-like cells, participating in initial bone formation [79]. Furthermore, chondrocyte-like and osteocytes-like cells are reported to co-exist in chondroid bone, with no clearly distinguishable boundary. According to the examination results of BMP-induced heterotopic bone tissue, chondrocyte-like cells demonstrate the two-phase function of the chondrocytes and osteocytes. These data also suggest that this BMP induced cartilage-like tissue should be classified as chondroid bone rather than as normal cartilage. As mentioned above, these examination data suggest that the cells involved in chondroid bone express cartilage phenotypes temporarily, then become bone-forming cells which survive in chondroid bone until tissue is resorped and remodeled by true bone tissue, although there is no direct evidence of this [38]. Using immunohistochemical and in situ hybridization examinations, there are some data that transforming growth factor-β peptide may be involved in the differentiation of chondrocytes into bone-forming cells [23,68]. BMP-induced bone tissue is formed by the third ossification mode, "transchondroid bone formation".

Concerning the "chondroid bone or chondroid bone forming cells", they also appear in the neoplastic lesions of bone. In osteochondromas, cartilage-capped bone tissue projections are typical histopathological features. The surface of the masses is covered with a cartilage tissue showing positive immunohistochemical reaction for type II collagen, and the deep region is composed of spongy bone, showing a positive immunohistochemical reaction for type II collagen and osteocalcin. Between the cartilage and spongy bone, which is a metaphysic-like region, a chondroidal pattern appeared in the matrix of hypertrophy cartilage. In these regions, both type I and II collagens and osteosarcoma are immunohistochemically detected. Furthermore, type I collagen mRNA in neoplastic chondrocytes is detected. Therefore, the cells involved in "chondroid bone" appear in osteochondromas and then change directly into bone-forming cells that survive in the "chondroid bone" until the tissue is resorbed and remodeled into true bone tissue. These features suggest that bone formation in osteochondromas, at least in some regions, occurs through transchondroid bone formation.

Conclusion

In the oral and craniofacial region, development is a complex and fascinating set of processes which require a sequential integration of numerous biological steps, and the differentiation in cells of neoplasms is also a complex and fascinating compound processes. In this region, there are many different types of neoplasms. It is known that proliferation, development and cytological-differentiation of the neoplastic cells reflect the normal physiological development of the outbreak mother cells and/or tissues. Therefore, some cell-differentiation, development and proliferation factors may also play some roles in the neoplastic cells, and therefore, their behavior is closely related in cytological differentiation and clinical behavior and/or grade.

References

[1] Andela VB, Siddiqui F, Groman A and Rosier RN (2005) An immunohistochemical analysis to evaluate an inverse correlation between Runx2/Cbfa1 and NF kappa B in human osteosarcoma. *J Clin Pathol 58*: 328-330.

[2] Barsky SH, Siegal GP, Jannotta F and Liotta LA (1983) Loss of basement membrane components by invasive tumors but not by their benign counterparts. *Lab Invest 49*: 140-147.

[3] Bernfield M, Gotte M, Park PW, Reizes O, Fitzgerald ML, Lincecum J and Zako M (1999) Functions of cell surface heparin sulfate proteoglycans. *Ann Rev Biochem 68*: 729-777.

[4] Blackhall FH, Merry CL, Davies EJ and Javson GC (2001) Heparan sulfate proteoglycans and cancer. *Br J Cancer 85*: 1094-1098.

[5] Blaumueller CM, Qi H, Zagouras P, and Artavanis-Tsakonas S (1997) Intracellular cleavage of Notch leads to a heterodimeric receptor on the plasma membrane. *Cell 90*: 281-291.

[6] Butler WT and Ritchie H (1995) The nature and functional significance of dentin extracellular matrix proteins. *Int J Dev Biol 39*: 169-179.

[7] Dehan P, Waltregny D, Baschin A, Noel A, Castronovo V, Tryggvason K, Leval DJ and Foidart JM (1997) Loss of type IV collagen alpha 5 and alpha 6 chains in human invasive prostate carcinomas. *Am J Pathol 151*: 1097-1104.

[8] D'errico JA, Macneil RL, Takata T, Berry J, Strayorn C and Somerman MJ (1997) Expression of bone associated makers by tooth root lining cells, in situ and in vitro. *Bone 20*: 117-126.

[9] Fujii H, Nagatsuka H, Lee YJ, Shinnou T, Tamamura R, Xiao J, Naitou I, Sado G, Nakagawa T and Nagai N (2004) Differential expression of type IV collagen alpha 1 to 6 chains in basement membrane of human tooth germ and odontogenic tumors. *J Hard Tissue Biol 13*: 103-109.

[10] Fujii H, Nagatsuka H, Lee YJ, Shinnou T, Tamamura R, Xiao J, Naitou I, Sado G, Nakagawa T and Nagai N (2004) Differential expression of type IV collagen alpha 1 to alpha 6 chains in basement membrane of malignant odontogenic tumors. *J Hard Tissue Biol 13*: 111-115.

[11] Gardner DG, Heikinheimo K, Shear M, Philipsen HP and Coleman H (2005) Ameloblastomas. In: Banes L, Eveson JW, Reichart P, Sidransky D ed. *World Health Organization Classification of Tumours Pathology and Genetics of Head and Neck Tumours*. IARC Press, Lyon, 296-300.

[12] Hamada T, Suda N and Takayuki K (1999) Immunohistochemical localization of fibroblast growth factor receptors in the rat mandibular condylar cartilage and tibial cartilage. *J Bone Miner Metab 17*: 274-282.

[13] Han PP, Nagatsuka H, Siar CH, Tsujigiwa H, Gunduz M, Tamamura R, Katase N, Nakajima M, Naomoto Y and Nagai N (2006) Immunodetection of heparin sulphate and heparanase molecules in benign and malignant odontogenic tumors. *Oral Med Pathol 11*: 49-54.

[14] Harada H, Mitsuyasu T and Toyono T (2002) Epithelial stem cells in teeth. *Odontol 90*: 1-6.

[15] Heikinheimo K, Jee KJ, Niini T, Aalto Y, Happonen RP, Leivo I and Knuutila S (2002) Gene expression profiling of ameloblastoma and human tooth germ by means of a cDNA microarray. *J Dent Res 81*: 525-30.

[16] Heikinheimo K, Morgan PR, Happonen RP, Stenman G and Virtanen I (1991) Distribution of extracellular matrix protein in odontogenic tumors and developing teeth. *Virchows Arch 61*: 101-109.

[17] Heikinheimo K and Salo T (1995) Expression of basement membrane type IV collagen and type IV collagenase (MMP-2 and MMP-9) in human fetal teeth. *J Dent Res 74*: 1226-1234.

[18] Hina M, Inoue M, Nagatsuka H, Takagi T and Nagai N (1993) Immunohistochemical demonstration of amelogenin, type IV collagen and fibronectin in human and rat tooth germs. *J Jpn Stomatol 42*: 659-664.

[19] Hudson BG, Reeders ST and Tryggvason K (1993) Type IV collagen: structure, gene organization, and role in human diseases. Molecular basis of Goodpasture and Alport syndromes and diffuse leiomyomatosis. *J Biol Chem 15*: 26033-26036.

[20] Ishida O, Inomata K and Nagai N (1992) Cytokeratin filaments in odontogenic epithelium of human fetus. *J Jpn Stomatol 41*: 19-29.

[21] Ishii M, Suda N, Toshimoto T, Shoichi S and Kuroda T (1998) Immunohistochemical findings type I and type II collagen in prenatal mouse mandibular condylar cartilage compared with the tibial anlage. *Arch Oral Biol 43*: 545-550

[22] Kawakami T (2001) Immunohistochemistry of BMP-induced hetorotopic osteogensis. *J Hard Tissue Biol 10*: 73-76.

[23] Kawakami T, Hiraoka BY, Kawai T, Takei N, Hasegawa H and Eda S (1999) Expression of transforming growth factor-beta peptide and its mRNA in chondrocytes in the early phase of BMP-induced heterotopic 'transchondroid bone formation'. *Med Sci Res 27*: 419-421.

[24] Kawakami T, Kawai T, Kimura A, Hasegawa H, Tsujigiwa H, Gunduz M, Nagatsuka H and Nagai N (2001) Characteristics of bone morphogenetic protein-induced chondroid bone: histochemical, immunohistochemical and in situ hybridization examinations. *J Int Med Res 29*: 480-487.

[25] Kawakami T, Kawai T, Kimura A, Hasegawa H, Yoshikawa Y and Eda S (1998) Transchondroid bone formation displayed in BMP-induced hetrotopic osteogenesis. *J Hard Tissue Biol 7*: 21-26.

[26] Kawakami T, Kawai T, Takei N, Kise T, Eda S and Urist MR (1997) Evaluation of heterotopic bone formation induced by squalane and bone morphogenetic protein composite. *Clin Orthop 337*: 261-266.

[27] Kawakami T, Kimura A, Yamada M, Matsuura S, Horio T, Hasegawa H and Kanda H (2002) Localization of matrix proteins of hard tissue in osteosarcomas. *Eur J Med Res 7*: 335-339.

[28] Kawakami T, Shimizu M and Shimizu T (2005) Immunohistochemical characteristics of developing mandibular angle in fetal mice. *Eur J Med Res 10*: 547-548.

[29] Kawakami T. Shimizu T, Kimura A, Hasegawa H, Siar CH, Ng KH, Nagatsuka H, Nagai N and Kanda H (2005) Immunohistochemical examination of cytological differentiation in osteosarcomas. *Eur J Med Res. 10*: 475-479.

[30] Kawakami T, Siar CH, Ng KH, Shimizu T, Okafuji N, Kurihara S, Hasegawa H, Tsujigiwa H, Nagatsuka H and Nagai N (2004) Expression of Notch in a case of osteosarcoma of the maxilla. *Eur J Med Res 9*: 533-535.

[31] Kawakami T, Uji H, Antoh M, Hasegawa, Kise T and Eda S (1993) Squalane as a possible carrier of bone morphogenetic protein. *Boimateials 14*: 575-577.

[32] Kimura A, Kawakami T, Matsuura S, Hasegawa H, Kanda H, Tsujigiwa H, Nagatsuka N and Nagai N (2003) Gene expression of type I collagen in neoplastic chondrocytes. *Eur J Med Res 8*: 165-167.

[33] Kuboki T, Kanyama M, Nakanishi T, Akiyama K, Nawachi K, Yatani H, Yamashita K, Takano-Yamamoto T and Takigawa M (2003) Cbfa1/Runx2 gene expression in articular chondrocytes of the mice temporomandibular and knee joints in vivo. *Arch Oral Biol. 48*(7): 519-525

[34] Kumamoto H, Ohki K and Ooya K (2004) Expression of sonic hedgehog (SHH) signaling molecules in ameloblastomas. *J Oral Pathol Med 33*: 185-190.

[35] Mitsiadis TA, Fried K and Goridis C (1995) Reactivation of Delta-Notch signaling after injury: complementary expression pattern of ligand and receptor in dental pulp. *Exp Cell Res 246*: 312-318.

[36] Mitsiadis TA, Henrique D, Thesleff I and Lendahl U (1997) Mouse Serrate-1 (Jagged-1): expression in the developing tooth is regulated by epithelial-mesenchymal interactions and fibroblast growth factor-4. *Development 124*: 1473-1483.

[37] Mizoguchi I, Nakamura M, Takahashi I, Kagayama M and Mitani H (1990) An immunohistochemical study of localization of type I and type II collagens in mandibular condylar cartilage compared with tibial growth plate. *Histochemistry 93* (6): 593-599

[38] Nagai N, Nagatsuka H, Murata M, Inoue M, Akagi T, Qin C-L, Nakano K, Ishiwari Y, Konouchi H, Tsujigiwa H, Chigono Y and Takagi T (1995) Gene expression of bone matrix protein mRNA during BMP induced chondrogenesis and osteogenesis by in situ hybridization. *J Hard Tissue Biol 4*: 15-23.

[39] Nagai N, Nakano K, Sado Y, Naito I, Gunduz M, Tsujigiwa H, Nagatsuka H, Ninomiya Y and Siar CH (2001) Localization of type IV collagen a1 to a6 chains in basement membrane during mouse molar germ development. *Int J Dev Biol 45*: 827-831.

[40] Nagai N, Yamachika E, Nishijima K, Inoue M, Shin HI, Suh MS, Nagatsuka H (1994) Immunohistochemical demonstration of tenascin and fibronectin in odontogenic tumors and human fetal tooth germs. *Oral Oncol, Eur J Cancer 30B*: 191-195.

[41] Nagatsuka H, Han PP, Tsujigiwa H, Siar CH, Gunduz M, Sugahara T, Sasaki A, Nakajima M, Naomoto Y and Nagai N (2005) Heparanase gene and protein expression in ameloblastoma: possible role in local invasion of tumor cells. *Oral Oncol 41*: 542-548.

[42] Nagatsuka H, Shin H-I, Park H-K, Ishiwari Y, Kuroda K, Nosaka Y, Song H, Qin C-L, Zhang S-Q, Nakano K, Chigono Y, Tsujigiwa H, Takagi T and Nagai N (1995) Immunohistochemical study of cytokeratin patterns in follicular and plexiform ameloblastoma. *J Hard Tissue Biol 4*: 50-56.

[43] Nagatsuka H, Siar CH, Kitamura Y, Tsujigiwa H, Lee Y-J, Gunduz M, Tamamura R, Komori T, Lefevre M and Nagai N (2004) Gene expression of matrix proteins in Cbfa1-knockout mice. *J Hard Tissue Biol 13*: 37-46.

[44] Nagatsuka H, Siar CH, Nakano K, Tsujigiwa H, Gunduz M, Choufuku H, Lee Y-J, Naito I, Sado Y and Nagai N (2002) Differential expression of collagen IV α1 to α6 chains in basement membranes of benign and malignant odontogenic tumors. *Virch Arch 441*: 392-399.

[45] Nakano K, Siar CH, Nagai N, Naito I, Sado Y, Nagatsuka H, Hor C, Kuroda K, Tsujigiwa H and Gunduz M (2002) Distribution of basement membrane type IV collagen a chains in ameloblastoma: an immunofluoressence study. *J Oral Pathol Med 31*: 494-499.

[46] Nakano S, Iyama K, Ogawa M, Yoshioka H, Sado Y, Oohashi T and Ninomiya Y (1999) Differential tissular expression and localization of type IV collagen alpha1 (IV),

alpha2 (IV), alpha5 (IV), and alpha6 (IV) chains and their mRNA in normal breast and in benign and malignant breast tumors. *Lab Invest 79*: 281-292.

[47] Ng KH and Siar CH (1990) Peripheral ameloblatoma with clear cell differentiation. *Oral Surg 70*: 210-213.

[48] Ninomiya Y, Kagawa M, Iyama K, Naito I, Kishiro Y, Seyer JM, Sugimoto M, Oohashi T and Sado Y (1995) Differential expression of two basement membrane collagen genes, COL4A6 and COL4A5, demonstrated by immunofluorescence staining using peptide-specific monoclonal antibodies. *J Cell Biol 130*: 1219-1229.

[49] Nobta M, Tsukazaki T, Shibata Y, Xin C, Moriishi T, Sakano S, Shindo H and Yamaguchi A (2005) Critical regulation of bone morphogenetic protein-induced osteoblastic differentiation by Delta1/Jagged1-activated Notch1 signaling. *J Biol Chem 280*: 15842-15848

[50] Parish CR, Freeman C and Hulett MD (2001) Heparanase: a key enzyme involved in cell invasion. *Biocem Biophys Acta 1471*: M99-108.

[51] Philipsen HP, Ormiston IW and Reichart PA (1992) The desmo and osteoplastic ameloblastoma. Histologic valiant or clinicopathoilogic entity? Case reports. *Int J Oral Maxillofac Surg 21*: 352-357.

[52] Rabie ABM and Hagg U (2002) Factors regulating mandibular condylar growth. *Am J Orthod Dentofacial Orthop 122*: 401-409.

[53] Reichart PA, Philipsen HP and Sonner S (1995) Ameloblastoma: biological profile of 3677 cases. *Eur J Cancer B Oral Oncol 31B*: 86-99.

[54] Sado Y, Kagawa M, Kishiro Y, Sugihara K, Naito I, Seyer JM, Sugimoto M and Oohashi T (1995) Establishment by the rat lymphnode method of epitope-defiened monoclonal antibodies recognizing the six different a chains of human type IV collagen. *Histochem Cell Biol 104*: 267-275.

[55] Sado Y, Kagawa M, Naito I, Ueki Y, Seki T, Momota R, Oohashi T and Ninomiya Y (1998) Organization and expression of basement membrane collagen IV genes and their roles in human disordcrs. *J Biochem 123*: 767-776.

[56] Schnabel M, Fichtel I, Gotzen L and Schlegel J (2002) Differential expression of Notch genes in human osteoblastic cells. *Int J Molecul Med 9*: 229-232

[57] Shibata S, Fukuda K, Suzuki S, Ogawa T and Yamashita Y. (2002) In situ hybridization and immunohistochemistry of bone sialoprotein and secreted phosphoprotein 1 (osteopontin) compared with limb bud cartilage. *J Anat 200*: 309-320

[58] Shibata S, Suda N, Suzuki S, Fukuoka H and Yamasita Y (2006) An in situ hybridization study of Runx2, Osterix, and Sox9 at the onset of condylar cartilage formation in fetal mouse mandible. *J Anat 208*: 169-177

[59] Shibata S, Suda N, Yoda S, Fukuoka H, Ohyama K, Yamashita Y, Komori T (2004) Runx2-deficient mice lack mandibular condylar cartilage. *Anat Embryol 208*(4): 273-80

[60] Shibata S, Suzuki S, Tengan T, Ishii M and Kuroda T (1996) A histological study of the developing condylar cartilage of the fetal mouse mandible using coronal sections. *Arch Oral Biol 41*(1): 47-54

[61] Shimizu M (2005) Histological and immunohistochemical observations of developing mandibular angle in mice. *J Hard Tissue Biol 14*: 346-350.

[62] Shimizu T (2005) Histological characteristics of mandible in fetal mice. *J Hard Tissue Biol 14*: 67-68.

[63] Shimizu T (2006) Participation of Runx2 in mandibular condylar cartilage development. *Eur J Med Res 11*: 455-461.

[64] Shimizu T, Tsujigiwa H, Nagatsuka H, Okafuji N, Kurihara S, Nagai N and Kawakami T (2005) Expression of Notch1 and Math1 in mandibular condyle cartilage in neonatal mice. *Angl Orthodont 75*: 993-995.

[65] Siar CH and Ng KH (1991) Calcifying and keratinizing ameloblastoma of the maxilla. *J Laryngol Otol 105*: 971-972.

[66] Siar CH, Ng KH, Ariff Z, Muraki E, Shimizu T, Tsujigiwa H, Nagatsuka H, Nagai N and Kawakami T (2006) A case report of ameloblastoma of the mandible with examination of Notch signaling. *Oral Med Pathol 11*: 35-39.

[67] Slater LJ (1999) Odontogenic sarcomas and carcinosarcomas. *Semin Diagn Pathol 16*: 325-332.

[68] Takei N, Kawakami T, Kawai T, Yoshikawa Y and Eda (1997) Expression of TGF-β in the course of BMP-induced heterotopic osteogenesis. *J Hard Tissue Biol 6*: 64-69.

[69] Tamamura R, Nagatsuka H, Lee Y-J, Xiao J, Naito I, Sado Y, Kawakami T and Nagai N (2004) Distribution of collagen type IV α1-6 chains and MMP in carcinogenesis in oral squamous cell carcinoma. *J Hard Tissue Biol 13*: 117-123.

[70] Tamamura R, Nagatsuka H, Lee Y-J, Xiao J, Naito I, Sado Y, Kawakami T and Nagai N (2004) Distribution of collagen type IV α1-6 chains and MMP in oral squamous cell carcinoma. *J Hard Tissue Biol 13*: 1125-130.

[71] Tanaka K, Iyama K, Kitaoka M, Ninomiya Y, Oohashi T, Sado Y and Ono T (1997) Differential expression of alpha 1 (IV), alpha 2 (IV), alpha 5 (IV) and alpha 6 (IV) collagen chains in the basement membrane of basal cell carcinoma. *Histochem J 29*: 563-570.

[72] Tezuka K, Yasuda M, Watanabe N, Morimura N, Kuroda K, Miyatani S and Hozumi N (2002) Stimulation of osteoblastic cell differentiation by Notch. *J Bone Miner Res 17*: 231-239.

[73] Thesleff I, Barrach HJ, Foidart JM, Vaheri A, Pratt RM and Martin GR (1981) Changes in the distribution of type IV collagen, laminin, proteoglycan, and fibronectin during mouse tooth development. *Dev Biol 81*: 182-192.

[74] Thesleff I and Ekblom P (1984) Distribution of keratin and laminin in ameloblastoma. Comparison with developing tooth and epidermoid carcinoma. *J Oral Pathol 13*: 85-96.

[75] Thomas DM, Johnson SA, Sims NA, Trivett MK, Slavin JL, Rubin BP, Waring P, MacArthur GA, Walkley CR, Holloway AJ, Diyagama D, Grim JE, Clurman BE, Bowtell DD, Lee JS, Gutierrez GM, Piscopo DM, Carty SA and Hinds PW (2004) Terminal osteoblast differentiation, mediated by Runx2 and p27LIP1, is disrupted in osteosarcoma. *J Cell Biol 167*: 925-934

[76] Tomich CE (1999) Benign mixed odontogenic tumors. *Semin Diagn Pathol 16*: 308-316.

[77] Tsujigiwa H, Nagatsuka H, Han PP, Gunduz M, Siar CH, Oida S and Nagai N (2005) Analysis of amelogenin gene (AMGX, AMGY) expression in ameloblastoma. *Oral Oncol 41*: 843-850.

[78] Watanabe N, Tezuka Y, Matsuno K, Miyatani S, Morimura N, Yasuda M, Fujimaki R, Kuroda K, Hiraki Y, Hozumi N, Tezuka K (2003) Suppression of differentiation and proliferation of early chondrogenic cells by Notch. *J Bone Miner Metab 21*: 344-352

[79] Yasui N, Sato M, Ochi T, Kimura T, Kawahara H, Kitamura M and Nomura S (1997) Tree modes of ossification during destruction osteogenesis in the rat. *J Bone Joint Surg 79-B*: 824-830.

[80] Yurchenco PD and O'Rear JJ (1994) Basal lamina assembly. *Curr Opin Cell Biol 6*: 674-681.

[81] Zagouras P, Stifani S, Blaumueller CM, Carcangiu ML and Artavanis-Tsakonas S (1995) Alterations in Notch signaling in neoplastic lesions of the human cervix. *Proc Natl Acad Sci USA 92*: 6414-6418.

[82] Zcharia E, Metzger S, Chajek-Shaul T, Aingorn H, Elkin M, Friedmann Y, Weinstein T, Li JP, Lindahl U and Vlodavsky I (2004) Ttansgenic expression of mammalian heparanase uncovers physiological functions of heparansulfate in tissue morphogenesis, vascularization, and feeding behavior. *FASEB J 18*: 252-263.

In: Cell Differentiation Research Developments
Editor: L. B. Ivanova, pp. 31-59
ISBN: 978-1-60021-939-9

Chapter II

DOES NEURAL PHENOTYPIC PLASTICITY FROM NON-NEURAL CELLS REALLY EXIST?

***S. Wislet-Gendebien*[1, 2,*]*, F. Wautier*[1]*, E. Laudet*[1] *and B. Rogister*[1, 3]**
[1]Center for Cellular and Molecular Neuroscience, University of Liège, B-4020 Liège, Belgium;
[2]Centre for Research in Neurodegenerative Diseases, University of Toronto, Canada;
[3]Department of Neurologie, CHU, Université de Liège, Belgium.

ABSTRACT

Cellular therapies are promising approaches in the treatment of several neurological diseases such as Parkinson's disease [Isacson et al., 2001] or Huntington's disease [Dunnett et al., 2000], but also for spinal cord injury [Hall et al., 2001]. One main problem concerns the origin and nature of the cells to be used for such procedures. In this context, recent studies suggest that somatic stem cells (stem cells from foetal or adult tissues) might be able to exhibit more plasticity than previously thought as they seem able to differentiate into many cell types, including cell types which are not encountered in their tissue origin. This last property, named phenotypic plasticity of somatic stem cells, is thus the capacity for a stem cell to develop in several phenotypes depending on their environment. Several recent reports suggest that bone marrow mesenchymal stem cells (MSC) could be a source of somatic stem cells suitable for cell replacement strategies in the treatment of central nervous system (CNS) disorders. MSC can differentiate into many types of mesenchymal cells, i.e. osteocytes, chondrocytes and adipocytes, but can also differentiate into non-mesenchymal cell, i.e. neural cells in appropriate *in vivo* and *in vitro* experimental conditions [Kopen et al., 1999; Brazelton et al, 2000; Mezey et al, 2000; Wislet-Gendebien et al., 2003, 2005]. Some works have attributed the neural phenotypic plasticity to "transdifferentiation" [Krause et al., 2001; Orlic et al., 2001; Priller et al., 2001; Wislet-Gendebien et al., 2005], while some other

* Correspondence concerning this article should be addressed to: S. Wislet-Gendebien, Telephone: +1416 978 0773; Fax: +1416 978 1878; e-mail: sabine.wislet@utoronto.ca.

works suggested that this neural plasticity could be explained by cell fusion [Terada et al., 2002; Ying et al., 2002; Vassilopoulos et al., 2003; Alvarez-Dolado et al., 2003]. These observations could suggest that mesenchymal cells are heterogeneous and there are two cell populations able to adopt a neural phenotype: one which is able to fuse with already-present neurons and a second one which is really able to differentiate in neurons. In the first part of this chapter, we will review the studies realized on the potential neural phenotypic plasticity of the mesenchymal stem cells. The second part of this chapter will focus on recent studies demonstrating that stem cells isolated from adipose, skin and umbilical cord cells have the ability to differentiate into neural cells [Nagase et al., 2007; McKenzie et al., 2006; Fallahi-Sichani et al., 2006]. This ability could be attributed to the presence of neural crest stem cells in those tissues [Fernandes et al., 2007; Crane and Trainor, 2006]. Consequently, we will address the question of the potential presence of neural crest stem cells in bone marrow.

INTRODUCTION

Embryologists and developmental biologists have introduced the stem cell concept several years ago. A stem cell, by definition, is an undifferentiated cell that can produce daughter cells that can either remain a stem cell (a process called self-renewal) or commit to a pathway leading to differentiation. The first stem cell (and probably the more symbolic) is the fertilized egg which results from the fusion of two haploid germinal stem cells. Once fertilized, the egg undergoes a series of divisions, yielding two, then four, then eight identical cells. These cells are *totipotent* (Figure 1), meaning that each one, if isolated and allowed to develop, can form a new viable embryo. This is how identical (homozygous) twins result from a single fertilized egg. Beside this particular type of stem cells, two other types of stem cells can be distinguished according to their origin and their potential of differentiation: embryonic stem cells (ES) and somatic stem cells (SSC) [Stanford et al., 1998]. After the eight-cell stage (in human, two or three days after fertilization), the cells continue to divide, but lose the ability to form a new embryo because trophoblastic differentiation is no more observed at this stage of development. The embryo takes the form of a hollow sphere, known as a blastocyst, containing an inner cell mass. ES are isolated for the inner mass of the blastocyte and are able to differentiate into the three germ layer cell types [Amit et al., 2000, Itskovitz-Eldor et al., 2000, Schuldiner et al., 2000]. This property is also call *pluripotentiality*. ES are not considered to be totipotent because they cannot produce all of the extra-embryonic tissues required for a full organism development. SSC are isolated from fetal (after gastrulation stage) or adult tissues and have a more restricted potential of differentiation also defined as *multipotentiality*. Currently, SSC have been isolated from various organs or tissues: brain [Davis and Temple, 1994], blood [Domen and Weissman, 1999], epidermis [Gandarillas and Watt, 1997], intestine [Potten et al., 2003], bone marrow [Bianco and Robey, 1999], pancreas [Lechner and Habener, 2003], liver [Sell, 1990], cornea [Wu et al., 2001] and skeletal muscles [Seale and Rudnicki, 2000]. It was classically admitted that SSC support tissue homeostasis by replacing lost cells. This function relies on the self-renewing capacity which, together with asymmetrical division, prevents the drying up of stem cell stocks and on the differentiation into mature cells to replace the lost cells. However, while a great number of organs or tissues contain somatic stem cells, it appears that, most of the time, those cells are quiescent or weakly active and, at least in some organs, are unable to efficiently repair the damaged tissue. This is especially true for neural stem cells which are found in the subependymal layer of the subventricular zone (SVZ) and the *dentate gyrus* (DG) of the hippocampus in adult mammals [Gage, 2002]. Those cells, while still able to proliferate and differentiate *in vitro* into neurons, astrocytes and oligodendrocytes [McKay, 1997], seem unable to sufficiently do so *in vivo* to ensure brain homeostasis as we define it above [Li et al., 2003]. These two neurogenic zones are only able to form a small microneurons population for olfactory bulb (SVZ) or CA1 (DG) [Hagg, 2005]. By contrast, mesenchymal stem cells (MSC) remain able to proliferate and differentiate into adipocytes, osteocytes, fibroblasts and chondrocytes during the whole life and thus guarantee such homeostasis for these cell types and tissues [Bianco and Robey, 1999].

Over the last decade, researchers have challenged the concept of tissue restriction of SSC and demonstrated that those stem cells have more differentiation abilities than previously thought. This ability, also refereed as stem cell *phenotypic plasticity*, can be defined as a cell property where a same genotype can express different phenotype in function of the environmental conditions. Such kind of plasticity can be adaptative or non-adaptative. Six different mechanisms of differentiation, described by the embryologists during the development could explain the phenotypic plasticity of SSC: de-differentiation, trans-differentiation, the fusion, the trans-determination, true pluripotent stem cell behaviour and the influence of trophic factors (Figure 2):

- *De-differentiation.* De-differentiation refers to the gain in differentiation potential that would occur if mature cells were pushed back up the hierarchical model of lineage restriction. After injury, some amphibians regenerate limbs, tail, and even brain and spinal cord by de-differentiation at the injury site [Tsonis, 2000; Stocum, 2003]. Normally, primordial germ cells transplanted into blastocysts are fate-restricted to germ cells. However, environmental manipulation allows these cells to contribute to all somatic tissues. For instance, oligodendrocyte precursor cells can be induced to generate neurospheres and subsequently neurons *in vitro* [Kondo and Raff, 2000, 2004]. Whether de-differentiation of mammalian cells occurs under normal circumstances or in repair remains unclear.
- *Fusion.* Investigators recognized in the 1960s that differentiated cell fate could be altered in the rather extreme experimental conditions used when a somatic cell nucleus was injected into an enucleated egg, and the cloning of mammals provides proof of principle [Hochedlinger and Jaenisch, 2006]. Nonetheless, much of our knowledge about cell fusion has been derived from studies involving heterokaryons-cells fused *in vitro*. Typically, specialized functions are lost in these cells, although they may be regained after chromosome loss. Silent genes may be activated; for example, synthesis of human muscle proteins occurs in human amniotic fibroblasts after fusion with mouse muscle cells [Clegg et Hauschka, 1987]. It has been postulated that gene dosage (the relative genetic contribution of the two fused cell types) is important both to novel gene activation in heterokaryons and to the suppression of malignancy in hybrids. The formation of (non-dividing)

heterokaryons indicates that differentiated cell phenotype can be altered without DNA replication and cell division.

- *Transdifferentiation.* In transdifferentiation, a mature cell assumes the phenotype and function of another fully differentiated cell. This mechanism occurs during normal oesophageal development, when smooth muscle cells switch to skeletal muscle [Reddy and Kablar, 2004]. Moreover, the transdifferentiation is encountered in various organs which are the target of a chronic inflammation leading to metaplasia [Slack, 2007]. Some reports of transdifferentiation without chronic inflammation have been based on morphological characteristics and lineage-specific markers alone. However, to fulfil the criteria for transdifferentiation, multilineage engraftment and functional activity must also be demonstrated.
- *Transdetermination.* Transdetermination is the redirection of lineage-committed stem cells or precursors to an alternative lineage [Johnston, 2005]. This happens during development in drosophila, but it is extremely difficult to establish definitively that cells are irreversibly committed to a lineage before transdetermination, either at single cell or population level.
- *True pluripotent cells.* True pluripotent cells may persist beyond embryogenesis and, if provided with the appropriate signals, differentiate into cells of multiple lineages. Presently, our knowledge of cell markers is inadequate to define cell populations accurately, so it is possible that cells with true pluripotency issued from adult organisms will have contaminated experiments previously reported as examples of transdifferentiation. However, if the presence of an adult reserve of stem cells is confirmed, the question of why such cells fail to effectively contribute to repair in disease states must be addressed. Stem cells of various origins, including bone marrow, do have tropism for inflammation suggesting that signals may be released from areas of tissue damage into the circulation.
- *Trophic factors.* Stem cells could also play a part in promoting functional recovery by means other than cell replacement. The production of trophic factors might confer resistance to disease, or promote the survival, migration, and differentiation of endogenous precursors. Certainly, bone-marrow cells are known to produce a wide variety of cytokines and exert paracrine effects [Wislet-Gendebien et al., 2004]. Moreover, these trophic factors could be concentrated in discrete functional localisations, the so-called "niche", where trophic factors should also include paracrine and extracellular influences [Scadden, 2006].

The finding of stem cell plasticity carries significant implications for potential cell therapy. For example, if differentiation can be redirected, stem cells of abundant source and easy access, such as bone marrow or umbilical cord blood, could be used to substitute stem cells in tissues that are difficult to isolate, such as heart and nervous system tissue. In this therapeutical objective, numerous studies have been performed on the plasticity of the mesenchymal stem cells. In this chapter, we will discuss the different studies of neural phenotypic plasticity of mesenchymal stem cells and the pathway used by those cells to express some neural characteristics.

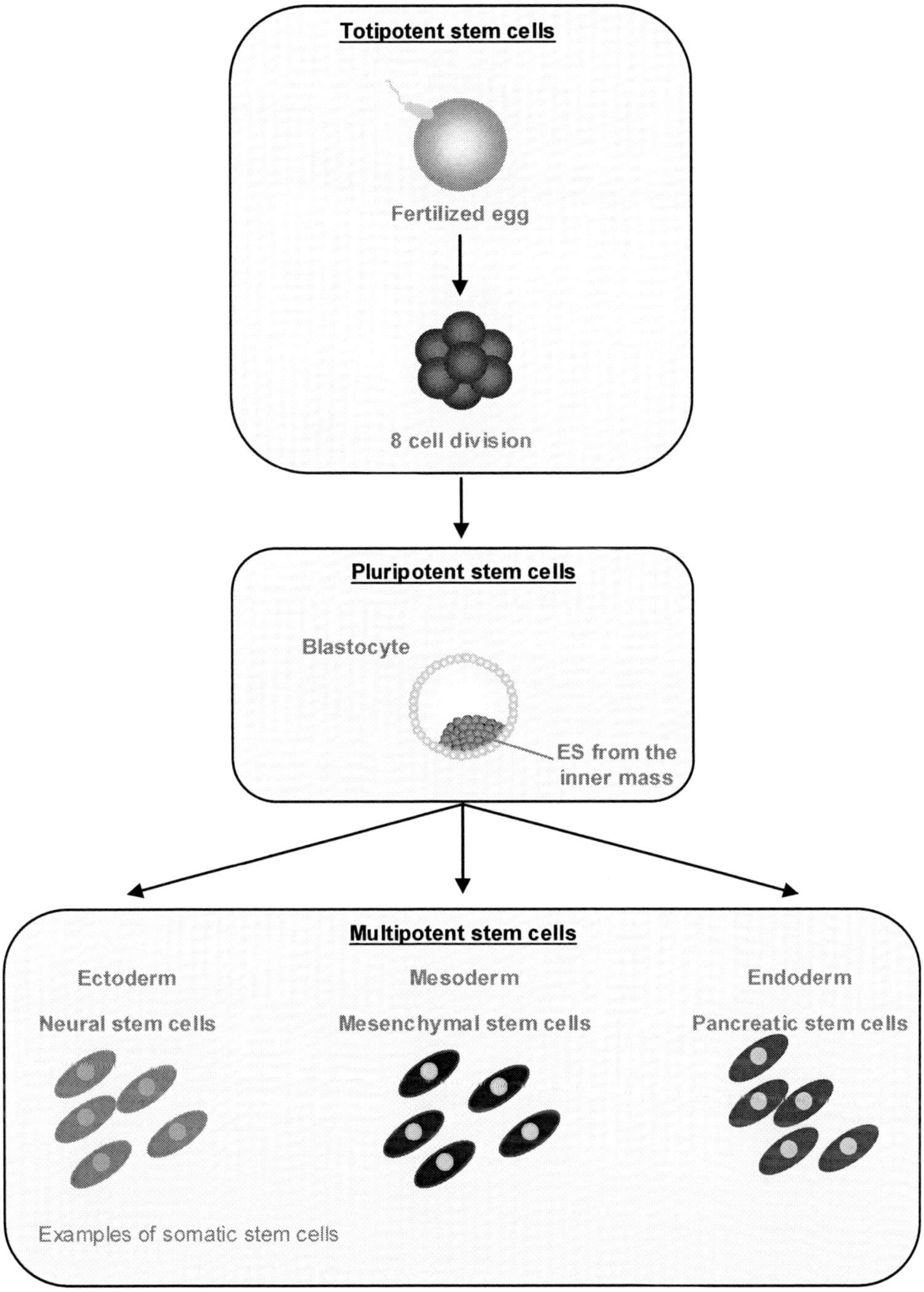

Figure 1. Stem cell types, origin and potential of differentiation. Totipotent stem cells results from the fusion of two haploid germinal stem cells and the first divisions of the zygote. Pluripotent stem cells are issued from the inner mass of the blastocyte, also named embryonic stem cells (ES). Multipotent stem cells also defined as somatic stem cells, have a more restricted potential of differentiation than the ES although this potential is not already well defined and seems to vary between somatic stem cells of various organs and tissues.

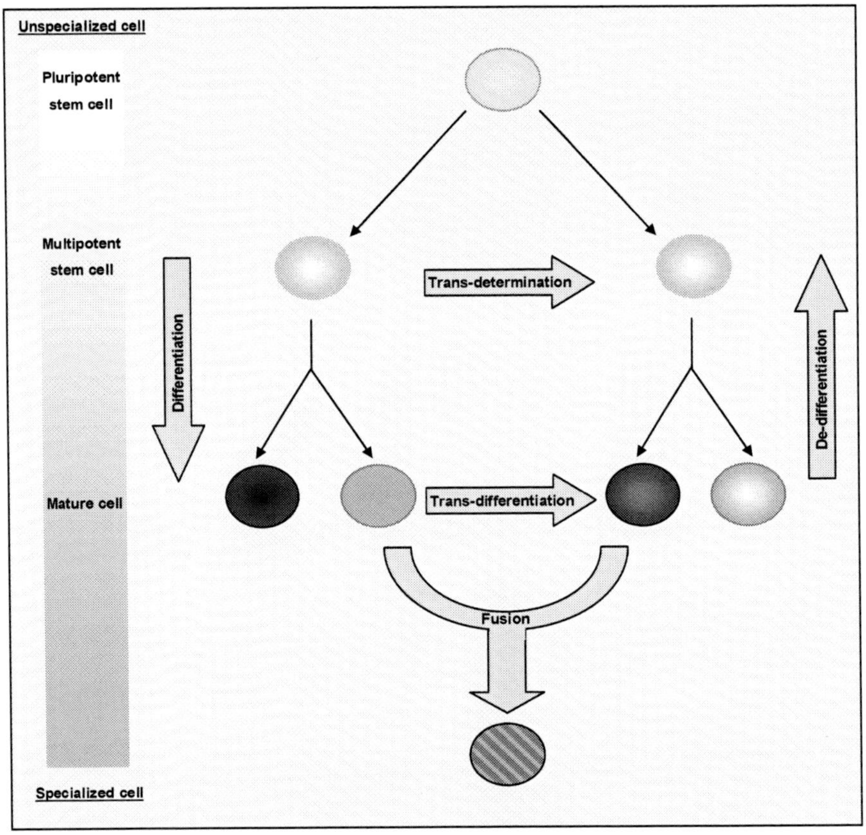

Figure 2. Phenotypic plasticity of somatic stem cells. The six different modes of phenotype acquisition can explain the unusual differentiation properties of the somatic stem cells: differentiation, trans-determination, trans-differentiation, fusion, real pluripotent behaviour and trophic factor influences.

Neural Phenotypic Plasticity of Mesenchymal Stem Cells: Mechanism(s) of Differentiation

Mesenchymal stem cells (MSC) are isolated from various tissues i.e. bone marrow, adipose tissue and muscles. Those cells are characterized by the expression of numerous surface antigens, but none of them appears to be exclusively expressed on MSC. It is classically admitted that those cells need to be positive for STRO-1, CD13, CD49a, CD29, CD44, CD90 (Thy1.1), VCAM-1 and p75-NGFr markers, but negative for CD11b, CD34 and CD45 [reviewed by Herzog, 2003]. As a high proliferate and accessible source of cells, MSC have been intensively studied for their potential use in restorative approaches for degenerative diseases and traumatic injuries. In the central nervous system (CNS), stem cell-based strategies have been proposed to replace lost neurons in degenerative diseases such as Parkinson's disease, Huntington's disease, and amyotrophic lateral sclerosis [Lou Gehrig's disease), or to replace lost oligodendrocytes in demyelinating diseases such as multiple sclerosis. However, even if some studies, which we will be described later in this chapter, demonstrated that MSC derived from adult tissues could be good candidates for cell replacement therapy, a better knowledge of the mechanisms underlying the phenotypic plasticity of somatic stem cells is a prerequisite before considering their use in the treatment of neurological diseases.

During the last decade, different studies have shown that MSC are able to adopt neural phenotypes. Those studies can be classified in 3 groups: A) *In vivo* studies; B) neural phenotypic plasticity induce *in vitro* using chemical agents and C) neural phenotypic plasticity *in vitro* using co-culture. Those studies have been summarized in the Table 1 and will be briefly described in the following paragraphs.

1. *In vivo* Studies

One of the first studies was realised by Kopen et al. [1999] who injected mouse MSC into the lateral ventricle of neonatal mice. By 12 days post-injection, grafted MSC migrated throughout the forebrain and cerebellum without disrupting the host brain architecture and some MSC labelled by BrdU incorporation before grafting were characterized within the striatum and the hippocampus by their expression of glial fibrillary acidic protein (GFAP) and, therefore, recognized as mature astrocytes. Brazelton et al. [2000] demonstrated that after intravascular delivery of genetically-marked adult mouse bone marrow into lethally irradiated normal adult hosts, donor-derived cells in the brain expressed neuronal and astroglial proteins. Lee et al. [2003] transplanted MSC stereotaxically into the brains of mice subjected to focal cerebral infarct. A large number of grafted cells survived after injection in the normal side of the brain 4 weeks after transplantation. MSC can then migrate into CNS lesions and differentiate there into neurons or astrocytes. After inducing the neuronal differentiation of MSC using cryptotanshinone, an AMP-kinase inhibitor extracted from the roots of *Salvia miltiorrhiza* [Kim et al, 2006], Deng et al. [2006] demonstrated that MSC-derived neuron-like cells are able to restore spinal cord function when grafted into injured monkey spinal cord.

2. *In vitro* Studies

Several studies were also performed *in vitro* and demonstrated in different ways the MSC ability to express neural markers. Sanchez-Ramos et al. [2000 and 2002] treated a subset of both human and murine bone marrow cells with retinoic acid and BDNF. This treatment allows MSC to express markers of immature neurons. Likewise, Woodbury et al. [2000 and 2002] found that β-mercaptoethanol (BME) added to cultivated adult rat MSC rapidly induced a transition into neuron-like cells, but not into glial cells. Deng et al. [2001] reported that compounds that increase intracellular cAMP levels such as isobutylmethylxanthine (IBMX) and dibutyryl cAMP (db-cAMP), stimulated cultured human MSC to exhibit neural cell morphology. Padovan et al. [2003] demonstrated that human MSC express an immature neuronal marker (β-III-tubulin) when they are stimulated with neurotrophins such as neurotrophin3 (NT3) or brain-derived neurotrophic factor (BDNF). Zhang et al. [2004]

Table 1. Summary of the *in vivo* / *in vitro* neural phenotypic plasticity of the MSC

Group	Experimental conditions	*in vivo*	*in vitro*	type of cell-like obtained	Immunological characteristics after experiment	References
A	Injection of MSC into the lateral ventricle of neonatal mice	+		Astrocytes	GFAP	Kopen et al., 1999
	Intravascular delivery of MSC into lethally irradiated mice	+		Neurons Astrocytes	NeuN, NF-H, βIII-tubulin GFAP	Brazelton et al., 2000
	Injection of MSC into injured spinal cord	+		Neurons Astrocytes	NSE, NF GFAP	Deng et al., 2006
B	Retinoic acid and BDNF treatment		+	Neurons Astrocytes	NeuN, Nestin, βIII-tubulin GFAP	Sanchez-Ramos et al.,2000
	Retinoic acid and BDNF treatment		+	Neurons Astrocytes	NeuN, Nestin, βIII-tubulin GFAP	Sanchez-Ramos et al., 2002
	Beta-mercaptoethanol treatment		+	Neurons	NeuN, NF-H, βIII-tubulin	Woodbury at al., 2000
	Beta-mercaptoethanol treatment		+	Neurons	NeuN, NF-H, βIII-tubulin	Woodbury at al., 2002
	Isobutylmethylxanthine and dibutyryl cAMP treatment		+	Neurons	NSE vimentin	Deng et al., 2001
	Neurotrophin NT3 treatment		+	Neurons	NeuN	Podovan et al., 2003
	Culture with bFGF and EGF		+	Neurons	Tuj1, NF, MAP2 and NeuN	Locatelli et al., 2003
	Induction with RA and Shh GDNF, dibutyryl AMPc and PACAP		+	Neurons	NF-L and synapsin	Tzeng et al., 2004
	bFGF and ganglioside GM1 treatment		+	Neurons Astrocytes	NSE GFAP	Zhang et al., 2004

Table 1. Summary of the *in vivo* / *in vitro* neural phenotypic plasticity of the MSC (Continued)

Group	Experimental conditions	*in vivo*	*in vitro*	type of cell-like obtained	Immunological characteristics after experiment	References
	BDNF transfection and retinoic acide treatment		+	Neurons Astrocytes	NeuN GFAP	Zhao et al., 2004
	GDNF, IL-1beta, mesencephalic glial-cell-conditioned medium and flash-frozen mesencephalic membrane fragments.		+	Dopaminergic neurons	NSE, MAP2ab and TH	Guo et al., 2005
	Culture with RA, IBMX, dAMPc and ascorbic acid		+	neurons	NSE and NeuN	Hellmann et al., 2006
	Conditionned medium from hippocampus and cerebellum		+	Neurons	GAP-43 and NF	Rivera et al., 2006
C	Co-culture of MSC with neural stem cells		+	astrocytes	GFAP	Wislet-gendebien et al. 2003
	Co-culture of MSC with atrocytes		+	Neurons astrocytes	NF-200 GFAP	Jiang et al., 2002 and 2003
	Co-culture with fetal midbrain cells		+	Neurons astrocytes	NeuN GFAP	Sanchez-Ramos et al., 2000
	Co-culture of MSC with cerebellar granule neurons		+	Neurons astrocytes	Tuj1, NF-M and NeuN GFAP	Wislet-gendebien et al. 2005

demonstrated that a combination of basic fibroblast growth factor (bFGF) with ganglioside GM1 may synergistically promote the transformation of adult rat MSC into neurons and astrocyte-like cells. More recently, Guo et al. [2005] demonstrated that MSC treated with IBMX for 2 days and replace in a differentiating medium containing GDNF, IL-1β, mesencephalic glial-cell-conditioned medium and flash-frozen mesencephalic membrane fragments are able to differentiate into dopaminergic neuron-like cells. Likewise, Rivera et al. [2006] demonstrated that soluble factors derived from adult hippocampus, cortex or cerebellum are sufficient to induce the expression of several neuronal markers like GAP-43 and neurofilament by MSC.

Most of those *in vitro* and *in vivo* studies have only characterized the morphological and immunological characteristic of the MSC-derived neural-like cells. Few studies, however, have demonstrated that beside their immunological similarities, MSC-derived neuron-like cells can also exhibit some neuron specific functionalities. Zhao et al. [2004] demonstrated that, after being transfected with the BDNF gene via a recombinant retroviral vector, human MSC become able to differentiate into neural cells when they are treated with all-trans-retinoic acid (RA). Those MSC-derived neural cells express neural-specific proteins such as NeuN, O4 and glial fibrillary acidic protein (GFAP). Electrophysiological analyses using the whole-cell patch-clamp technique recorded voltage-dependent K^+/Ca^{2+} currents with a resting membrane potential of -65.4±6.07mV. Similarly, we recently demonstrated that adult rats MSC are able to differentiate into excitable neuron-like cells when they are co-cultivated with mouse cerebellar granule neurons [Wislet-Gendebien et al., 2005]. First we demonstrated that those cells express several neuronal markers (NeuN and β-III-tubulin), an axonal marker (neurofilament protein recognized by the monoclonal antibody, SMI31) and a dendritic marker (MAP2ab). Electrophysiological recordings of these nestin-positive MSC-derived neuron-like cells (MDN) were performed and three maturation stages were observed. At 4–6 days of co-culture, MDN showed some neurotransmitter responsiveness (GABA, glycine, serotonin and glutamate) and voltage-gated K^+ currents inhibited by TEA (tetraethylammonium). At that stage, MDN do not express functional sodium voltage-gated channels and have a low membrane potential (V_{rest}) (-37.6±3mV, $n = 61$). During the second week of co-culture, MDN started to display Na^+ currents reversely inhibited by TTX (tetrodotoxin) and were able to fire single action potential spikes. In those older co-cultures, the V_{rest} reaches a more negative value which is closer to the value usually measured in neurons (7–9 days, -50.3±2mV, $n = 76$ and 10–15 days, -56.7±2.3mV, $n = 97$).

Altogether those studies suggest that MSC have some real potential to mimic immunological and functional characteristic of neural cells. Different ways of differentiation have been proposed to explain the unusual abilities of MSC: fusion, real pluripotency and trophic factors.

- *Fusion:* Alvarez-Dolado et al. used an elegant approach based on Cre/lox recombination methodology to detect cell fusion events and demonstrated that after being intravenously injected in X-irradiated animals, bone marrow-derived cells fuse spontaneously *in vivo* with hepatocytes in liver, Purkinje neurons in the cerebellum and cardiac muscle in the heart, resulting in the formation of bi-nucleated cells. For this study, they used as donor mice expressing ubiquitously the Cre-recombinase gene under the control of a hybrid cytomegalovirus enhancer/β-actin (ACTB) promoter and the conditional Cre reporter mouse line R26R12 as a host animal. In these mice, the LacZ reporter gene is exclusively expressed after the excision of a lox P-flanked (floxed) stop cassette by Cre-mediated recombination. When Cre-expressing cells fuse with R26R12 cells, Cre recombinase excises the floxed stop cassette of the reporter gene in the R26R12 nuclei, resulting in expression of LacZ in the fused cells. Consequently, fused cells can be detected easily by 5-bromo-4-chloro-3-indolyl-beta-d-galactoside (X-gal) staining. To study cell fusion *in vivo*, R26R12 reporter mice were lethally irradiated, and two days later were grafted with bone marrow from mice constitutively expressing Cre recombinase and green fluorescent protein (GFP) under the control of the β-actin promoter. In all animals, cells labelled with X-gal were only found in brain, heart and liver (animals examined 10 months after grafting: in brain 5 cells/1.5×10^6 Purkinje neurons; in liver 59 cells/5.5×10^5 hepatocytes; in heart 71 cells/7500 cardiomyocytes). In the same study, they co-cultured bone marrow stromal cells from R26R12 reporter mice with Cre-positive multipotent progenitor cells isolated from postnatal brain and grown as neurospheres. After 4 days *in vitro*, a small proportion of LacZ-positive cells (1–2 cells per 80,000 cells) were found in these co-cultures. Weimann et al. injected intravenously GFP-expressing MSC in X-irradiated mice and observed after at least 6 weeks, that some Purkinje cells in the host animal were GFP-positive [2003]. However, these GFP-positive Purkinje neurons are characterized by the presence of two nuclei and when only male GFP-MSC donor cells are injected into female receiver animals, one nucleus is recognized by a Y-chromosome probe by in situ hybridisation. The frequency of cell fusion events in these experiments seems to be roughly the same as described above.

- *True pluripotent cells:* We recently analysed the mechanism underlying the neural phenotypic plasticity of nestin-positive MSC when these cells are co-cultivated with cerebellar granule neurons (CG) [Wislet-Gendebien et al., 2005]. Three sets of experiments were performed. The first group involved the analysis of the DNA content of nestin-positive MSC and CG before and after co-culture. In all cases, we never observed a significant difference in ploidy which could explain how 60% of the MSC show some neural differentiation (40% as astrocytes and 20% as neuron-like cells). Although the sensitivity of this technique does not allow us to rule out that some fusion events can take place, we performed two other experiments. Rats MSC were co-cultivated with granule cerebellar neurons cultivated from green-mice expressing the Green Fluorescent Protein (GFP) under the control of the actin promoter [Okabe et al., 1997]. Thereafter, double labellings with GFAP and M2 antibodies on one hand, and Tuj1 (anti-β-III-tubulin) and M6 antibodies on the other hand were performed. M2 and M6 recognize respectively and specifically mouse astrocytes and mouse neurons [Lagenaur et al., 1981]. We could observe that: (1) all GFAPpositive/ GFP-negative and Tuj1-positive/GFP-negative cells are also negative for M2 and M6, respectively, allowing us to conclude that those cells are of rat origin and thus likely derive from the MSC population, (2) all the cells which were recognized either by the M2 or by the M6 antibodies were also GFP-positive, ruling out a down-regulation of GFP expression during the co-culture period. Finally, nestin-positive MSC were cultivated for 5 days on paraformaldehyde-fixed GFP-positive granule cerebellar neurons in the presence of CG-conditioned medium (which has been centrifuged and filtered). In such conditions, we observed that 16.1±2.6% (*n* = 8) nestin-positive MSC-derived cells were NeuN-positive and 23.1±2.1% (*n* = 7) nestin-positive MSC-derived cells were GFAP-positive. Note that the level of GFP fluorescence in the fixed cells maintained for 5 days in culture remains stable. Although some rare fusion events cannot be excluded in these three sets of experiments using a co-culture paradigm, we can conclude that most MSC-derived neuron-like cells appear as a consequence of a differentiation process of nestin-positive MSC.

- *Trophic factors:* As described above, Podovan et al. [2003], Zhao et al. [2004], Zhang et al. [2004] and Rivera et al. [2006] demonstrated that several factors like GDNF, bFGF, NT3 or unknown factors contain in neural cell conditioned medium are able to orientate the MSC toward a neural fate, although most of these studies are only based on an immunological approaches except for Zhao et al. which infect cells with retroviruses (see above).

All of these results suggest that the mechanisms underlying the neural phenotypic plasticity of MSC could vary as a function of the environment and/or the cell status. We have also to consider that bone marrow mesenchymal stem cells are an heterogeneous population suggesting that some cells would be more subjected for cell fusion while the other would respond to environmental factors (like trophic factors) and/or adopt a true pluripotent stem cell behaviour. Cell fusion could be a hallmark of precursors or progenitors of monocytes and/or osteoclasts, two well-known cell types which usually fused with other cells or with themselves.

Could Neural Crest Stem Cells be Present in the Bone Marrow?

Beside MSC, other cell types form different tissues were studied for their potential ability to differentiate into neural cells, i.e. adipose tissue, umbilical cord cells, and skin cells. Adipose tissue, like bone marrow, is derived from the embryonic mesoderm. While available literature conflicts about the presence of certain cell receptor antigens, there is general agreement that adipose-derived stem cells and bone marrow stromal cells share adhesion and receptor molecules but are distinct for several adhesion markers with known function in hematopoietic stem cell homing and mobilization that is unique to the bone marrow function [Gronthos et al., 2001; De Ugarte et al., 2003]. Other similarities of adipose-derived stem cells to bone marrow stromal cells is that both stem cell types are capable of differentiating into chondrocytes, osteoblasts, and myocytes [Sanchez-Ramos et al., 2000; Zuk et al., 2001; Safford et al., 2004]. To date, there have been several publications that demonstrate neural cell differentiation of adipose-derived stem cells both *in vivo* and *in vitro* [reviewed by Kokai et al. 2005]. According to those studies, it appears that adipose-derived stem cells can express, under specific conditions, neural markers like nestin, NF-M, NeuN, NSE, Map2, S100 and GFAP.

The mononuclear fraction of umbilical cord blood cells (UCBmf) is rich in stem/progenitor cells. Like bone marrow stem cells, UCBmf cells are capable of self-renewal [Kim et al., 2005], proliferation, subsequent lineage commitment for multiple differentiated cell types [Goodwin et al., 2001]. Moreover those cells can be used to reconstitute the blood and immune systems [reviewed by Broxmeyer, 2005] in various haematological diseases. Some of these pluripotent mesodermal cells have recently been shown to differentiate into cells derived from other germ layers both *in vitro* and *in vivo* [Kong et al., 2004]. It has been show that UCBmf are able to express numerous markers of either stemness or neural fate *in vitro* such as nestin, Musashi1, Oct-4, TuJ1, NCAM, vimentin, GFAP, S100, GalC and MAP2 [McGuckin et al., 2005; Chen et al., 2005]. Further, these cells express neurotrophic receptors trkB, trkC and p75NTR and cytokine receptor CXCR4 [Sanchez-Ramos et al., 2006; Chen et al., 2005].

Similarly, recent evidence indicates that multipotent stem cell populations are present in the mammalian dermis. These cells, defined skin-derived stem cells (SKS), were isolated from the embryo and adult mouse dermis, expanded, and differentiated into cells of various lineages when cultured in optimized media [Toma et al., 2001 and Jahoda et al., 2003]. Like bone marrow and UCBmf cells, the SKS are able to express some neural markers, i.e., nestin, NF-H, NF-M, beta-III-tubulin, Map2, GAP43 and GFAP [Toma et al., 2001, Gorio et al., 2004, Kawase et al., 2004 and Collo et al., 2006].

Beside the ability to express some neural marker that adipose, umbilical cord and skin cells share, recent studies reported that neural crest stem cells can be isolated from those tissue [Reviewed by Crane and Trainor, 2006; Fernandes et al., 2007]. Therefore, in the second part of this chapter, we will analyse the possibility that neural crest stem cells can be isolated from adult bone marrow.

Origin of the Neural Crest Cells

In early vertebrate development, the neural crest is specified in the embryonic ectoderm at the boundary of the neural plate and the ectoderm. Once specified, the neural crest cells (NCCs) undergo a process of epithelium to mesenchyme transition (EMT) that will confer them the ability to migrate. The EMT involves different molecular and cellular machineries and implies deep changes in cell morphology and in the type of cell surface adhesion and recognition molecules. When the EMT is complete, they delaminate from the neural folds/neural tube and migrate along characteristic pathways to differentiate into a wide variety of derivates (Figure 3) [reviewed by Kalcheim, 2000]. Classically, the neural crest can be divided in four main domains, each with characteristic derivatives and functions: the *cranial* (*cephalic*), the *trunk*, the *vagal* and *sacral*, and the *cardiac* neural crest cells [Le Lièvre et Le Douarin, 1975].

- The *cranial neural crest* arises in the anterior and populates the face and the pharyngeal arches giving rise to cartilage, bone, cranial neurons of the peripheral nervous system, glial and connective tissue of the face. Cells that migrated into the pharyngeal arches play an inductive role in the thymus and thyroid gland development.

- The *trunk neural crest* is a transient structure which disappears soon after the neural tube closes, lies between the vagal and sacral neural crest and gives rise to two groups of cells. One group migrates dorso-lateral and populates the skin, forming pigment cells and the other migrates ventro-lateral through the anterior sclerotome to become the epinephrine-producing cells of the adrenal gland and the neurons of the sympathetic nervous system. Some cells remain in the sclerotome to form the dorsal root ganglia.
- The *vagal* and *sacral neural crest* arises in the neck and tail and populates the gut, forming the parasympathetic neurons that regulate peristalsis and control blood vessel dilation.
- *Cardiac neural crest* cells as the cranial neural crest cells can differentiate into melanocytes, neurons, cartilage and connective tissue, but can also more specifically produce the entire muscular-connective tissue wall of the large arteries as they arise from the heart as well as contributing to the septum that separate pulmonary circulation from the aorta.

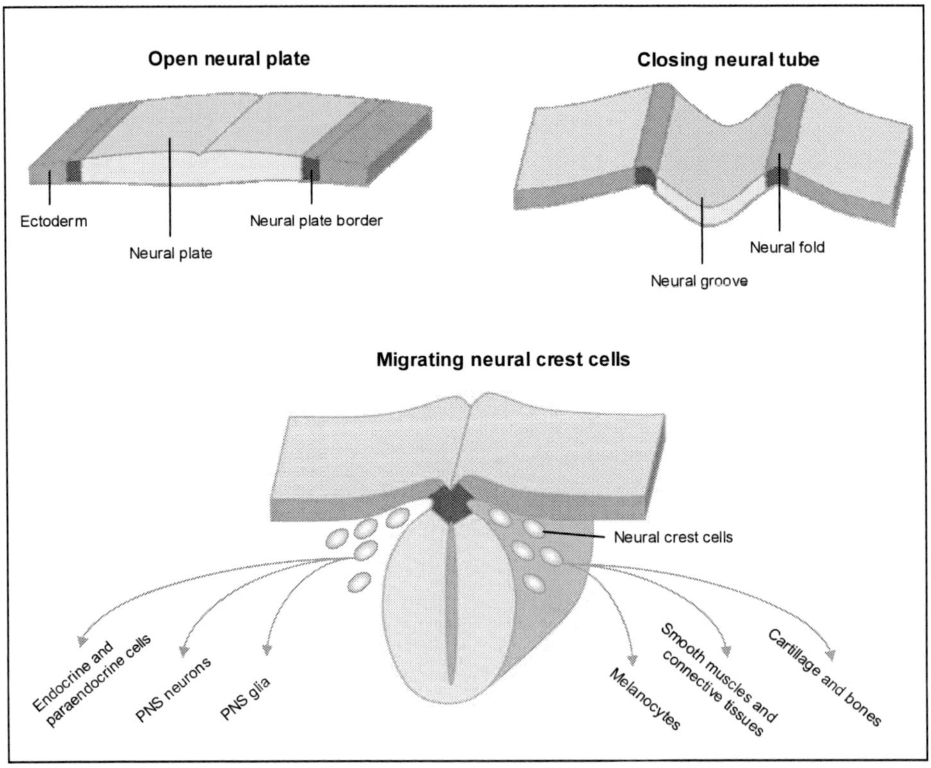

Figure 3. Neurulation and neural crest migration. As neurulation proceeds, the neural plate rolls up and the neural plate border becomes the neural folds. Near the time of neural tube closure (depending on the species), the neural crest cells go through an epithelial to mesenchymal transition (EMT) and delaminate from the neural folds or dorsal neural tube and migrate along defined pathways.

Molecular Pathways Underlying the Formation of the Neural Crest Stem Cells

A large body of *in vitro* studies as well as analyses of gene mutations in mice has led to the identification of several growth factor/receptor signalling pathways implicated in the development of specific neural crest derivative subtypes [reviewed by Anderson, 1997; Le Kalcheim, 2000; Sieber-Blum, 1997]. In mice, neural crest stem cells are defined by expression of two markers: p75NGFR (low-affinity nerve growth factor receptor) and the transcription factor Sox10 [Stemple et Anderson, 1992; Rao et Anderson, 1997; Paratore et al., 2001; Kim et al., 2003].

A number of signals have been implicated in the formation of the neural crest, including members of the *Wnt*, *FGF* and *BMP* families. These secreted proteins (extrinsic factors) regulate early expression of transcription factors, cell adhesion molecules, extracellular glycoproteins, etc. while intrinsic factors act to stabilize the competence of the epithelium to form neural crest [Knecht and Bronner-Fraser, 2002] and, moreover, some also regulate subsequent developmental events, such as delamination and initiation of migration [Kalcheim, 2005, LaBonne and Bronner- Fraser, 2000].

1) Extrinsic Factors Regulation the Neural Crest Formation

The formation of neural crest has traditionally been considered as a classic example of induction, in which signals from one tissue elicit differentiation in a responding competent tissue [Liem et al., 1995; Selleck and Bronner-Fraser, 1995; Mancilla and Mayor, 1996]. As described in the Figure 4, signals inducing the formation of the neural crest formation are sent by the mesoderm as well as the epidermis. *A) Signals from the mesoderm*: a graded signal from the mesoderm is responsible for neural crest induction. Members of the *Wnt* (wingless/INT) family of secreted glycoproteins may mediate the neural crest inducing ability of paraxial mesoderm [Bang et al., 1999]. Likewise, a member of the fibroblast growth factor (FGF) family, *FGF-8*, mediates the inductive effects of paraxial mesoderm. Several studies suggested that FGF's ability to induce neural crest is dependent on Wnt signaling [LaBonne and Bronner-Fraser, 1998]. *B) Signals from ectoderm*: a gradient of *BMP* signalling acts initially to specify epidermal, neural, and border (prospective neural crest) fates in the ectoderm. Both Wnt and FGF signals have been proposed to play a role in this process [Mayor et al., 2000; LaBonne and Bronner-Fraser, 1998; Marchant et al., 1998; Bastidas et al., 2004]. Wnt proteins play significant roles in neural crest cell development at different developmental times [reviewed in Wu *et al.*, 2003]. *Wnt6* is synthesized in the epidermal ectoderm and might mediate crest specification [Garcia-Castro *et al.*, 2002]. Slightly later, *Wnt1* and *Wnt3a* are present in the dorsal neural tube following initial specification of crest cells [Dickinson *et al.*, 1995]. Wnt3a is intense already opposite the segmental plate while Wnt1 becomes apparent slightly later, opposite epithelial somites and concomitant with BMP relief from noggin inhibition [Burstyn-Cohen *et al.*, 2004]. Whereas Wnt1 is likely to be directly regulated by BMPs [Marcelle *et al.*, 1999, Sela-Donenfeld and Kalcheim, 2002, Burstyn-Cohen *et al.*, 2004], the transcription of Wnt3a is not; suggesting that Wnt1 better fits to be involved in crest delamination [Burstyn-Cohen *et al.*, 2004]. Wnt signaling act through the transmembrane receptor Frizzled which is required to modulate the distribution and function of β-catenin [Miller and Moon, 1997]. β-catenin associate directly with the highly conserved cytoplasmic domain of cadherins. The so formed cadherin-catenin complex links to the actin filament network [Ozawa *et al.*, 1989; Hinck *et al.*, 1994; Knudsen *et al.*, 1995; Weiss *et al.*, 1998]. BMP/Wnt-mediated signals could induce changes in the actin cytoskeleton via rhoB. A role for rhoB in crest delamination has been already suggested based on inhibition experiments in culture [Liu and Jessell, 1998]. A molecular pathway for the activation of Rho by Wnt/frizzled was suggested, which involves the formation of a complex between Rho, dishevelled and Daam1 in the plasma membrane, resulting in the generation of a polarized cytoskeleton [Habas *et al.*, 2002]. Thus, the dynamic association of the catenin-cadherin complex and that of rhoB with the cytoskeleton may be essential for regulating cell-cell interactions leading to neural crest delamination. Notably, Rho GTPases could also be effectors of Wnt signals in this pathway as they were shown to affect morphogenesis by interfering with cell proliferation [Wei *et al.*, 2002].

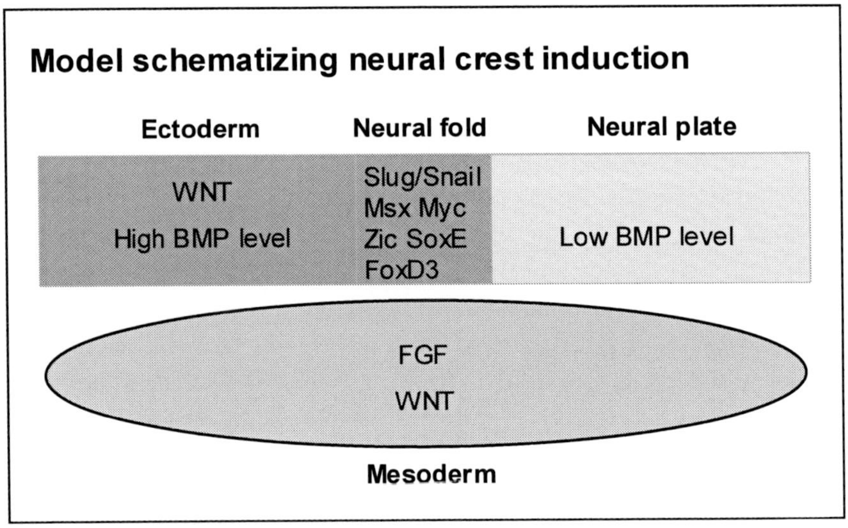

Figure 4. Neural crest induction and its relation to neural plate and neural plate border. Members of the BMP, FGF, and Wnts families of signalling molecules are involved in the formation of the neural plate, the establishment of the neural plate border, and in neural crest induction. These signals can originate from the ectoderm, or the mesoderm.

Later during the development, the induced cells interact to complete neural crest induction by a process that requires *Notch/Delta* signalling. Notch signalling has two roles during neural crest development: first in establishing the neural crest domain within the ectoderm via lateral induction and subsequently in diversifying the fates of cells that arise from the neural crest via lateral inhibition. The first of these roles, specification of neural crest via lateral induction, has been explored primarily in the cranial neural folds from which the cranial neural crest arises. Evidence for such a role has thus far only been obtained from chick and frog; results from these two species differ, but share the feature that Notch signalling regulates genes that are expressed by cranial neural crest through effects on expression of Bmp family members. The second of these roles, diversification of neural crest progeny via lateral inhibition, has been identified thus far only in trunk neural crest. Evidence from several species suggests that Notch-mediated lateral inhibition functions in multiple episodes in this context, in each case inhibiting neurogenesis. In the 'standard' mode of lateral inhibition, Notch promotes proliferation and in the 'instructive' mode, it promotes specific secondary fates, including cell death or glial differentiation [reviewed by Cornell et Eisen, 2005].

Once neural crest formation has been induced, some extracellular factors have been reported to induce the ultimate steps of neural crest differentiation. Bone morphogenetic protein 2 (*BMP2*) was shown to induce neurogenesis in culture, glial growth factor (*GGF*) drives the cells into glial, transforming growth factor-β (*TGFβ*) into smooth muscle differentiation [Shah et al., 1994 and 1996]. Several groups published data on melanocyte differentiation of NC cells [Maxwell et al., 1996; Takano et al., 2002] and demonstrated that Endothelin 3 is an essential factor in that pathway [Lahav et al., 1996 and 1998]. Recent work showed, that chondrogenesis can also be induced in trunk NC cells [McGonnell et Graham, 2002; Oka et Ito, 2007; Maurer et al., 2006].

2) Intrinsic Factors Regulating the Neural Crest Formation

The boundary region between the neural plate and the epidermis is defined by expression of a variety of specific markers, including transcription factors such as *Slug* or *Snail*, *AP-2, Foxd3, PAX3, twist, Sox9, Zic5*, etc.

Group E Sox genes (Sox8, Sox9 and Sox10) are expressed in the prospective neural crest and *Sox9* expression precedes expression of premigratory neural crest markers. Group E Sox genes act at two distinct steps in neural crest differentiation. Forced expression of Sox9 promotes neural-crest-like properties in neural tube progenitors at the expense of central nervous system neuronal differentiation. Subsequently, in migratory neural crest cells, SoxE gene expression biases cells towards glial cell and melanocyte fate, and away from neuronal lineages. Although SoxE genes are sufficient to initiate neural crest development, they do not efficiently induce the delamination of ectopic neural crest cells from the neural tube consistent with the idea that this event is independently controlled [Reviewed by Cheung et Briscoe 2003]. *Sox10* is expressed by migrating neural crest cells and is required for the survival of neural crest cells before lineage segregation [Southard-Smith et al., 1998; Kapur, 2000; Paratore et al., 2001]. Furthermore, in developing dorsal root ganglia, Sox10 is also required for glial fate acquisition, and Sox10 expression appears to be maintained by differentiating glial cells but not by differentiating neurons [Britsch et al., 2001, Paratore et al., 2001]. Finally, Sox10 heterozygous mutant neural crest cells, survival appears to be normal, while fate specifications are drastically affected. Thereby, the fate chosen by a mutant neural crest cell is context dependent. Several studies indicate that combinatorial signalling by Sox10, extracellular factors such as neuregulin 1, and local cell-cell interactions is involved in fine-tuning lineage decisions by neural crest stem cells [reviewed by Paratore, 2001]. Likewise, the expression of *Sox2* has been proposed to act by maintaining a neural progenitor identity [Graham et al., 2003]. It becomes down-regulated in the neural plate when neural crest cells segregate from the dorsal neural tube, and it remains low during neural crest cell migration. *Sox2* expression is subsequently up-regulated in some crest-derived cells in the developing peripheral nervous system and is later restricted to glial sublineages [Wakamatsu et al., 2004].

Similarities between Neural Plasticity of MSC and Neural Crest Stem Cells

We recently demonstrated that MSC express the Neural stem cells protein (nestin) [Wislet-Gendebien et al., 2003], an intermediate filament protein predominantly expressed by neural progenitors [Lendahl et al., 1990]. During embryogenesis, nestin is expressed in migrating and proliferating cells (including neural crest stem cells), whereas in adult tissues, nestin is mainly restricted to areas of regeneration [Wiese et al., 2004]. As MSC are able to adopt some neural phenotypes, it appears that the nestin expression is a pre-requisite for the neural differentiation: only nestin-positive MSC are able to differentiate into neuron- and glial-like cells [Wislet-Gendebien et al., 2005]. Moreover, the number of nestin-positive MSC present in culture increase with the number of cell passages suggesting a higher proliferation rate for the MSC with neural abilities. Interestingly, several factors like Sox transcription factors (Sox2 and Sox10); Pax6 transcription factor; ErBb2 and ErBb4 neuregulin receptors; as well as Frizzled Wnt receptors (FZD1, FZD2 and FZD5) are over-expressed by nestin-positive MSC compare to nestin-negative MSC [Wislet-Gendebien et al., 2005].

As previously described, neural crest stem cells are defined by expression of two markers: p75NGFR (low-affinity nerve growth factor receptor) and the transcription factor Sox10 [Stemple et Anderson, 1992; Rao et Anderson, 1997; Paratore et al., 2001; Kim et al., 2003]. Once again, MSC express also those two factors [Wislet-Gendebien et al., 2005].

During the embryonic development, migration and differentiation of neural crest stem cells involved several factors like neuregulins and their receptors ErbB. The *neuregulin* gene encodes various isoforms of a recently identified growth and differentiation factor, which all contain an EGF-like domain. So far, four genes has been described to encode for neuregulin *NRG1*, *NRG2*, *NRG3* and *NRG4* [Burden et Yarden, 1997; Busfield et al., 1997; Carraway et al ., 1995; Higashiyama et al., 1997; Harari et al., 1999]. NRG1 gene is the more studied and contains numerous exons that after alternative splicing allow the expression of 14 different proteins that can be classified in 3 groups: (I) type I isoforms (originally identified as *NDF* Neu Differentiation Factor/ *HER* heregulin or *ARIA*, Acetylcholine Receptor Inducing Activity) contain an Ig-like domain, an EGF-like domain that suffices to elicit biological responses [Holmes et al., 1992], a proteolysis site, a hydrophobic domain suggested to act as internal signal sequence for secretion of the factor and additional C-terminal sequences [Holmes et al., 1992]; (II) type II isoforms (originally identified as *GGF* Glial Growth Factor) contain a signal peptide, a kringle-like sequence plus Ig and EGF-like domains [Marchionni et al.,1993]; (III) type III isoforms (originally identified as *SMDF* Sensory and Motor Neuron-derived Factor) share only the EGF-like domain with other isoforms; notable in the N-terminal part is a hydrophobic domain within a cysteine-rich sequence [Ho et al., 1995].

Neuregulin-induced cellular responses are mediated by tyrosine kinase receptors of the erbB family: neuregulin binds erbB3 and erbB4 with high affinity, but not the erbB2 (HER2) and erbB1 (EGF) receptors [Peles et al., 1993; Plowman et al., 1993; Carraway et al., 1995]. Binding affinity of neuregulin to erbB3 is increased by the presence of erbB2 [Sliwkowski et al., 1994]. Moreover, when co-expressed with erbB3 or erbB4, neuregulin induces tyrosine phosphorylation of the erbB2 receptor [Holmes et al., 1992; Wen et al., 1992; Carraway and

Cantley, 1994; Beerli et al., 1995]. This is the result of heterodimerization of erbB2 with erbB3 or erbB4 and subsequent receptor cross-phosphorylation.

Meyer et al. [1997] demonstrated that type I neuregulin is expressed in cephalic mesenchyme and cranial ganglia as well as the endocardium and governs development of neural crest-derived neurons. In contrast, type III neuregulin is expressed in differentiating sensory and motor neurons and acts on the Schwann cell precursors, driving their initial development. Concerning the receptors ErbB, distinct receptor combinations are essential in different developmental events: the ErbB2 and ErbB4 receptors cooperate in transmission of neuregulin-1 signals in the heart, whereas ErbB2 and ErbB3 cooperate in neural crest cells [Gassmann et al. 1995; Lee et al. 1995; Erickson et al. 1997; Riethmacher et al. 1997], however, precise developmental event that requires the neuregulin-1 signal and ErbB receptors has not been elucidated. At this point, the lack of ErbB3 expression by the mesenchymal-derived stem cells would not be enough to rule out those cells as potential neural crest stem cells. Indeed, when we look at the functional side, type I neuregulin can interact with both ErbB2/ErbB3 and ErbB2/ErbB4 receptors. Likewise, it seems that ErbB3 expression by neural crest cells is observed early in the development (at the migration stage) or later (at the neural specification stage). Potential neural crest stem cells located in the bone marrow would not express the ErbB3 receptor because they at a post-migration but pre-differentiation stage. Finally, neural crest cells located in the heart express ErbB4 receptor but not ErbB3. However, heart-derived neural crest stem cells are also able to differentiate into mature neural cells under specific conditions [Tomita et al., 2005].

Upon binding to cell surface receptors, Wnts initiate an intracellular cascade that, via several intermediate steps, leads to the translocation of beta-catenin to the nucleus (see above). Ikeya et al. [1997] demonstrated that wnt1 and wnt3a play an important role in the specification of the neural crest derivates and Giarre et al, [1999] demonstrated that the responsiveness to Wnt-1 at the biochemical level is a common property of both epithelial and mesenchymal cells. More recently, Jackson et al. [2005] demonstrated that mesenchymal stem cells express different wnt1- and wnt3a-induced gene that activated the β-catenin dependent pathway. Moreover, wnt3a can activated frizzled receptor Fz1 and Fz2 [Kennell et MacDougald, 2005] which are overexpressed by nestin-positive MSC [Wislet-Gendebien et al., 2005].

Conclusion

According to the different studies realised on bone marrow-derived stem cells, it seems that those cells share numerous similarities with the neural crest stem cells. Beside their abilities to differentiate into mesenchymal and neural lineages, bone marrow-derived stem cells are p75NGFr- and Sox10-positives and express some wnt receptors (Frizzled) as well as neuregulin receptors (erbB2 and erbB4), which could activate the intracellular pathways underlying the neural differentiation.

However, if those evidences strongly suggest that neural crest stem cells can be present in the adult bone marrow, the "ultimate" demonstration seems to be missing. This demonstration would involve the tracking of the neural crest stem cells form the early stage of embryonic development to the colonisation of various tissues. Interestingly, Jiang et al. in 2000 developed a two-component genetic system based on Cre/lox recombination to label indelibly the entire mouse neural crest population at the time of its formation, and to detect it at any time thereafter. Briefly, the fate of neural crest cells was mapped *in vivo* by mating *ROSA26* Cre reporter (*R26R*) mice, which express β-galactosidase upon Cre-mediated recombination, with mice expressing Cre recombinase under the control of the *Wnt1* promoter. In *Wnt1-Cre/R26R* double transgenic mice, virtually all neural crest stem cells express β-galactosidase. Using this transgenic model, Sieber-Blum and Grim [2004] demonstrated the presence of pluripotent neural crest stem cells in adult follicle hairs and Wong et al. [2006] recently demonstrated the presence of neural crest cells in the mouse adult skin. This approach could also be useful to look for a possible presence of neural crest-derived cells in bone marrow. These cells keep the ability to differentiation into neurons or astrocytes and as the facility of getting mesenchymal stem cells from bone marrow, could be useful for a cell therapy based on auto-graft. The physiological role of these neural crest derived cells in bone marrow should also be addressed, especially regarding the regulation of haematopoiesis [Katayama et al., 2006].

References

Alvarez-Dolado M, Pardal R, Garcia-Verdugo JM, Fike JR, Lee HO, Pfeffer K, Lois C, Morrison SJ, Alvarez-Buylla A: Fusion of bone-marrow-derived cells with Purkinje neurons, cardiomyocytes and hepatocytes. *Nature* 2003, 425:968-973.

Amit M, Carpenter MK, Inokuma MS, Chiu CP, Harris CP, Waknitz MA, Itskovitz-Eldor J, Thomson JA: Clonally derived human embryonic stem cell lines maintain pluripotency and proliferative potential for prolonged periods of culture. *Dev.Biol.* 2000, 227:271-278.

Bang AG, Papalopulu N, Goulding MD, Kintner C: Expression of Pax-3 in the lateral neural plate is dependent on a Wnt-mediated signal from posterior nonaxial mesoderm. *Dev.Biol.* 1999, 212:366-380.

Bastidas F, De Calisto J, Mayor R: Identification of neural crest competence territory: role of Wnt signaling. *Dev.Dyn.* 2004, 229:109-117.

Beerli RR, Graus-Porta D, Woods-Cook K, Chen X, Yarden Y, Hynes NE: Neu differentiation factor activation of ErbB-3 and ErbB-4 is cell specific and displays a differential requirement for ErbB-2. *Mol.Cell Biol.* 1995, 15:6496-6505.

Bianco P, Robey P: Diseases of bone and the stromal cell lineage. *J.Bone Miner.Res.* 1999, 14:336-341.

Brazelton TR, Rossi FM, Keshet GI, Blau HM: From marrow to brain: expression of neuronal phenotypes in adult mice. *Science* 2000, 290:1775-1779.

Britsch S, Goerich DE, Riethmacher D, Peirano RI, Rossner M, Nave KA, Birchmeier C, Wegner M: The transcription factor Sox10 is a key regulator of peripheral glial development. *Genes Dev.* 2001, 15:66-78.

Broxmeyer HE: Biology of cord blood cells and future prospects for enhanced clinical benefit. *Cytotherapy.* 2005, 7:209-218.

Burden S, Yarden Y: Neuregulins and their receptors: a versatile signaling module in organogenesis and oncogenesis. *Neuron* 1997, 18:847-855.

Burstyn-Cohen T, Stanleigh J, Sela-Donenfeld D, Kalcheim C: Canonical Wnt activity regulates trunk neural crest delamination linking BMP/noggin signaling with G1/S transition. *Development* 2004, 131:5327-5339.

Busfield SJ, Michnick DA, Chickering TW, Revett TL, Ma J, Woolf EA, Comrack CA, Dussault BJ, Woolf J, Goodearl AD, Gearing DP: Characterization of a neuregulin-related gene, Don-1, that is highly expressed in restricted regions of the cerebellum and hippocampus. *Mol.Cell Biol.* 1997, 17:4007-4014.

Carraway KL, III, Cantley LC: A neu acquaintance for erbB3 and erbB4: a role for receptor heterodimerization in growth signaling. *Cell* 1994, 78:5-8.

Carraway KL, III, Burden SJ: Neuregulins and their receptors. *Curr.Opin.Neurobiol.* 1995, 5:606-612.

Chen N, Hudson JE, Walczak P, Misiuta I, Garbuzova-Davis S, Jiang L, Sanchez-Ramos J, Sanberg PR, Zigova T, Willing AE: Human umbilical cord blood progenitors: the potential of these hematopoietic cells to become neural. *Stem Cells* 2005, 23:1560-1570.

Cheung M, Briscoe J: Neural crest development is regulated by the transcription factor Sox9. *Development* 2003, 130:5681-5693.

Clegg CH, Hauschka SD: Heterokaryon analysis of muscle differentiation: regulation of the postmitotic state. *J.Cell Biol.* 1987, 105:937-947.

Collo G, Goffi F, Merlo PE, Baldelli P, Benfenati F, Spano P: Immature neuronal phenotype derived from mouse skin precursor cells differentiated in vitro. *Brain Res.* 2006, 1109:32-36.

Crane JF, Trainor PA: Neural crest stem and progenitor cells. *Annu.Rev.Cell Dev.Biol.* 2006, 22:267-286.

Davis AA, Temple S: A self-renewing multipotential stem cell in embryonic rat cerebral cortex. *Nature* 1994, 372:263-266.

De Ugarte DA, Alfonso Z, Zuk PA, Elbarbary A, Zhu M, Ashjian P, Benhaim P, Hedrick MH, Fraser JK: Differential expression of stem cell mobilization-associated molecules on multi-lineage cells from adipose tissue and bone marrow. *Immunol.Lett.* 2003, 89:267-270.

Deng YB, Liu XG, Liu ZG, Liu XL, Liu Y, Zhou GQ: Implantation of BM mesenchymal stem cells into injured spinal cord elicits de novo neurogenesis and functional recovery: evidence from a study in rhesus monkeys. *Cytotherapy.* 2006, 8:210-214.

Dickinson ME, Selleck MA, McMahon AP, Bronner-Fraser M: Dorsalization of the neural tube by the non-neural ectoderm. *Development* 1995, 121:2099-2106.

Domen J, Weissman IL: Self-renewal, differentiation or death: regulation and manipulation of hematopoietic stem cell fate. *Mol.Med.Today* 1999, 5:201-208.

Dunnett SB: Functional analysis of fronto-striatal reconstruction by striatal grafts. *Novartis.Found.Symp.* 2000, 231:21-41.

Erickson SL, O'Shea KS, Ghaboosi N, Loverro L, Frantz G, Bauer M, Lu LH, Moore MW: ErbB3 is required for normal cerebellar and cardiac development: a comparison with ErbB2-and heregulin-deficient mice. *Development* 1997, 124:4999-5011.

Fallahi-Sichani M, Soleimani M, Najafi SM, Kiani J, Arefian E, Atashi A: In vitro differentiation of cord blood unrestricted somatic stem cells expressing dopamine-associated genes into neuron-like cells. *Cell Biol.Int.* 2007, 31:299-303.

Fernandes KJ, Toma JG, Miller FD: Multipotent skin-derived precursors: adult neural crest-related precursors with therapeutic potential. *Philos.Trans.R.Soc.Lond B Biol.Sci.* 2007.

Fernandez CI, Alberti E, Mendoza Y, Martinez L, Collazo J, Rosillo JC, Bauza JY: Motor and cognitive recovery induced by bone marrow stem cells grafted to striatum and hippocampus of impaired aged rats: functional and therapeutic considerations. *Ann.N.Y.Acad.Sci.* 2004, 1019:48-52.

Gage FH: Neurogenesis in the adult brain. *J.Neurosci.* 2002, 22:612-613.

Gandarillas A, Watt FM: c-Myc promotes differentiation of human epidermal stem cells. *Genes Dev.* 1997, 11:2869-2882.

Garcia-Castro MI, Marcelle C, Bronner-Fraser M: Ectodermal Wnt function as a neural crest inducer. *Science* 2002, 297:848-851.

Gassmann M, Casagranda F, Orioli D, Simon H, Lai C, Klein R, Lemke G: Aberrant neural and cardiac development in mice lacking the ErbB4 neuregulin receptor. *Nature* 1995, 378:390-394.

Giarre M, Semenov MV, Brown AM: Wnt signaling stabilizes the dual-function protein beta-catenin in diverse cell types. *Ann.N.Y.Acad.Sci.* 1998, 857:43-55.

Goodwin NC, Ishida Y, Hartford S, Wnek C, Bergstrom RA, Leder P, Schimenti JC: DelBank: a mouse ES-cell resource for generating deletions. *Nat.Genet.* 2001, 28: 310-311.

Gorio A, Torrente Y, Madaschi L, Di Stefano AB, Pisati F, Marchesi C, Belicchi M, Di Giulio AM, Bresolin N: Fate of autologous dermal stem cells transplanted into the spinal cord after traumatic injury (TSCI). *Neuroscience* 2004, 125:179-189.

Graham V, Khudyakov J, Ellis P, Pevny L: SOX2 functions to maintain neural progenitor identity. *Neuron* 2003, 39:749-765.

Gronthos S, Franklin DM, Leddy HA, Robey PG, Storms RW, Gimble JM: Surface protein characterization of human adipose tissue-derived stromal cells. *J.Cell Physiol* 2001, 189:54-63.

Guo L, Yin F, Meng HQ, Ling L, Hu-He TN, Li P, Zhang CX, Yu S, Duan DS, Fan HX: Differentiation of mesenchymal stem cells into dopaminergic neuron-like cells in vitro. *Biomed.Environ.Sci.* 2005, 18:36-42.

Habas R, Kato Y, He X: Wnt/Frizzled activation of Rho regulates vertebrate gastrulation and requires a novel Formin homology protein Daam1. *Cell* 2001, 107:843-854.

Hagg T: Molecular regulation of adult CNS neurogenesis: an integrated view. *Trends Neurosci.* 2005, 28:589-595.

Hall ED: Pharmacological treatment of acute spinal cord injury: how do we build on past success? *J.Spinal Cord.Med.* 2001, 24:142-146.

Harari D, Tzahar E, Romano J, Shelly M, Pierce JH, Andrews GC, Yarden Y: Neuregulin-4: a novel growth factor that acts through the ErbB-4 receptor tyrosine kinase. *Oncogene* 1999, 18:2681-2689.

Herzog EL, Chai L, Krause DS: Plasticity of marrow-derived stem cells. *Blood* 2003, 102:3483-3493.

Higashiyama S, Horikawa M, Yamada K, Ichino N, Nakano N, Nakagawa T, Miyagawa J, Matsushita N, Nagatsu T, Taniguchi N, Ishiguro H: A novel brain-derived member of the epidermal growth factor family that interacts with ErbB3 and ErbB4. *J.Biochem.(Tokyo)* 1997, 122:675-680.

Hinck L, Nelson WJ, Papkoff J: Wnt-1 modulates cell-cell adhesion in mammalian cells by stabilizing beta-catenin binding to the cell adhesion protein cadherin. *J.Cell Biol.* 1994, 124:729-741.

Hochedlinger K, Jaenisch R: Nuclear reprogramming and pluripotency. *Nature* 2006, 441:1061-1067.

Holmes WE, Sliwkowski MX, Akita RW, Henzel WJ, Lee J, Park JW, Yansura D, Abadi N, Raab H, Lewis GD, .: Identification of heregulin, a specific activator of p185erbB2. *Science* 1992, 256:1205-1210.

Ikeya M, Lee SM, Johnson JE, McMahon AP, Takada S: Wnt signalling required for expansion of neural crest and CNS progenitors. *Nature* 1997, 389:966-970.

Isacson O, Bjorklund L, Pernaute RS: Parkinson's disease: interpretations of transplantation study are erroneous. *Nat.Neurosci.* 2001, 4:553.

Itskovitz-Eldor J, Schuldiner M, Karsenti D, Eden A, Yanuka O, Amit M, Soreq H, Benvenisty N: Differentiation of human embryonic stem cells into embryoid bodies compromising the three embryonic germ layers. *Mol.Med.* 2000, 6:88-95.

Jackson A, Vayssiere B, Garcia T, Newell W, Baron R, Roman-Roman S, Rawadi G: Gene array analysis of Wnt-regulated genes in C3H10T1/2 cells. *Bone* 2005, 36:585-598.

Jahoda CA: Cell movement in the hair follicle dermis - more than a two-way street? *J.Invest Dermatol.* 2003, 121:ix-xi.

Johnston LA: Regeneration and transdetermination: new tricks from old cells. *Cell* 2005, 120:288-290.

Kalcheim C: Mechanisms of early neural crest development: from cell specification to migration. *Int.Rev.Cytol.* 2000, 200:143-196.

Kalcheim C, Burstyn-Cohen T: Early stages of neural crest ontogeny: formation and regulation of cell delamination. *Int.J.Dev.Biol.* 2005, 49:105-116.

Kapur RP: Colonization of the murine hindgut by sacral crest-derived neural precursors: experimental support for an evolutionarily conserved model. *Dev.Biol.* 2000, 227:146-155.

Katayama Y, Battista M, Kao WM, Hidalgo A, Peired AJ, Thomas SA, Frenette PS: Signals from the sympathetic nervous system regulate hematopoietic stem cell egress from bone marrow. *Cell* 2006, 124:407-421.

Kawase Y, Yanagi Y, Takato T, Fujimoto M, Okochi H: Characterization of multipotent adult stem cells from the skin: transforming growth factor-beta (TGF-beta) facilitates cell growth. *Exp.Cell Res.* 2004, 295:194-203.

Kennell JA, MacDougald OA: Wnt signaling inhibits adipogenesis through beta-catenin-dependent and -independent mechanisms. *J.Biol.Chem.* 2005, 280:24004-24010.

Kim J, Lo L, Dormand E, Anderson DJ: SOX10 maintains multipotency and inhibits neuronal differentiation of neural crest stem cells. *Neuron* 2003, 38:17-31.

Kim S, Honmou O, Kato K, Nonaka T, Houkin K, Hamada H, Kocsis JD: Neural differentiation potential of peripheral blood- and bone-marrow-derived precursor cells. *Brain Res.* 2006, 1123:27-33.

Kim YM, Jung MH, Song HY, Yang HO, Lee ST, Kim JH, Kim YT, Nam JH, Mok JE: Ex vivo expansion of human umbilical cord blood-derived T-lymphocytes with homologous cord blood plasma. *Tohoku J.Exp.Med.* 2005, 205:115-122.

Knecht AK, Bronner-Fraser M: Induction of the neural crest: a multigene process. *Nat.Rev.Genet.* 2002, 3:453-461.

Knudsen KA, Soler AP, Johnson KR, Wheelock MJ: Interaction of alpha-actinin with the cadherin/catenin cell-cell adhesion complex via alpha-catenin. *J.Cell Biol.* 1995, 130:67-77.

Kokai LE, Rubin JP, Marra KG: The potential of adipose-derived adult stem cells as a source of neuronal progenitor cells. *Plast.Reconstr.Surg.* 2005, 116:1453-1460.

Kondo T, Raff M: Oligodendrocyte precursor cells reprogrammed to become multipotential CNS stem cells. *Science* 2000, 289:1754-1757.

Kondo T, Raff M: Chromatin remodeling and histone modification in the conversion of oligodendrocyte precursors to neural stem cells. *Genes Dev.* 2004, 18:2963-2972.

Kong PY, Luo CJ, Zhou YH, Guo CH: [Promoting effects of stromal cells on hematopoietic reconstitution capability of bone marrow cells expanded under different conditions]. *Zhongguo Shi Yan.Xue.Ye.Xue.Za Zhi.* 2004, 12:265-269.

Kopen GC, Prockop DJ, Phinney DG: Marrow stromal cells migrate throughout forebrain and cerebellum, and they differentiate into astrocytes after injection into neonatal mouse brains. *Proc.Natl.Acad.Sci.U.S A* 1999, 96:10711-10716.

Krause DS, Theise ND, Collector MI, Henegariu O, Hwang S, Gardner R, Neutzel S, Sharkis SJ: Multi-organ, multi-lineage engraftment by a single bone marrow-derived stem cell. *Cell* 2001, 105:369-377.

LaBonne C, Bronner-Fraser M: Neural crest induction in Xenopus: evidence for a two-signal model. *Development* 1998, 125:2403-2414.

LaBonne C, Bronner-Fraser M: Snail-related transcriptional repressors are required in Xenopus for both the induction of the neural crest and its subsequent migration. *Dev.Biol.* 2000, 221:195-205.

Lagenaur C, Schachner M: Monoclonal antibody (M2) to glial and neuronal cell surfaces. *J.Supramol.Struct.Cell Biochem.* 1981, 15:335-346.

Lahav R, Ziller C, Dupin E, Le Douarin NM: Endothelin 3 promotes neural crest cell proliferation and mediates a vast increase in melanocyte number in culture. *Proc.Natl.Acad.Sci.U.S A* 1996, 93:3892-3897.

Lahav R, Dupin E, Lecoin L, Glavieux C, Champeval D, Ziller C, Le Douarin NM: Endothelin 3 selectively promotes survival and proliferation of neural crest-derived glial and melanocytic precursors in vitro. *Proc.Natl.Acad.Sci.U.S A* 1998, 95:14214-14219.

Le Lievre CS, Le Douarin NM: Mesenchymal derivatives of the neural crest: analysis of chimaeric quail and chick embryos. *J.Embryol.Exp.Morphol.* 1975, 34:125-154.

Lechner A, Habener JF: Bone marrow stem cells find a path to the pancreas. *Nat.Biotechnol.* 2003, 21:755-756.

Lee J, Kuroda S, Shichinohe H, Ikeda J, Seki T, Hida K, Tada M, Sawada K, Iwasaki Y: Migration and differentiation of nuclear fluorescence-labeled bone marrow stromal cells after transplantation into cerebral infarct and spinal cord injury in mice. *Neuropathology* 2003, 23:169-180.

Lendahl U, Zimmerman LB, McKay RD: CNS stem cells express a new class of intermediate filament protein. *Cell* 1990, 60:585-595.

Liem KF, Jr., Tremml G, Roelink H, Jessell TM: Dorsal differentiation of neural plate cells induced by BMP-mediated signals from epidermal ectoderm. *Cell* 1995, 82:969-979.

Liu JP, Jessell TM: A role for rhoB in the delamination of neural crest cells from the dorsal neural tube. *Development* 1998, 125:5055-5067.

Mancilla A, Mayor R: Neural crest formation in Xenopus laevis: mechanisms of Xslug induction. *Dev.Biol.* 1996, 177:580-589.

Marcelle C, Ahlgren S, Bronner-Fraser M: In vivo regulation of somite differentiation and proliferation by Sonic Hedgehog. *Dev.Biol.* 1999, 214:277-287.

Marchant L, Linker C, Ruiz P, Guerrero N, Mayor R: The inductive properties of mesoderm suggest that the neural crest cells are specified by a BMP gradient. *Dev.Biol.* 1998, 198:319-329.

Maurer J, Fuchs S, Jager R, Kurz B, Sommer L, Schorle H: Establishment and controlled differentiation of neural crest stem cell lines using conditional transgenesis. *Differentiation* 2007.

Maxwell GD, Reid K, Elefanty A, Bartlett PF, Murphy M: Glial cell line-derived neurotrophic factor promotes the development of adrenergic neurons in mouse neural crest cultures. *Proc.Natl.Acad.Sci.U.S A* 1996, 93:13274-13279.

Mayor R, Guerrero N, Young RM, Gomez-Skarmeta JL, Cuellar C: A novel function for the Xslug gene: control of dorsal mesendoderm development by repressing BMP-4. *Mech.Dev.* 2000, 97:47-56.

McGonnell IM, Graham A: Trunk neural crest has skeletogenic potential. *Curr.Biol.* 2002, 12:767-771.

McGuckin C, Forraz N, Baradez MO, Basford C, Dickinson AM, Navran S, Hartgerink JD: Embryonic-like stem cells from umbilical cord blood and potential for neural modeling. *Acta Neurobiol.Exp.(Wars.)* 2006, 66:321-329.

McKay R: Stem cells in the central nervous system. *Science* 1997, 276:66-71.

McKenzie IA, Biernaskie J, Toma JG, Midha R, Miller FD: Skin-derived precursors generate myelinating Schwann cells for the injured and dysmyelinated nervous system. *J.Neurosci.* 2006, 26:6651-6660.

Meyer D, Yamaai T, Garratt A, Riethmacher-Sonnenberg E, Kane D, Theill LE, Birchmeier C: Isoform-specific expression and function of neuregulin. *Development* 1997, 124:3575-3586.

Mezey E, Chandross KJ: Bone marrow: a possible alternative source of cells in the adult nervous system. *Eur.J.Pharmacol.* 2000, 405:297-302.

Moon RT, Brown JD, Yang-Snyder JA, Miller JR: Structurally related receptors and antagonists compete for secreted Wnt ligands. *Cell* 1997, 88:725-728.

Nagase T, Matsumoto D, Nagase M, Yoshimura K, Shigeura T, Inoue M, Hasegawa M, Yamagishi M, Machida M: Neurospheres from human adipose tissue transplanted into cultured mouse embryos can contribute to craniofacial morphogenesis: a preliminary report. *J.Craniofac.Surg.* 2007, 18:49-53.

Oka K, Oka S, Sasaki T, Ito Y, Bringas P, Jr., Nonaka K, Chai Y: The role of TGF-beta signaling in regulating chondrogenesis and osteogenesis during mandibular development. *Dev.Biol.* 2007, 303:391-404.

Okabe M, Ikawa M, Kominami K, Nakanishi T, Nishimune Y: 'Green mice' as a source of ubiquitous green cells. *FEBS Lett.* 1997, 407:313-319.

Orlic D, Kajstura J, Chimenti S, Limana F, Jakoniuk I, Quaini F, Nadal-Ginard B, Bodine DM, Leri A, Anversa P: Mobilized bone marrow cells repair the infarcted heart, improving function and survival. *Proc.Natl.Acad.Sci.U.S A* 2001, 98:10344-10349.

Ozawa M, Baribault H, Kemler R: The cytoplasmic domain of the cell adhesion molecule uvomorulin associates with three independent proteins structurally related in different species. *EMBO J.* 1989, 8:1711-1717.

Padovan CS, Jahn K, Birnbaum T, Reich P, Sostak P, Strupp M, Straube A: Expression of neuronal markers in differentiated marrow stromal cells and CD133+ stem-like cells. *Cell Transplant.* 2003, 12:839-848.

Paratore C, Goerich DE, Suter U, Wegner M, Sommer L: Survival and glial fate acquisition of neural crest cells are regulated by an interplay between the transcription factor Sox10 and extrinsic combinatorial signaling. *Development* 2001, 128:3949-3961.

Peles E, Yarden Y: Neu and its ligands: from an oncogene to neural factors. *Bioessays* 1993, 15:815-824.

Plowman GD, Green JM, Culouscou JM, Carlton GW, Rothwell VM, Buckley S: Heregulin induces tyrosine phosphorylation of HER4/p180erbB4. *Nature* 1993, 366:473-475.

Potten CS, Booth C, Tudor GL, Booth D, Brady G, Hurley P, Ashton G, Clarke R, Sakakibara S, Okano H: Identification of a putative intestinal stem cell and early lineage marker; musashi-1. *Differentiation* 2003, 71:28-41.

Priller J, Flugel A, Wehner T, Boentert M, Haas CA, Prinz M, Fernandez-Klett F, Prass K, Bechmann I, de Boer BA, Frotscher M, Kreutzberg GW, Persons DA, Dirnagl U: Targeting gene-modified hematopoietic cells to the central nervous system: use of green fluorescent protein uncovers microglial engraftment. *Nat.Med.* 2001, 7:1356-1361.

Rao MS, Anderson DJ: Immortalization and controlled in vitro differentiation of murine multipotent neural crest stem cells. *J.Neurobiol.* 1997, 32:722-746.

Reddy T, Kablar B: Evidence for the involvement of neurotrophins in muscle transdifferentiation and acetylcholine receptor transformation in the esophagus of Myf5(-/-):MyoD(-/-) and NT-3(-/-) embryos. *Dev.Dyn.* 2004, 231:683-692.

Riethmacher D, Sonnenberg-Riethmacher E, Brinkmann V, Yamaai T, Lewin GR, Birchmeier C: Severe neuropathies in mice with targeted mutations in the ErbB3 receptor. *Nature* 1997, 389:725-730.

Rivera FJ, Sierralta WD, Minguell JJ, Aigner L: Adult hippocampus derived soluble factors induce a neuronal-like phenotype in mesenchymal stem cells. *Neurosci.Lett.* 2006, 406:49-54.

Safford KM, Safford SD, Gimble JM, Shetty AK, Rice HE: Characterization of neuronal/glial differentiation of murine adipose-derived adult stromal cells. *Exp.Neurol.* 2004, 187:319-328.

Sanchez-Ramos J, Song S, Cardozo-Pelaez F, Hazzi C, Stedeford T, Willing A, Freeman TB, Saporta S, Janssen W, Patel N, Cooper DR, Sanberg PR: Adult bone marrow stromal cells differentiate into neural cells in vitro. *Exp.Neurol.* 2000, 164:247-256.

Sanchez-Ramos J: Stem cells from umbilical cord blood. *Semin.Reprod.Med.* 2006, 24:358-369.

Sanchez-Ramos JR: Neural cells derived from adult bone marrow and umbilical cord blood. *J.Neurosci.Res.* 2002, 69:880-893.

Scadden DT: The stem-cell niche as an entity of action. *Nature* 2006, 441:1075-1079.

Schuldiner M, Yanuka O, Itskovitz-Eldor J, Melton DA, Benvenisty N: Effects of eight growth factors on the differentiation of cells derived from human embryonic stem cells. *Proc.Natl.Acad.Sci.U.S A* 2000, 97:11307-11312.

Seale P, Rudnicki MA: A new look at the origin, function, and "stem-cell" status of muscle satellite cells. *Dev.Biol.* 2000, 218:115-124.

Sela-Donenfeld D, Kalcheim C: Localized BMP4-noggin interactions generate the dynamic patterning of noggin expression in somites. *Dev.Biol.* 2002, 246:311-328.

Sell S: Is there a liver stem cell? *Cancer Res.* 1990, 50:3811-3815.

Selleck MA, Bronner-Fraser M: Origins of the avian neural crest: the role of neural plate-epidermal interactions. *Development* 1995, 121:525-538.

Shah NM, Marchionni MA, Isaacs I, Stroobant P, Anderson DJ: Glial growth factor restricts mammalian neural crest stem cells to a glial fate. *Cell* 1994, 77:349-360.

Shah NM, Groves AK, Anderson DJ: Alternative neural crest cell fates are instructively promoted by TGFbeta superfamily members. *Cell* 1996, 85:331-343.

Sieber-Blum M, Zhang JM: Growth factor action in neural crest cell diversification. *J.Anat.* 1997, 191 (Pt 4):493-499.

Sieber-Blum M, Grim M, Hu YF, Szeder V: Pluripotent neural crest stem cells in the adult hair follicle. *Dev.Dyn.* 2004, 231:258-269.

Slack WV: Cybermedicine for the patient. *Am.J.Prev.Med.* 2007, 32:S135-S136.

Sliwkowski MX, Schaefer G, Akita RW, Lofgren JA, Fitzpatrick VD, Nuijens A, Fendly BM, Cerione RA, Vandlen RL, Carraway KL, III: Coexpression of erbB2 and erbB3 proteins reconstitutes a high affinity receptor for heregulin. *J.Biol.Chem.* 1994, 269:14661-14665.

Southard-Smith EM, Kos L, Pavan WJ: Sox10 mutation disrupts neural crest development in Dom Hirschsprung mouse model. *Nat.Genet.* 1998, 18:60-64.

Stanford WL, Caruana G, Vallis KA, Inamdar M, Hidaka M, Bautch VL, Bernstein A: Expression trapping: identification of novel genes expressed in hematopoietic and endothelial lineages by gene trapping in ES cells. *Blood* 1998, 92:4622-4631.

Stemple DL, Anderson DJ: Isolation of a stem cell for neurons and glia from the mammalian neural crest. *Cell* 1992, 71:973-985.

Stocum DL: Amphibian regeneration and stem cells. *Curr.Top.Microbiol.Immunol.* 2004, 280:1-70.

Takano N, Kawakami T, Kawa Y, Asano M, Watabe H, Ito M, Soma Y, Kubota Y, Mizoguchi M: Fibronectin combined with stem cell factor plays an important role in melanocyte proliferation, differentiation and migration in cultured mouse neural crest cells. *Pigment Cell Res.* 2002, 15:192-200.

Terada N, Hamazaki T, Oka M, Hoki M, Mastalerz DM, Nakano Y, Meyer EM, Morel L, Petersen BE, Scott EW: Bone marrow cells adopt the phenotype of other cells by spontaneous cell fusion. *Nature* 2002, 416:542-545.

Toma JG, Akhavan M, Fernandes KJ, Barnabe-Heider F, Sadikot A, Kaplan DR, Miller FD: Isolation of multipotent adult stem cells from the dermis of mammalian skin. *Nat.Cell Biol.* 2001, 3:778-784.

Tomita Y, Matsumura K, Wakamatsu Y, Matsuzaki Y, Shibuya I, Kawaguchi H, Ieda M, Kanakubo S, Shimazaki T, Ogawa S, Osumi N, Okano H, Fukuda K: Cardiac neural crest cells contribute to the dormant multipotent stem cell in the mammalian heart. *J.Cell Biol.* 2005, 170:1135-1146.

Tsonis PA: Regeneration of the lens in amphibians. *Results Probl.Cell Differ.* 2000, 31:179-196.

Vassilopoulos G, Russell DW: Cell fusion: an alternative to stem cell plasticity and its therapeutic implications. *Curr.Opin.Genet.Dev.* 2003, 13:480-485.

Wakamatsu Y, Endo Y, Osumi N, Weston JA: Multiple roles of Sox2, an HMG-box transcription factor in avian neural crest development. *Dev.Dyn.* 2004, 229:74-86.

Wei Y, Renard CA, Labalette C, Wu Y, Levy L, Neuveut C, Prieur X, Flajolet M, Prigent S, Buendia MA: Identification of the LIM protein FHL2 as a coactivator of beta-catenin. *J.Biol.Chem.* 2003, 278:5188-5194.

Weimann JM, Charlton CA, Brazelton TR, Hackman RC, Blau HM: Contribution of transplanted bone marrow cells to Purkinje neurons in human adult brains. *Proc.Natl.Acad.Sci.U.S A* 2003, 100:2088-2093.

Weimann JM, Johansson CB, Trejo A, Blau HM: Stable reprogrammed heterokaryons form spontaneously in Purkinje neurons after bone marrow transplant. *Nat.Cell Biol.* 2003, 5:959-966.

Weiss EE, Kroemker M, Rudiger AH, Jockusch BM, Rudiger M: Vinculin is part of the cadherin-catenin junctional complex: complex formation between alpha-catenin and vinculin. *J.Cell Biol.* 1998, 141:755-764.

Wen D, Peles E, Cupples R, Suggs SV, Bacus SS, Luo Y, Trail G, Hu S, Silbiger SM, Levy RB, .: Neu differentiation factor: a transmembrane glycoprotein containing an EGF domain and an immunoglobulin homology unit. *Cell* 1992, 69:559-572.

Wiese C, Rolletschek A, Kania G, Blyszczuk P, Tarasov KV, Tarasova Y, Wersto RP, Boheler KR, Wobus AM: Nestin expression--a property of multi-lineage progenitor cells? *Cell Mol.Life Sci.* 2004, 61:2510-2522.

Wislet-Gendebien S, Leprince P, Moonen G, Rogister B: Regulation of neural markers nestin and GFAP expression by cultivated bone marrow stromal cells. *J.Cell Sci.* 2003, 116:3295-3302.

Wislet-Gendebien S, Bruyere F, Hans G, Leprince P, Moonen G, Rogister B: Nestin-positive mesenchymal stem cells favour the astroglial lineage in neural progenitors and stem cells by releasing active BMP4. *BMC Neurosci.* 2004, 5:33.

Wislet-Gendebien S, Hans G, Leprince P, Rigo JM, Moonen G, Rogister B: Plasticity of cultured mesenchymal stem cells: switch from nestin-positive to excitable neuron-like phenotype. *Stem Cells* 2005, 23:392-402.

Wong CE, Paratore C, Dours-Zimmermann MT, Rochat A, Pietri T, Suter U, Zimmermann DR, Dufour S, Thiery JP, Meijer D, Beermann F, Barrandon Y, Sommer L: Neural crest-derived cells with stem cell features can be traced back to multiple lineages in the adult skin. *J.Cell Biol.* 2006, 175:1005-1015.

Woodbury D, Schwarz EJ, Prockop DJ, Black IB: Adult rat and human bone marrow stromal cells differentiate into neurons. *J.Neurosci.Res.* 2000, 61:364-370.

Woodbury D, Reynolds K, Black IB: Adult bone marrow stromal stem cells express germline, ectodermal, endodermal, and mesodermal genes prior to neurogenesis. *J.Neurosci.Res.* 2002, 69:908-917.

Wu J, Saint-Jeannet JP, Klein PS: Wnt-frizzled signaling in neural crest formation. *Trends Neurosci.* 2003, 26:40-45.

Wu KY, Hong SJ, Lin CP, Lai YH, Wang HZ: Endothelin-induced changes of secondary messengers in cultured corneal endothelial cells. *J.Ocul.Pharmacol.Ther.* 2001, 17:351-361.

Ying QL, Nichols J, Evans EP, Smith AG: Changing potency by spontaneous fusion. *Nature* 2002, 416:545-548.

Zhang H, Wang JZ, Sun HY, Zhang JN, Yang SY: The effects of GM1 and bFGF synergistically inducing adult rat bone marrow stromal cells to form neural progenitor cells and their differentiation. *Chin J.Traumatol.* 2004, 7:3-6.

Zhao LX, Zhang J, Cao F, Meng L, Wang DM, Li YH, Nan X, Jiao WC, Zheng M, Xu XH, Pei XT: Modification of the brain-derived neurotrophic factor gene: a portal to transform mesenchymal stem cells into advantageous engineering cells for neuroregeneration and neuroprotection. *Exp.Neurol.* 2004, 190:396-406.

Zuk PA, Zhu M, Mizuno H, Huang J, Futrell JW, Katz AJ, Benhaim P, Lorenz HP, Hedrick MH: Multilineage cells from human adipose tissue: implications for cell-based therapies. *Tissue Eng* 2001, 7:211-228.

In: Cell Differentiation Research Developments
Editor: L. B. Ivanova, pp. 61-95
ISBN: 978-1-60021-939-9

Chapter III

Immunolocalization and Biosynthesis of Acid Glycohydrolases in Spermatogenic Cells Isolated from Rodent Testes by Unit Gravity Sedimentation

***Daulat R.P. Tulsiani*[1] *and Aïda Abou-Haila*[2]**
[1]Departments of Obstetrics & Gynecology and Cell & Developmental Biology, Vanderbilt University Medical Center, Nashville, TN 37232, USA;
[2]UFR Biomédicale, Université Paris Descartes, 75270 Paris cedex 06, France.

Abstract

Spermatozoa are formed in the mammalian testes by a complex differentiation process collectively referred to as spermatogenesis. The process involves multiple molecular events during mitotic cell division, meiosis and spermiogenesis. The last event is the final phase of the complex process when a non-dividing ordinary looking round spermatid is transformed into a uniquely shaped spermatozoon containing a well-developed sperm head (with a nucleus and a new organelle, the acrosome) and a fully formed flagellum. It is important to emphasize that the constituent proteins/glycoproteins, present within the acrosome of a testicular spermatozoon, are synthesized in testicular germ cells during spermatogenesis. However, our knowledge on the origin and molecular processes that regulate the expression and processing of acrosomal enzymes in the germ cells is very limited. In this article, we have described multiple biochemical and immunohistochemical approaches to examine the synthesis, processing and localization of two acid glycohydrolases in spermatogenic cells from rodent testes. Mixed germ cells, prepared from the rat or mouse testis by enzymatic digestion, were separated by unit gravity sedimentation using a linear gradient of 2-4% bovine serum albumin. Fractions rich in spermatocytes, round spermatids, and condensed/elongated spermatids (>95% pure cell populations) were pooled separately and the enriched cells were assayed for several acid glycohydrolase activities. All these cell populations as well as the cauda epididymal spermatozoa were found to contain several acid glycohydrolase activities.

The presence of two enzymes, namely β-D-galactosidase and β-D-glucuronidase in the germ cells was further confirmed using immunocytochemical and biochemical approaches. Immunocytochemical approaches at the electron microscopic level revealed that the two enzymes were present in the Golgi membranes, Golgi-associated vesicles, and lysosome-like structures in the late spermatocytes and round spermatids. Indirect immunofluorescence studies at the light microscopic level revealed intense immunopositive reaction in the forming/formed acrosome of the round spermatids, elongated spermatids and the cauda epididymal spermatozoa. A combination of radiolabeling of spermatocytes and round spermatids in cell culture, immunoprecipitation of the [^{35}S]methionine-labeled (newly synthesized) β-D-galactosidase, SDS-PAGE and autoradiography procedures were used to examine the synthesis and processing of the enzyme in rat spermatogenic cells. Data from these approaches demonstrate that the β-D-galactosidase is synthesized in the spermatocytes and round spermatids as high molecular weight precursor forms of 90/88-kD that undergo processing to the lower molecular weight mature forms in a cell-specific manner. The net result is the formation of predominantly 64-kD and 62-kD forms in spermatocytes and round spermatids, respectively. Taken together, our studies demonstrate that the two acid glycohydrolases present in the acrosome of the epididymal spermatozoa are first expressed in spermatocytes. Our final intention is to briefly discuss the significance of acrosomal glycohydrolases in sperm function and fertilization. We hope that various studies reported in this article will contribute to a better understanding of the sperm acrosome and its potential role in fertilization.

Keywords: Spermatogenesis; sperm acrosome; acrosomal enzymes; glycohydrolases; sperm formation; acrosomal reaction.

Introduction

Mammalian spermatozoa undergo many fascinating changes during their formation in the testes. The cells are formed throughout postpubertal male reproductive life from spermatogonial stem cells by a highly orchestrated process referred to as spermatogenesis. The entire process consists of three sequential phases of cell proliferation and differentiation [1-3]. First, spermatogonial stem cells undergo extensive multiplication and proliferation to produce an optimal number of spermatogonia that give rise to primary spermatocytes and also to maintain a pool of stem cells. Second, the primary spermatocytes undergo a lengthy meiotic prophase, followed by the first meiotic cell division that results in the formation of two secondary spermatocytes. Each diploid cell undergoes the second meiotic cell division to produce four haploid round spermatids. Finally, the haploid cells undergo gradual remodeling of cellular components during transformation into sperm cells by a process referred to as spermiogenesis [3,4].

The haploid round spermatids have a spherical nucleus with dispersed chromatin and absence of nucleolii [5,6]. These cells are incapable of further cell division. During spermiogenesis, the cell undergoes nuclear and cytoplasmic changes that result in the formation of a hydrodynamically shaped spermatozoon with a head containing the condensed nucleus and acrosome, and a flagellum [1-4]. During this transformation, several cellular organelles (nucleus, endoplasmic reticulum, Golgi apparatus, mitochondria, and centriole etc.) simultaneously undergo biochemical and structural changes that result in a gradual shape change from round to elongated as spermatid metamorphoses into a developing spermatozoon.

Spermatozoa released from the seminiferous tubules are terminally differentiated cells with no known synthetic activity. Their constituent proteins/glycoproteins are synthesized in the spermatogenic cells during spermatogenesis. Although the sperm acrosome and its contents (acid glycohydrolases, proteinases etc.) play an important role in sperm function, our understanding of how these molecules are synthesized and processed is very limited. There are multiple reasons for the lack of this information. First, the methods available for the preparation of enriched populations of spermatogenic cells are highly technical and have discouraged many investigators from approaching these studies. Second, the male germ cells are viable in culture for only a short period of time [7] making it difficult to carry out detailed *in vitro* biosynthetic studies. Finally, very few high titer antibodies have been prepared against acrosomal antigens to attempt these studies. The availability of high titer antibodies against rat preputial gland β-D-glucuronidase and rat epididymal luminal fluid β-D-galactosidase that cross-reacted with these enzymes in the germ cells and spermatozoa has allowed us to attempt these studies.

In this article, we have used multiple approaches to: i) chemically characterize the two glycohydrolases in spermatogenic cells and spermatozoa; ii) immunolocalize these glycohydrolases in the mouse and rat testicular germ cells and spermatozoa; and iii) examine synthesis and processing of β-D-galactosidase in spermatocytes and round spermatids in cell culture. Data from these studies collectively demonstrate that the two enzymes are first expressed in 4N diploid cells (pachytene spermatocytes). Both enzymes are localized in the Golgi apparatus, Golgi-associated vesicles, and lysosome-like vesicles in the spermatocytes and round spermatids. Biosynthetic studies with β-D-galactosidase indicated that the enzyme is first synthesized in pachytene spermatocytes and round spermatids where it undergoes processing before it is packaged into the sperm acrosome. In addition, using indirect immunofluorescence approaches, we have examined the progressive formation of the sperm acrosome during spermiogenesis. Interestingly, our data demonstrate that the formation of the acrosome coincides with the decrease in the number of lysosome-like multivesicular bodies in spermatids. The procedures and results outlined in this article are based on several reports from our laboratories [8-12].

The Sperm Acrosome: Formation and Contents

The acrosome and its contents play an important role at the site of sperm-egg (zona) binding during the fertilization process. Clinical studies have identified a distinct group of men whose infertility is associated with abnormal acrosome reaction [13], a complex process involving exocytosis of acrosomal contents [3,14]. The acrosome is formed during spermiogenesis and represents one of the defining features of the sperm development in the testis. After meiosis is completed, the nondividing haploid round spermatids begin to transfer into a uniquely shaped spermatozoon, a process that is completed in 12-14 days. The Golgi complex at an early stage of spermiogenesis (round spermatid) consists of a prominent system of closely packed tubules and vesicles localized close to one of the poles of the nucleus. Some of these vesicles enlarge and become filled with newly synthesized glycoprotein components. These vesicles migrate and empty their glycoprotein-rich contents into the forming acrosome [15]. Glycoprotein synthesis and transport to the acrosome is quite active during early stages of spermiogenesis and continues until late spermiogenesis. There are several published reports demonstrating that the acrosomal glycohydrolases are first expressed in late spermatocytes [9,11,16]. Whether the antigens destined for various regions of the acrosome are tagged with specific signals for transportation and packaging is not yet known.

The formation and evolution of the acrosomic system of the rat take place during four distinct phases of spermiogenesis [for review see reference 3]. First, during the Golgi phase, numerous proacrosomic granules are formed from trans-Golgi stacks and accumulate in the medullary region. The small proacrosomic granules fuse with each other to form a single large acrosomic granule that establishes close contact with the nuclear envelope. Second, during the cap phase, the spherical acrosomic granule enlarges by the addition of newly synthesized incoming glycoproteins from the Golgi apparatus. The developing acrosomic granule grows and flattens over the surface of nucleus. The region of the nucleus that is in contact with the enlarging acrosome vesicle is characterized by a thin layer of condensed chromatin just beneath the nuclear membrane. Third, during the acrosomal phase, the dense acrosomic granule applies itself to the inner acrosomal membrane and becomes hemispherical. This structure remains distinct throughout the fourth and final maturation phase of spermiogenesis and represents the acrosome proper [3]. The acrosome membrane close to the nuclear envelope is termed the inner acrosomal membrane, whereas the membrane surrounding the acrosomal contents and underlying the sperm plasma membrane is termed the outer acrosomal membrane. The acrosome is quite variable among species, ranging from an skull-cap (spatulate) shape in man to a sickle-shape in rodents [3,14,17].

Table 1. Glycohydrolase Activities Present in Mammalian Sperm Acrosome

Enzyme	Sugar linkage [a]
Hyaluronidase	(GlcNAc-Gluc)n
α -L-Fucosidase	Fuc-GlcNAc
α -D-Galactosidase	Gal-Gal
β -D-Galactosidase	Gal-GlcNAc
β -D-Glucuronidase	Gluc-GlcNAc
β -*N*-Hexosaminidase	GlcNAc-(Gal)GalNAc
α -D-Mannosidase	Man-Man
β -D-Mannosidase	Man-GlcNAc
Neuraminidase	NANA-(Gal)GalNAc
Aryl sulfatases A, B and C	Sulfates

[a] GlcNAc, *N*-acetylglucosamine; Gluc, glucuronide; Fuc, fucose; Gal, galactose GalNAc, *N*-acetylgalactosamine; Man, mannose, NANA, *N*-acetylneuraminic acid.

In earlier studies, our group reported immunolocalization of β-D-galactosidase and β-D-glucuronidase, two of the several glycohydrolases thought to be present in the sperm acrosome (Table 1). In addition to glycohydrolases, the acrosome contains proteinases, esterases, sulfatases, phosphatases, and phospholipases. These enzymes have been described in sufficient details in earlier publications [14,18] and will not be repeated here. Thus, the acrosome, formed during the early stage of spermiogenesis, resembles the cellular lysosome, a bag-like structure which normally functions in intracellular digestive and defensive mechanisms [19]. However, the two organelles are considered different. Because of its exocytotic properties, the acrosome is considered more like a secretory granule. Its important features as reported by Burgess and Kelly [20] are: (i) the secretory contents are stored over an extended period of time and are present in a concentrated form; (ii) the contents form a dense structure surrounded by an inner and outer acrosomal membrane; and (iii) the organelle undergoes exocytosis as a result of an extracellular stimulus.

After several decades of intensive research, investigators are beginning to understand the complex nature of the fully developed acrosome [for review see reference 3]. Accumulated evidence from several laboratories suggests the involvement of cytoskeletal domains such as actin, calmodulin, and α-spectrin-like antigens in the organization of the acrosome. In addition, the organelle contains filamentous structures primarily associated with the outer acrosomal membrane. However, the biological significance of the filamentous structures in the acrosomal organization or its function is not yet known.

Potential Role of Acrosomal Enzymes in Sperm Function

It is noteworthy that over four decades ago, de Duve proposed that penetration of vestments surrounding the egg may be mediated by the release of hydrolytic enzymes from the sperm acrosome during the acrosome reaction [19]. Although the precise role of the acrosomal enzymes is not yet known, the hydrolytic action of glycohydrolases and proteinases released at the site of sperm-egg (zona pellucida) binding, along with the enhanced thrust generated by the hyperactivated beat pattern of the spermatozoon bound to the egg, are important factors that regulate the penetration of the egg [21]. These enzymes could also be functional in capacitated spermatozoa (see below).

Spermatozoa penetrate the ovulated egg consisting of an oocyte surrounded by cumulus cells in an extracellular matrix (oocyte-cumulus complex). The matrix is primarily composed of hyaluronic acid covalently attached to a protein backbone [14]. The hyaluronic acid is a polymer composed of repeat disaccharide units containing glucuronic acid and N-acetylglucosamine in β1,3-linkage [14]. Each disaccharide unit in the polymer is attached to the next by β1,4-linkage forming alternate β1,3- and β1,4-linkages. The polymer, covalently linked to the protein backbone, participates in maintaining the level of hydration within the cumulus cells. The carboxyl groups in hyaluronic acid are completely ionized, and the polymer has a net negative charge at physiological pH. This property enables the acid to become soluble in aqueous medium forming an extremely viscous solution.

The enzyme hyaluronidase catalyzes the hydrolysis of β1,4-linkages in the hyaluronic acid [14]. Other hydrolytic enzymes, such as β-D-glucuronidase and N-acetylglucosaminidase may aid in the process of dispersion of the cumulus mass by further hydrolyzing disaccharide chains into monomeric units. Several other acrosomal glycohydrolases and proteinases have been implicated in the passage through the cumulus matrix. These include β-D-galactosidase [22], arylsulfatases [23,24], and acrosin [25]. Any reasonable hypothesis suggesting a role for the acrosomal enzymes in the hydrolysis and dispersion of the cumulus mass must include a possible mechanism(s) by which the hydrolytic enzymes could be present on the surface of the acrosome-intact spermatozoa so that they can react with their substrates on the cumulus oophorus. Interestingly, epididymal luminal fluid and seminal fluid are rich in most of the hydrolytic enzymes [26,27]. Thus, one likely possibility is that these enzymes tightly bind to the sperm surface during epididymal transit and ejaculation and remain bound during interaction of spermatozoa with the cumulus mass in the oviduct. A second possibility is that the hydrolytic enzymes present in the acrosome diffuse during capacitation and are exposed on the sperm surface during sperm-cumulus interaction (see below). Alternatively, some of the capacitated spermatozoa may undergo spontaneous acrosome reaction releasing the acrosomal contents (hydrolytic enzymes, proteinases, etc.). These powerful enzymes could disperse the cumulus cells allowing the acrosome-intact spermatozoa a clear passage through the cells.

Characteristics of Acid Glycohydrolases

Since acrosomal contents are thought to aid in the penetration of vestments surrounding the egg [19], an understanding of these contents is important and may lead to new strategies to regulate the sperm function. The acrosomal glycohydrolases, like the lysosomal glycohydrolases, have a high substrate specificity and will not hydrolyze even a closely related glycosidic linkage(s). These enzymes catalyze hydrolytic cleavage of terminal sugar residues from the glycan portion of glycoproteins and glycolipids [28], and are named on the basis of the sugar residue they cleave. Thus, a galactosidase will hydrolyze only a galactosyl residue(s), whereas a mannosidase will cleave only a mannosyl residue(s). Under *in vivo* circumstances, the hydrolytic enzymes function sequentially in such a way that the product of the first enzymatic cleavage becomes the substrate for the next enzyme [3].

The mechanism of action of glycohydrolases is thought to follow the model advanced for the lysozyme. In this model, there are two carboxylic acid moieties in the active site: one ionized and the other proteinated [29]. The former moiety is thought to stabilize the resulting oxocarbonium ions, either by ion pair interaction or by covalent binding, whereas the latter moiety facilitates departure of the cleaving group. Thus, the catalytic mechanism of all hydrolytic enzymes is the formation of an enzyme: substrate (i.e., enzyme: sugar) intermediate before the cleavage of the glycosidic bond and the release of the terminal sugar residue.

The sperm acrosome and cellular lysosomes contain several acid glycohydrolases with similar catalytic and immunological properties. There is a vast literature on the biosynthesis and targeting of glycohydrolases in the somatic cell lysosomes. The series of studies leading to the search for both the recognition markers on lysosomal glycohydrolases and the receptor for these enzymes have been reviewed before [3,30-33] and will be discussed briefly. The lysosomal glycohydrolases have both the anionic and carbohydrate recognition marker demonstrated to be mannose 6-phosphate (Man6-P). The formation of Man6-P on glycohydrolases involves an α-N-acetylglucosamine 1-phospho-6-mannose diester generated by direct transfer of GlcNAc-1-phosphate from UDP-GlcNAc to mannose(s) on protein-linked oligosaccharide, without involvement of a lipid-linked intermediate. As the glycohydrolases pass through the Golgi membranes, the GlcNAc-1-phosphate on high mannose oligosaccharide is hydrolyzed by Golgi N-acetylglucosamine 1-phosphodiester α-N-acetylglucosaminidase, exposing Man6-P residues [31,32]. Enzymatic removal of the GlcNAc residues from glycohydrolases involves a glycosidase rather than a phosphodiesterase.

The newly synthesized glycohydrolases are transported (targeted) to lysosomes by virtue of the exposed Man6-P ligand which is recognized by two types of Man6-P receptors; one a 215-kD cation-independent (CI) and the other a 46-kD cation-dependent (CD). The two receptors are transmembrane glycoproteins localized in the trans-Golgi network. The Man6-P ligand(s) provides recognizable signal for the CI/CD receptors which bind to the newly synthesized glycohydrolases, package them and segregate them into transport vesicles. The specific transport vesicles bud off from the trans-Golgi-associated membranes (transport vesicles), fuse with the late endosomes (prelysosomal structures) and deliver their newly synthesized enzymes to the lumen of these organelles. The affinity of the receptors for the phosphorylated oligosaccharides on the glycohydrolases is higher at pH 7.0 in the trans-Golgi network, and low at pH below 5.0, which is the pH in the lumen of the late endosomes. Thus, when transport vesicles fuse with the acidic prelysosomal structures, the glycohydrolases dissociate from the receptor, leaving them in the acidic vesicles while the released Man6-P receptor(s) recycles back to the trans-Golgi network for reuse [33].

A published report from our laboratory demonstrated the presence of β-D-glucuronidase, a glycohydrolase present in the lysosomes and microsomes of several somatic cells [9], in the Golgi membranes, Golgi-associated vesicles, and lysosome-like multivesicular structures in the late spermatocytes and early spermatids (see below). This distribution was not surprising since the germ cell enzyme, like somatic cell glycohydrolases, is likely phosphorylated and transported to the lysosome-like structures in the late diploid (pachytene spermatocytes) and early haploid (round spermatids) cells. Furthermore, the enzyme was localized in the forming/formed acrosome during spermiogenesis and exhibited similar lysosomal (L) form in spermatocytes, round spermatids, and condensed/elongated spermatids [9]. These data, and the fact that mouse germ cells have been reported to contain CD and CI Man6-P receptors [34], strongly suggest that spermatocytes and spermatids synthesize β-D-glucuronidase (and other glycohydrolases) which undergoes similar post-translational processing before being transported to either the lysosome or the forming acrosome by virtue of these receptors, as has been suggested [34]. This suggestion is consistent with other studies demonstrating that α-L-fucosidase [35] and β-D-galactosidase [11], two other glycohydrolases, synthesized in the spermatocytes and round spermatids have identical molecular mass. Additional studies on other glycohydrolases as well as Man6-P receptors in spermatogenic cells are needed and will provide insights into the targeting of these enzymes and biogenesis of the acrosome.

Accumulated evidence also suggests that some of the acrosomal antigens may not use the typical Golgi apparatus pathway described above to reach the organelle [36,37]. It should also be noted that several intra-acrosomal antigens have been reported to undergo modifications inside the acrosome. The growing list includes β-D-galactosidase [38], α-L-fucosidase [35], proacrosin-acrosin [39], SP-10 [40] and acrogranin [41]. Proacrosin-acrosin, a trypsin-like serine protease, is an acrosomal component. Proacrosin, the enzymatically inactive high molecular weight precursor form undergoes proteolytic processing into acrosin, an enzymatically active low molecular weight mature form in the spermatozoa of several species including rabbit [42], boar [43], human [44], and rat [39]. In the guinea pig spermatozoa, the reduction in the size of proacrosin has been demonstrated to be due to the processing of glycan moieties [41]. These data imply that most acrosomal antigens are synthesized as precursors that undergo modifications in the acrosome during sperm development in the testes and/or sperm maturation in the epididymis. These modifications are essential part of the maturation process before spermatozoa can capacitate, bind to the zona-intact egg, undergo the acrosome reaction, and fertilize an egg.

β-D-Glucuronidase: Biochemical Characterization and Immunolocalization in Mouse Testicular Germ Cells and Spermatozoa

β-D-glucuronidase is one of the most extensively studied glycohydrolases. It is present in all mammalian tissues examined, including male reproductive tissues [45] and spermatozoa [9]. The sperm-associated enzyme, along with other glycohydrolases, is thought to be localized in the acrosome and cytoplasmic droplets. The enzyme has long been of particular interest because of its dual localization, being present in the lysosomes and endoplasmic reticulum of several rodent tissues [46-48]. The lysosomal form (L form) of the enzyme purified from the rat [46] and mouse [47] liver and rat preputial gland [48] is a homotetrameric glycoprotein of an apparent molecular mass 280-kD. The microsomal form (M form) in mouse tissues consists of four forms (M1-M4) that correspond to non-covalent complexes formed between the proenzyme (X form) and one to four molecules of egasyn, a 64-kD glycoprotein with carboxyl esterase activity. The esterase active site of the egasyn interacts with the exposed carboxyl terminus of the X form of β-D-glucuronidase, forming the enzyme: egasyn complexes that are retained in the lumen of the endoplasmic reticulum [49].

Although β-D-glucuronidase has been extensively studied in somatic cells, our understanding of its expression, properties, and the localization of various molecular forms in testicular germ cells and spermatozoa is very limited. Since the enzyme, along with other glycohydrolases, may be important in sperm function, it was of interest to characterize and immunolocalize the enzyme in spermatogenic cells and spermatozoa. In the first part of this article, we describe procedures to: 1) characterize the molecular form(s) of the enzyme present in mouse testicular germ cells; and 2) examine its localization in the spermatogenic cells and cauda epididymal spermatozoa.

Isolation of Mouse Testicular Germ Cells and Cauda Spermatozoa: Preparation of their Extracts for Biochemical Studies

Adult male mice (C57BL/6J strain) from Harlan Industries, Inc. were sacrificed by CO_2 asphyxiation and perfused through the left ventricle with Enriched Kreb-Ringers Bicarbonate (EKRB) buffer to clear blood from the testes. Twenty testes, free of epididymis and fat pads, were detunicated and the seminiferous tubules were dispersed by enzymatic digestions as described [9]. Briefly, testes were mixed with 60 ml of collagenase solution (0.5 mg collagenase/ml EKRB), and the mixture was incubated for 20 min at 33°C in the presence of 5% CO_2 in air. The exposed seminiferous tubules were washed with 15 ml of EKRB buffer (3 washes) and subjected to trypsin (0.25 mg/ml EKRB) and DNase (0.5 µg/ml EKRB) treatment for 15 min at 33°C as above. The resulting cell suspension was made homogeneous by gentle pipetting for 3-5 min and the enzyme treatment was stopped by the addition of trypsin inhibitor solution (0.25 mg/ml EKRB). The released cells were filtered through a 70-µm nylon mesh and the filtered cells centrifuged (400 g for 5 min). The pelleted cells were washed by gentle suspension in 10 ml EKRB buffer and centrifugation as above (2 washes), and the washed cells were suspended in 20-25 ml of EKRB buffer containing 0.5% bovine serum albumin (BSA). The cells were counted using a hemocytometer and subjected to separation by unit gravity sedimentation on a Staput sedimentation chamber as below.

The cell separation was carried out at 4°C for a total of 4 hours beginning from the loading of the cells to the sedimentation chamber and the collection of the last fraction. The sedimentation chamber was initially loaded from the bottom through the cell-loading syringe with ~50 ml of EKRB buffer to a level just above the baffle. The EKRB buffer should completely fill the tubing leading from the syringe without any air bubbles. An aliquot containing 3-5 x 10^8 germ cells, suspended in the EKRB buffer supplemented with 0.5% BSA, was loaded through the syringe in the sedimentation chamber at a flow rate of 10 ml/min. Immediately after introducing the cell suspension, the chamber was filled with 1200 ml of a linear BSA gradient (2-4%) in the EKRB buffer at a flow rate of 10 ml/min adjusted with the T-valve and later increased to 40 ml/min. After the gradient loading was completed, the micrometering valve was closed and the cells were allowed to sediment. After 2 h and 40 min at 4°C, fractions (10 ml) were collected in disposable polystyrene tubes (17 x 100 mm) at a rate of 40-42 sec per tube. The tubes were centrifuged (400 g for 5 min) and the top 9 ml of the supernatants were aspirated, leaving ~1 ml in each tube. The presence of cells was confirmed by the appearance of the cell pellet and their quality and quantity assessed by examining cells in every fourth or fifth fraction using phase-contrast microscopy (Figures 1 and 2). Spermatocytes, which are the largest spermatogenic cells, were collected first followed by round spermatids of intermediate size and then condensed/elongated spermatids. The residual bodies were found only in the last few fractions. The fractions rich in spermatocytes, round spermatids, and condensed/elongated spermatids were pooled separately (Figure 1), washed with EKRB buffer and photographed using Nomarski differential interference contrast optics (Figure 2).

Three critical steps during cell separation are important that will have an effect on the purity of cells. First, loading of the mixed cells to the sedimentation chamber and the formation of a linear BSA gradient are important steps, and should be followed carefully. Second, the sedimentation time is very critical and should be established in a few preliminary runs. Finally, any vibrations to the sedimentation chamber during cell separation should be avoided to obtain reproducible results.

Mouse cauda epididymal spermatozoa were obtained by retrograde flushing through the vas deferens with 10 mM potassium phosphate buffer (pH 7.4) containing 0.15 M NaCl (PBS) and 25 mM benzamidine, an inhibitor of serine protease [9]. The collected fluid was centrifuged at 500 g for 10 min and the pelleted spermatozoa were suspended in the desired buffer for biochemical or immunocytochemical studies.

For the preparation of germ cell or sperm extract, the pelleted cells were suspended in PBS containing 25 mM benzamidine, and 0.2% Triton X-100 (v/v). The cell suspension was sonicated (2 x 15s) in a Fisher sonicator set at position 40, and centrifuged at 105,000 g for 30 min in a Beckman Ultracentrifuge at 4°C. The supernatant was removed by aspiration and the residue was extracted one more time by suspending in a small volume of the extraction buffer, followed by sonication and centrifugation as above. The combined supernatant, containing over 95% of the total β-D-glucuronidase activity present in a given cell suspension, was designated as cell extract and used for biochemical studies.

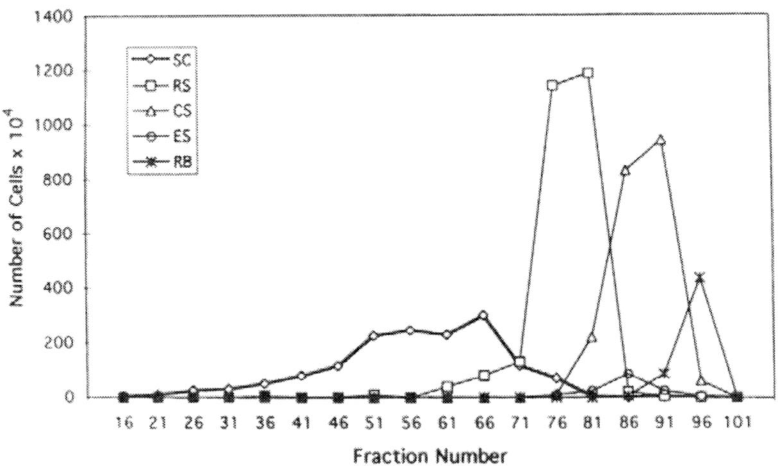

Figure 1. The sedimentation profile of mouse testicular germ cells following unit gravity sedimentation on a linear BSA gradient. Fraction (10ml) were collected from the bottom and their composition was examined using phase-contrast microscopy. The fractions were pooled as follows: spermatocytes (SC), fractions 35-57; round spermatids (RS), fractions 73-83; condensed spermatids (CS)/ elongated spermatids (ES), fractions 84-92; residual bodies (RB). Other details are in the text. Reproduced from Abou-Haila et al.[9] with permission.

Characterization of β-D-Glucuronidase Activity in Germ Cells

Highly enriched germ cell (>90% pure spermatocytes, round spermatids and condensed/elongated spermatids) populations were assayed for β-D-glucuronidase activity using published procedure [9]. Data presented in Table 2 demonstrate the presence of β-D-glucuronidase activity in the germ cell preparations. Interestingly, the spermatocyte-enriched fraction contained five times more enzyme activity per cell than the spermatid-enriched fractions. These data are consistent with our published studies in rat testicular germ cells (see below) which show that most glycohydrolase activities are 2 to 3 -folds higher in diploid cells than haploid cells [11]. Combined, the studies provided evidence suggesting that sperm-associated β-D-glucuronidase and other glycohydrolases are first expressed in the diploid spermatogenic cells.

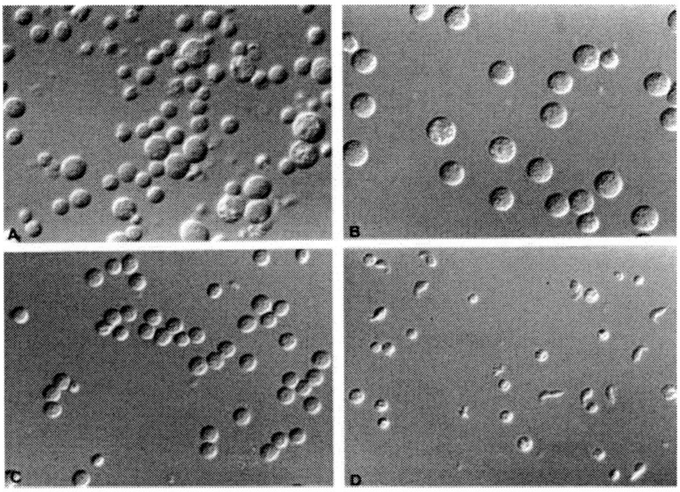

Figure 2. Microscopic appearance of spermatogenic cells prepared from mouse testes. A, mixed cells obtained after enzymatic disruption of the testes; B, spermatocytes; C, round spermatids; and D, condensed/elongated spermatids. Aliquots from the mixed cells (A) or enriched germ cells (B-D) were photographed using Nomarski differential interference contrast optics. Other details are in the legend to Figure 1. Reproduced from Abou-Haila et al. [9] with permission.

Table 2. β -D-Glucuronidase Activity in Mouse Testicular Germ Cells [a]

Cells [b]	Enzyme activity [c] (units/10^6 cells)
Spermatocytes	2.32
Round spermatids	0.45
Condensed/elongated spermatids	0.49

[a] The testicular germ cells released by enzymatic disruption of mouse testis were separated on a Staput cell separator. Fractions rich in spermatocytes, round spermatids and condensed/elongated spermatids were separately pooled.

[b] The purity of pooled fractions was examined by phase contrast microscopy (see Fig. 2). The cell purity was spermatocytes > 95%, round spermatids > 87% and condensed/elongated cell composition was condensed 94% and elongated 6%.

[c] Values are averages of two separate experiments.

Since some mouse tissues contain several molecular forms of β-D-glucuronidase (see above), attempts were made to characterize various forms present in the enriched germ cell extracts using 2D-gel electrophoresis carried out under non-denaturing conditions. The extracts were resolved by isoelectrofocusing in the first dimension, followed by gel electrophoresis in the second dimension using the Phast System (Pharmacia LKB Biotechnology, Inc.) as described [9,45]. The gels when stained for β-D-glucuronidase activity showed that the three cell preparations (spermatocytes, round spermatids, and condensed/elongated spermatids) contained a similar form of the enzyme with a broad isoelectrofocusing point (pI 5.5-6.0), which migrated between the two standard markers of the molecular mass 440- and 232-kD [9]. Furthermore, when the enzyme was immunoprecipitated from the pooled spermatogenic cell extract and resolved by sodium dodecyl sulfate (SDS)-PAGE under denaturing conditions (Laemmli's buffer), it revealed the presence of a single diffuse band with an apparent molecular mass of 74- kD. Taken together, these data are consistent with our suggestion that the spermatocytes (diploid cells) and spermatids (haploid cells) contain only one form of β-D-glucuronidase.

Two lines of evidence strongly suggested that this form is the L form. First, when the purified germ cell extracts were resolved by 2D-gel electrophoresis, the enzyme, like the L form in other tissues, had a broad isoelectrofocusing point in the pH range 5.5 to 6.0 and separated between standard proteins of apparent molecular mass 232- and 440-kD [9]. Second, when the germ cell β-D-glucuronidase was immunoprecipitated and resolved on SDS-PAGE, the subunit resolved as a diffuse (broad) band of an apparent molecular mass 74-kD, a value identical to the molecular mass of the L form present in the proximal epididymis [45]. Since the lysosomal form in other tissues is a glycoprotein [46-48], it is reasonable to suggest that an extensive microheterogeneity in sugar residues contributes to the broad isoelectric point and the presence of diffuse subunit of the germ cell β-D-glucuronidase.

The presence of only the lysosomal "L" form of β-D-glucuronidase in the spermatocytes, round spermatids, and condensed/elongated spermatids suggests that the diploid and haploid cells synthesize and process only the L form of the enzyme that undergoes similar post-translational processing before being transported to the forming acrosome. This suggestion is consistent with an earlier study demonstrating that the α-L-fucosidase, an exo-glycohydrolase that cleaves α-linked fucosyl residues from glycopeptides/glycoproteins, synthesized in rat spermatocytes or spermatids had an identical molecular mass [35].

In summary, the biochemical approaches demonstrate that β-D-glucuronidase is expressed in primary spermatocytes (diploid) and round spermatids (haploid). These cells express only the lysosomal "L" form of the enzyme, a homotetramer with an apparent subunit molecular weight of 74-kD. The enzyme present in the diploid and haploid cells has similar kinetic properties. These similarities allow us to suggest that these cells synthesize only the "L" form of the enzyme which undergoes similar post-translational processing before being transported to the forming acrosome [9].

Immunolocalization of β-D-Glucuronidase in the Mouse Spermatogenic Cells and Spermatozoa

β-D-glucuronidase is an acid glycohydrolase and is thought to be present in the sperm acrosome. However, the enzyme has never been localized. In fact, very few acid glycohydrolases have been studied in germ cells and spermatozoa using immunohistochemical procedures because very few high titer antibodies are available to approach these studies. In preliminary studies, we examined the cross-reactivity of the β-D-glucuronidase activity present in spermatogenic cells and spermatozoa using affinity purified or monospecific anti-rat preputial gland β-D-glucuronidase IgG [9]. Data from these studies demonstrated that the enzyme present in these cells is antigenically similar to that of the rat preputial gland β-D-glucuronidase. The cross-reactivity of the antibody with the mouse spermatogenic cells and sperm β-D-glucuronidase allowed us to use the antibody for immunolocalization studies. These preliminary studies also indicated that only the IgG purified on an immobilized β-D-glucuronidase column (monospecific IgG) was effective in immunolocalizing the enzyme on the methanol-permeabilized spermatozoa. Thus, all immunolocalization studies reported here were done using the monospecific IgG.

First, we examined the binding of the preimmune or monospecific IgG to the mouse cauda epididymal spermatozoa using indirect immunofluorescence microscopy. This was done with washed spermatozoa before (live sperm) or after permeabilization with -20°C methanol. Since data from these studies have been presented in our original report [9], they will be described briefly. The mouse sperm treated with preimmune IgG (9 μg IgG protein/ml) showed very light staining in the postacrosomal region and the midpiece (non-specific staining) before or after permeabilization. Similarly, the live sperm treated with immune IgG showed intense staining on the midpiece (non-specific). In contrast, the permeabilized sperm incubated with the immune IgG fraction followed by diluted secondary antibody (FITC-labeled anti-rabbit goat IgG) showed positive staining in the acrosomal (dorsal side of the sperm head) region. The cytoplasmic droplets revealed light immunopositive reaction only in the permeabilized sperm.

The acrosomal localization was confirmed by electron microscopy carried out in the mouse cauda epididydmal sections. Immunopositive reaction was revealed in the epithelial cells (data not included) as well as the intra-acrosomal region of the spermatozoa as evident by the presence of gold particles only when the immune IgG was used (compare Figure 3A vs Figure 3B). The gold particles were present in the acrosomal head cap, but not in the equatorial segment. In addition, very few particles were visible within the contents of the cytoplasmic droplets (Table 3). The last result is in agreement with the biochemical studies indicating the presence of glycohydrolase activities in the cytoplasmic droplets from various species [50,51].

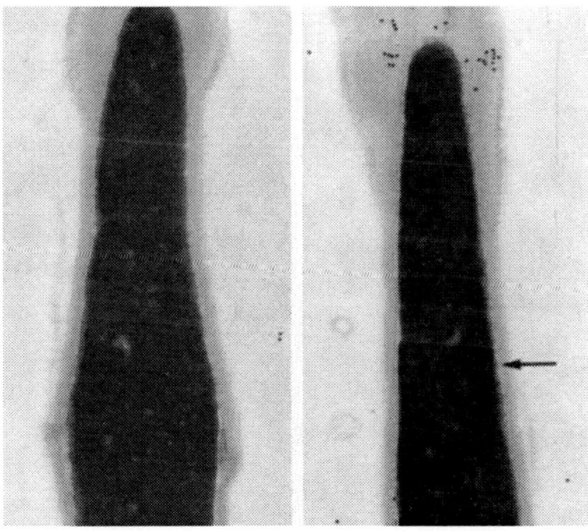

Figure 3. Distribution of β-D-glucuronidase in mouse cauda epididymal spermatozoa incubated with preimmune (A) or monospecific (B) IgG. Note the presence of gold particles in the acrosomal head cap (convex dorsal face of the sperm head) in B. The equatorial segment (arrow) has no gold particles. Original magnification x 40,000. Reproduced from Abou-Haila et al [9] with permission.

As stated above, the β-D-glucuronidase seen in the sperm acrosome is synthesized and transported to the organelle during spermatogenesis. It was, therefore, of interest to examine its localization during sperm development in the testis. This was done using preimmune or monospecific (immune) IgG in three different protocols. In preliminary studies, we examined the binding of preimmune/immune IgG (9 or 12 μg IgG protein/ml) to the somatic and germ cells present in paraffin sections of the mouse testis using immunoperoxidase method [9]. Somatic cells (Leydig and Sertoli cells) revealed immunopositive reaction with both concentrations of the immune IgG. In contrast, the late spermatocytes (pachytene spermatocytes) and early round spermatids showed immunopositive reaction in the Golgi region only when the higher concentration of the immune IgG and longer incubation conditions were used. However, the nonspecific background staining under these conditions was higher, which made interpretation rather difficult.

In the next set of experiments, we examined the binding of preimmune/immune IgG to the mixed or purified germ cell populations using indirect immunofluorescence microscopy. The cells revealed immunopositive reaction only when they were permeabilized and immune IgG was used as the primary antibody. Details from this approach are presented in our original article [9] and will be summarized in this report. Briefly, the spermatocytes (diploid cells) and early round spermatids (haploid cells stages 2-4) revealed strong immunopositive fluorescence in small granules disseminated within the cytoplasm. In addition, some spermatids displayed intense immunofluorescent spherical granules that likely correspond to the formation of the acrosome. At the later stages of spermiogenesis (stages 5-8), the number of fluorescent granules decreased and an intensely fluorescent cap-like structure appeared at one pole of the round spermatids. In some of these cells, depending on their orientation, the acrosomal cap was visible as two fluorescent rings. As the spermiogenesis continued (stages 9-16), the intense immunofluorescent reaction became confined to a distinct sickle-shaped structure, the forming/formed acrosome. The residual bodies were also intensely immunopositive (Table 3).

Immunoelectron microscopy was the final approach to localize β-D-glucuronidase within the germ cell organelles. Data from this approach described in our original report [9] are summarized below. The Golgi stacks and associated vesicles as well as the lysosome-like structures in the pachytene spermatocytes were immunopositive as evident by the presence of gold particles. The immunolabeling was more intense in the Golgi-associated vesicles of round spermatids. The acrosomal vesicles and the acrosomal cap showed significant immunolabeling. The reaction was uniformly distributed within the acrosome of condensed/elongated spermatids. In the later stage of sperm formation, the immunopositive reaction was confined to the acrosomal head cap (see Figures 9F, G in reference 9).

Our data demonstrate that β-D-glucuronidase is present in the Golgi apparatus and lysosome-like structures of the late spermatocytes and early round spermatids (Table 3). As spermatogenesis continued, the lysosome-like structures likely corresponding to the multivesicular bodies described by Clermont and Tang [15] disappear. Their disappearance coincided with the appearance (development) of the acrosome, an observation consistent with our suggestion that the lysosome-like structures (multivesicular bodies) either fuse or empty their contents into the forming acrosome. There was no evidence of compartmentalization of the enzyme in the acrosomal cap of the forming/formed acrosome since the gold particles were seen uniformly distributed throughout the region (see Figure 9 in reference 9). However, in the final stage of the testicular sperm formation, the enzyme was confined to the head cap of the sickle-shaped acrosome. There was no immunopositive reaction in the equatorial segment of spermatozoa, a result consistent with an earlier report indicating that acrosin, a trypsin-like proteinase, is localized in the rostral (but not the equatorial) segment of spermatozoa [52].

Table 3. Immunolocalization of β -D-Glucuronidase and β -D-Galactosidase in Spermatogenic Cells and Spermatozoa

Cell-Type	Structures	Procedures		
		IP	IF	IG
Spermatocytes	Golgi stacks	+	+	+
	Golgi-associated vesicles	NI	NI	+
	Lysosome-like-structures	+	+	+
Round spermatids	Golgi stacks	+	+	+
	Multivesicular bodies	NI	+	+
	Acrosomal vesicles	+	+	+
Elongated spermatids	Sickle-shaped acrosome	+	+	+
	Residual bodies	+	+	+
Spermatozoa	Acrosome	+	+	+
	Cytoplasmic droplets	ND	+	+

Details are described in our published reports [9-11].
IP: immunoperoxidase; IF: immunofluorescence; IG: immunogold;
NI: not identified; ND: not determined.

Characterization, Immunolocalization, and Biosynthesis of β-D-Galactosidase in Rat Spermatogenic Cells and Spermatozoa

β-D-galactosidase is another exo-glycohydrolase which cleaves terminal β-galactosyl residues from glycoproteins and glycolipids [19,22,27,28,38]. The enzyme is present in all mammalian tissues examined, including male reproductive tissues [27,38,53,54] and spermatozoa [10,11,22,38]. It is present in a latent form in lysosomes [55,56] and in soluble forms in body fluids, including serum [57] and epididymal luminal fluid [27,58]. The soluble enzyme in the luminal fluid occurs in two molecular forms of 97-kD and 84-kD [58]. The two forms have identical or very similar polypeptide backbone of 70-kD; however, the 97-kD form was shown by our group to contain much more total carbohydrate and sialic acid than the 84-kD form [58]. Both isoforms optimally cleave a synthetic substrate (*p*-nitrophenyl [PNP] β-D-galactopyranoside) at acidic pH and two glycoprotein substrates ([^{3}H]gal-fetuin and [^{3}H]gal-ovomucoid) at neutral pH [58]. The last finding suggests that the two isoforms will be functional within the physiological pH of the epididymal lumen believed to be 6.6-6.8 [59]. Evidence presented in earlier reports strongly suggests that the epididymal luminal fluid β-D-galactosidases have a role in the modification of sperm plasma membrane glycoproteins during epididymal maturation of spermatozoa [27,58].

Sperm-associated glycohydrolases are thought to be present in the acrosome and cytoplasmic droplet. Relatively few antibodies have been generated against glycohydrolases that are suitable for examining low levels of enzymes present in spermatogenic cells and spermatozoa. Thus, our understanding on the biosynthesis of acrosomal glycohydrolases is very limited. Since testicular germ cells are not suitable for long-term culture, studies on transport of the acrosomal components have relied on ultrastructural studies. Data have provided evidence suggesting that, like the lysosome formation, the assembly of an acrosome during spermiogenesis involves transport of newly synthesized antigens through the Golgi apparatus via coated vesicles before they are delivered to the forming acrosome [60].

Although β-D-galactosidase has been extensively studied, our understanding of its expression, properties, and the localization of various molecular forms in rat spermatogenic cells and spermatozoa is very limited. Since the enzyme, along with other glycohydrolases, may be important in mammalian fertilization [22], the aims of the studies presented here were to: 1) use immunocytochemical approaches to examine its localization in spermatogenic cells and spermatozoa; and 2) examine its synthesis and processing in the purified spermatogenic cells. Immunocytochemical approaches indicate that β-D-galactosidase is localized in the sperm acrosome and cytoplasmic droplets. Biochemical studies included in this article indicate that the enzyme, like other glycohydrolases studied, is first expressed in late spermatocytes. Multiple biochemical as well as biosynthetic studies indicate that spermatocytes and round spermatids contain several molecular forms of the enzyme, some of them appear to be processed in a cell-specific manner.

Purification of β-D-Galactosidase, Production of Antiserum and Preparation of Affinity Purified and Monospecific γ-Globulin (IgG)

β-D-galactosidase (97-kD form) was purified to apparent homogeneity from rat epididymal luminal fluid and used for the production of antiserum by immunizing a female white rabbit by our published procedure [58]. Affinity purified polyclonal antibody (IgG fraction) was prepared on a column of immobilized protein G (Pharmacia LKB Biotechnology, Piscataway, NJ) using manufacturer's protocol as previously described [58]. The IgG fraction was dialyzed against 50 mM Tris-HCl buffer, pH 7.5 containing 0.15 M NaCl and concentrated (3-4 mg protein/ml) using Centricon 10 micro-concentrator from Amicon (Beverly, MA). The monospecific IgG fraction was prepared by applying the anti-serum to a column of immobilized β-D-galactosidase (immobilized on Sepharose 4B) equilibrated with 50 mM Tris-HC1 buffer, pH 7.5 containing 0.15 M NaCl. The bound γ-globulin was eluted with the above buffer containing 7 M urea, followed by dialysis and concentration as above. The affinity purified or monospecific antibodies were aliquoted, and stored frozen at -75°C [58].

Rat epididymal β-D-galactosidase used as a carrier (unlabeled) enzyme during biosynthetic studies was partially purified from rat epididymis by acid and heat treatment [61]. Briefly, epididymal homogenates (10%, w/v) were prepared at 4°C in a solution containing 0.1 M Tris-HCl buffer, pH 7.5, and 0.15 M NaCl by homogenizing for 50s with a polytron homogenizer (Type PT 1020 3500; Brinkman Instruments, Westbury, NY) set at position 5. Sodium deoxycholate was added to a final concentration of 0.5% to the vigorously stirred homogenate to solubilize the lysosomal β-D-galactosidase. The pH of the homogenate was lowered to 4.8 by adding 1 M acetic acid (0.06 ml/ml), and the mixture was heat treated by incubation at 56°C for 30 min. Deoxycholate (which precipitated due to the acidic pH) and denatured proteins were removed by centrifugation (25,000 g for 30 min). The supernatant was removed by aspiration, assayed for β-D-galactosidase activity using PNP-β-D-galactoside substrate at pH 3.5. The partially purified enzyme (specific activity, 65 units/mg protein) was concentrated to a small volume (~2 ml) using a microconcentrator, aliquoted (2 units/aliquot), stored frozen at -20°C, and used as a source of carrier (unlabeled) enzyme [10,11].

Isolation of Spermatogenic (Germ) Cells from Rat Testes

The protocol consists of the: i) sequential enzymatic dissociation of seminiferous tubules and epithelial cells of the testis, followed by ii) separation of germ cells by unit gravity sedimentation on a continuous BSA gradient as above. The cells sedimented on the basis of their size were removed by collecting fractions from the bottom of the gradient. The procedure our group used was adopted from a published procedure for mice and rats [7], and is similar to the one discussed above for mice. Since the method used for rats has been discussed in two earlier publications [8,12], it will be summarized in this article. In brief, rats (~ 8 weeks old) were killed by CO_2 asphyxiation. The testis along with the epididymis and fat pad, was excised and perfused with the EKRB medium through the testicular artery. The EKRB solution was prepared just before use by mixing stock solutions and the pH was adjusted to 7.0-7.2 by bubbling with 5% CO_2 in air for 5-10 min as described [7,8]. The perfused testes were excised and their contents (seminiferous tubules) released by a small incision in the tunica albuginea. Mechanical dispersion of the seminiferous tubules is not recommended since it causes the formation of multinucleated cells.

The seminiferous tubules were dispersed by enzymatic digestion (collagenase solution 1mg/ml EKRB) under 5% CO_2 in air for 20 min at 33°C in a shaking water bath operated at 120 cycles/min. The reaction was stopped by washing the exposed tubules with EKRB (3 washes). The washed tubules were subjected to trypsin (0.5 mg/ml EKRB) and DNase (1µg/ml EKRB) treatment for 15 min at 33°C as described [8]. Cell aggregates that remained after these treatments were made homogeneous by gently pipetting for 2-5 min. The reaction was stopped by the addition of trypsin inhibitor (0.5 mg/ml EKRB) and the cell suspension was filtered through a 70-µm nylon mesh. The filtered cells were pelleted by centrifugation at 400 g for 5 min and the pelleted cells were washed by suspending in EKRB buffer containing 0.5% BSA and centrifugation as above (2 washes). The washed cells were suspended in 20-25 ml EKRB containing 0.5% BSA. The yield of spermatogenic cells prepared from two testes was usually ~ 7.5 to 9 x 10^8 as counted by a hemocytometer on a phase-contrast microscope. Over 95% of the cells were viable as assessed by the exclusion of trypan blue. Finally, the cells were allowed to sediment by gravity on a linear BSA gradient according to the method of O'Brien et al. [7] as we have described in our earlier publications [8,11,12]. Fractions (10 ml) were collected, cells were sedimented by centrifugation (400 g for 5 min), and their composition monitored by phase-contrast microscopy. The fractions rich in spermatoytes (fractions 25-56); round spermatids (fractions 65-72); condensed/elongated spermatids (fractions 84-99); and residual bodies (fractions 105-110) were pooled separately, washed by suspending gently in EKRB and used for further studies. Other details with the figure of the sedimentation profile of rat testicular germ cells following unit gravity sedimentation can be found in our earlier publications [8,12].

The unit gravity sedimentation procedure allowed us to obtain highly enriched (> 90%) germ cell populations and residual bodies. The isolated germ cells and residual bodies were assayed for β-D-galactosidase and four other acid glycohydrolase activities by our published procedure [57] in the presence of 0.2% Triton X-100 to ensure that the cells were lyzed and the PNP-glycoside substrates were in contact with the active site of the intracellular enzymes. Data reported in Table 4 show the presence of all five enzyme activities. In the absence of the detergent, the enzyme activities were 50-70% of these values. With the exception of α-L-fucosidase activity which was similar in spermatocytes and round spermatids, the other four glycohydrolase activities were two to three-fold higher in diploid cells compared to the haploid cells. Furthermore, the sum of glycohydrolase activities present in condensed/elongated spermatids and residual bodies nearly equals the amount present in the round spermatids (Table 4). These data provided evidence indicating that sperm-associated glycohydrolase activities are first expressed in the diploid (spermatocyte) germ cells.

Table 4. Glycohydrolase Activities in Rat Testicular Germ cells And Spermatozoa [a]

	Cells (milli units/10^6 Cells)			
Enzyme [b]	SC	RS	C/ES	SP
β -N-Acetylglucosaminidase	31.0 ± 2.7	7.4 ± 0.2	4.3 ± 0.6	220.8 ± 12.4
β -D-Galactosidase	12.4 ± 2.9	6.6 ± 0.4	3.7 ± 0.6	43.5 ± 4.4
β -D-Glucuronidase	1.6 ± 0.2	0.8 ± 0.1	0.5 ± 0.1	2.4 ± 0.2
α -D-Mannosidase	15.3 ± 1.1	7.5 ± 0.5	4.3 ± 0.2	18.1 ± 1.0
α -L-Fucosidase	44.8 ± 2.7	51.1 ± 3.0	51.1 ± 3.0	ND

[a]The germ cells were prepared from the rat testes and spermatozoa from the distal cauda epididymidis as described in our earlier publication [8]. Fractions rich in spermatocytes (SC), round spermatids (RS), condensed/elongated spermatids (C/ES) were pooled (see Figure 1) and centrifuged (400 g/5min). The pelleted cells were suspended in PBS (approximately 100 x 10^6 cells/ml) and used for enzyme assay. Values are average of four separate experiments in triplicate with ± standard deviation (SD).

[b]All enzymes were assayed using PNP-glycoside substrates.

ND, not determined.

Table 5. Apparent Km for Rat Spermatogenic Cell and Cauda Sperm β -D-Galactosidase

Germ cell [a]	Km (mM) [b]
Spermatocytes	0.80
Round spermatids	0.82
Condensed/elongated spermatids	0.83
Residual bodies	1.02
Cauda spermatozoa	0.52

[a] Rat germ cells and residual bodies were prepared as described in the text. The purity of germ cells was similar to the values reported in Table 4.

[b] Values reported were calculated from a linear double-reciprocal plot generated following substrate concentration vs enzyme activity using PNP-galactoside as substrate at pH 3.5.

β-D-galactosidase activities present in the spermatogenic cells and spermatozoa were found to optimally cleave p-nitrophenyl-β-D-galactopyranoside (PNP-galactoside) at an acidic pH of 3.5. The PNP-substrate concentration vs enzyme activity using enriched populations of germ cells and the cauda epididymal spermatozoa generated a linear reciprocal plot for all cells. An apparent K*m* calculated from various plots is presented in Table 5.

Localization of β-D-galactosidase in Rat spermatogenic cells and cauda epididymal spermatozoa

β-D-galactosidase is an acidic glycohydrolase and was shown by our group to be localized in the sperm acrosome and cytoplasmic droplets. Data from these approaches have been published in two earlier articles [10,11] and are briefly summarized here: i) the cauda epididymal spermatozoa treated with the preimmune IgG showed immunostaining on the midpiece before or after permeabilization with methanol; ii) the non-permeabilized sperm treated with the immune IgG showed nonspecific fluorescence on the midpiece; iii) in contrast, permeabilized sperm treated with the immune IgG revealed intense immunopositive staining (fluorescence) mostly on the dorsal side (acrosome region) of the sperm head; iv) most of the cytoplasmic droplets were detached from the spermatozoa and displayed intense immunostaining; however, the very few cytoplasmic droplets still attached to spermatozoa revealed light staining only when the cells were permeabilized. The presence of enzyme within the sperm acrosome and cytoplasmic droplets was in agreement with biochemical and morphological studies from several laboratories [10,11,50,51].

The acrosomal localization of β-D-galactosidase was further confirmed by electron microscopy carried out on the rat cauda epididymal sections. The immunopositive reaction was seen in the epididymal epithelial cells and in the intra-acrosomal region of the sperm head as evident by the presence of gold particles only when the immune IgG was used [11]. The gold particles were present in the acrosomal head cap, but not in the equatorial region, a result similar to the localization of β-D-glucuronidase (see above).

As stated above, β-D-galactosidase and other glycohydrolases present in the acrosome are synthesized and packaged in the organelle during sperm development in the testis. We, therefore, examined the stage-specific localization of the enzyme during spermatogenesis. The localization studies were attempted using light and electron microscopic approaches in multiple ways. First, we examined the binding of the preimmune or immune IgG to the somatic and germ cells present in paraffin sections of the rat testis using the immunoperoxidase approach. Experimental details have been described in earlier reports [11] and will not be repeated here. Data from this approach, summarized here show the following: i) no immunopositive reaction was seen in the somatic (Leydig or Sertoli) or spermatogenic cells when preimmune IgG was used as the primary antibody; ii) however, an intense immunopositive reaction was revealed in the Leydig cells and a moderate immunopositive reaction was seen in the cytoplasm of Sertoli cells when the immune IgG was used as the primary antibody. In the spermatogenic cells, immunopositive reaction was confined to granules dispersed in the cytoplasm (around the nucleus) of the late spermatocytes as well as the round spermatids; iii) the forming acrosome in the elongated spermatids was immunopositive. In addition, an intense immunopositive reaction was seen in the residual bodies present near the lumen of the seminiferous tubules (Table 3).

Next, we used the postembedding immunogold labeling protocol to localize the β-D-galactosidase in the testicular sections, and the enzyme was visualized by the presence of gold particles using electron microscopy. As expected, an intense immunopositive reaction was seen in the lysosomes of the somatic cells. In spermatogenic cells, the enzyme was seen in the late spermatocytes (pachytene spermatocytes) and round spermatids as evident by the presence of gold particles in the Golgi apparatus and lysosome-like structures (Table 3). The immunolabeling was obvious in the Golgi-associated vesicles of the early Golgi phase (stages 2-3) and in the floculent material surrounding the acrosomal granule [11]. The round spermatids during the cap phase (stage 5) had intense immunolabeling in the head cap extending over the nucleus and light labeling of the acrosomal granule. In addition, the immunopositive reaction was seen in small vesicles dispersed over the outer acrosomal membrane in the Golgi region. The gold particles were also present in the head cap of elongating/elongated spermatids (stages 8-16) and became essentially confined to the acrosomal cap of mature spermatids and cauda epididymal spermatozoa. This localization is similar to β-D-glucuronidase (see above), another exoglycohydrolase which is localized in the pachytene spermatocytes and round spermatids [9].

It is important to emphasize that immunocytochemical approaches are qualitative and detect the antigen only when its local concentration is reasonably high. However, when the antigen is present in low local concentration throughout the cell cytoplasm, the immunopositive reaction may not be intense enough for the microscopic evaluation. In addition, the preservation of the antigenic site(s) during tissue fixation is an important factor that may have an effect on the intensity of the immunopositive reaction. Thus, it is reasonable to suggest that a high titer antibody is needed to detect even low levels of glycohydrolases (and other antigens) present in the testicular germ cells. We highly recommend to first establish optimal conditions of antibody concentration and incubation time to determine non-specific background staining. Also, samples (tissue sections or isolated germ cells) should be incubated in the presence of immune IgG preabsorbed with an excess of the purified enzyme to establish negative staining (negative controls).

Whether the germ cell glycohydrolases use Man6-P ligand pathway to be targeted to the lysosomes and ultimately to the forming acrosome is not yet known. The presence of β-D-glucuronidase and β-D-galactosidase in the Golgi membranes, Golgi-associated vesicles, and lysosome-like structures in spermatocytes and round spermatids suggests that the enzymes are likely phosphorylated in the Golgi membranes and are recognized by two classes of Man6-P receptors localized in trans-Golgi cisternae [31-33]. The facts that mouse germ cells and Sertoli cells contain both the 215-kD CD and 46-kD CI Man 6-P receptors [34] and bovine testes contain two glycosylation isoforms of the cation-dependent Man6-P receptor suggest the potential involvement of these receptors in the segregation and targeting of newly synthesized β-D-glucuronidase and β-D-galactosidase (and other glycohydrolases) to the lysosomes/multivesicular bodies or the acrosome seen in the germ cells. To the best of our knowledge, data included here and in the original reports [9,11] provided first evidence suggesting that two acid glycosidases are present in the Golgi apparatus and Golgi-associated vesicles in spermatocytes and spermatids. Studies are needed to examine the synthesis and distribution of other glycohydrolases as well as Man6-P receptors in germ cells and may provide additional insights into the biogenesis of the sperm acrosome.

Characterization of Various Molecular Forms of β-D-galactosidase in the Spermatogenic Cells

Since β-D-galactosidase occurs in several forms, attempts were first made to identify various molecular forms of the enzyme in spermatocytes and round spermatids. This was done by first partially purifying the enzyme from the germ cell extracts by affinity chromatography on a column of immobilized p-aminophenyl-β-D-thiogalactopyranoside by our published protocol [58]. Briefly, the spermatocyte or round spermatid extract in 2 ml of ice cold 20 mM phosphate-citrate buffer, pH 4.3, containing 0.1 M NaCl, 1% Triton X-100 (v/v), and 25 mM benzamidine was applied to the affinity column (0.5 x 6.0 cm) at 0-4°C, equilibrated with the above phosphate-citrate buffer. The extracts were separately applied to the column at a flow rate of 2 ml/h followed by washing with 25 ml of the above buffer at a flow rate of 4-5 ml/h. The bound enzyme was eluted with the above buffer containing 7M urea [58]. The enzymatically active fractions were pooled, concentrated to a small volume using a microconcentrator, dialysed against 10 mM Tris-HCl buffer containing 0.1M NaCl and 0.02% sodium azide, and assayed for PNP-β-D-galactosidase activity at pH 3.5. Aliquots containing 0.05-0.08 units/tube were dried in a Speed Vac and stored frozen at –20°C.

The β-D-galactosidase activities purified from spermatocytes and round spermatids were resolved by SDS-PAGE and transferred to a nitrocellulose sheet. The immunoreactive polypeptides were detected by Western blot analysis. Data from this approach shown in our original publication [11] demonstrated that under steady-state conditions, both spermatocytes and round spermatids, contained several molecular forms (103- , 88- , 70- , 64- and 62-kD). Interestingly, the 103- and 62-kD immunoreactive polypeptides present in the round spermatids were not seen in the spermatocytes.

The 88-kD is likely the glycosylated form of β-D-galactosidase, whereas the 70- kD monomer appears to be the de-N-glycosylated form. The latter form was demonstrated by our group to be generated following N-glycanase treatment of the newly synthesized β-D-galactosidase in germ cells [10], and the enzyme present in the rat spermatozoa [38] and epididymal luminal fluid [58]. Interestingly, the 62-kD form is expressed in a cell-specific manner, being present only in round spermatids. It is not known at the present time whether the 103-kD form seen in the immunoblots of the round spermatids (see below) is related to the β-D-galactosidase. The fact that this polypeptide was not seen in biosynthetic studies (Figure 4) suggests that it is not related to the β-D-galactosidase. A less likely possibility would be that the half-time (t ½) for the appearance of 103-kD form is longer than the chase time of 10-h. Studies using longer chase times were not attempted since the germ cells are viable in culture for a short period of time [7,34].

The enzyme activities present in the spermatocytes, round spermatids and cauda epididymal sperm extracts were found to optimally cleave PNP-β-D-galactoside substrate at an acidic pH of 3.5. The substrate concentration vs enzyme activity studies generated a linear double-reciprocal plot for all cells. An apparent K*m* calculated from these plots is presented in Table 5.

Our published studies have demonstrated that anti-β-D-galactosidase IgG (monospecific) effectively immunoprecipitated the enzyme present in the detergent solubilized spermatocytes, round spermatids and cauda epididymal spermatozoa [10,11]. These results, and the fact that β-D-galactosidase activities present in the detergent solubilized spermatogenic cells and spermatozoa have similar pH optima, allowed us to suggest that the germ cells and spermatozoa contain immunologically and kinetically similar β-D-galactosidase activities. It should, however, be noted that the K*m* for the germ cell β-D-galactosidase is somewhat higher than the enzyme present in the cauda epididymal spermatozoa (Table 5). A likely explanation is that whereas the germ cells possess precursor and mature forms of the enzyme (see below), the epididymal spermatozoa possess only the mature form of β-D-galactosidase. The mature form in spermatozoa is expected to have high affinity (and low K*m*) for the PNP-β-D-galactoside substrate.

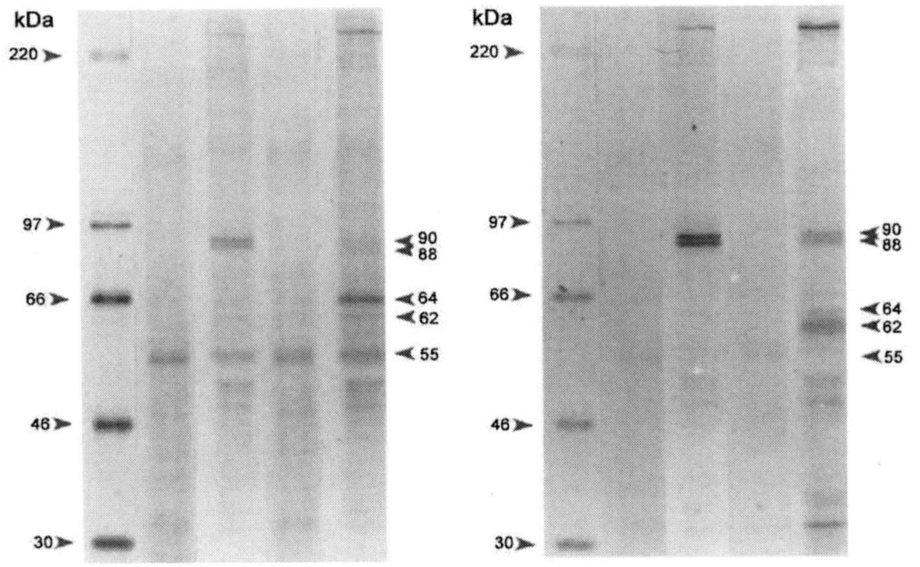

Figure 4. Synthesis of β-D-galactosidase and conversion of the newly synthesized precursor forms to the mature forms in rat testicular germ cells. Enriched populations of spermatocytes (left) or round spermatids (rignt) were pooled (see Figure 1 in reference 8). Cells (1.5 x 10^7 cells/ml) were incubated in 3 ml of methionine free MEM for 30 min, followed by incubation in methionine-free MEM supplemented with [^{35}S] methionine (400 μCi/ml) as described (11). Following 30 min of pulse labelling (lanes 2 and 3), the cells were washed and incubated (chase) in non-radioactive medium for 10-h (lanes 4 and 5). After this incubation, the cells were washed, sonicated and the radiolabeled β-D-galactosidase was immunoprecipitated. Immunoprecipitates were electrophoresed on 7% polyacrylamide gels containing SDS (SDS-PAGE) at 30 mA constant current and prepared for fluorography [11]. Lane 1, [^{14}C]-labeled standard marker proteins (Amersham); lanes 2 and 4, preimmune IgG; and lanes 3 and 5, immune IgG. Molecular weight x 10^{-3}. Reproduced from Skudlarek et al. [11] with permission.

Synthesis and turnover of β-D-Galactosidase in Spermatocytes and Round Spermatids

Radiolabeling of spermatocytes and round spermatids. Enriched populations of spermatocytes (>95% pure, impurities being 1.2% Sertoli cells; 2.4% round spermatids; and 1.2% condensed/elongated spermatids) or round spermatids (> 95% pure, impurities being 4.5% condensed/elongated spermatids) were pooled and 1.5 x 10^7 cells/ml were incubated in 3 ml of methionine-free MEM [11]. After 30 min at 34°C, the methionine-depleted cells were centrifuged at 400g for 5 min, and the pelleted cells were suspended in the above medium containing [^{35}S] methionine (400 μCi/ml) and incubated at 34°C under 5% CO_2 in air. After 30 min of labeling (pulse), the cells were pelleted and washed four times by suspending each time in 2 ml of PBS and centrifugation as above. The washed cells were either frozen (pulse) or suspended in 3 ml of medium containing non-radioactive methionine and cultured for pulse/chase studies as described in each experiment.

Extraction, immunoprecipitation, and SDS-PAGE of ^{35}S-labeled β-D-galactosidase. The frozen cells were suspended in ice-cold 0.1M Tris-HCl buffer, pH 7.5, containing 0.15M NaCl, 1% Triton X-100, 0.5% sodium deoxycholate, and a protease inhibitor cocktail (10, 11). The cells were sonicated and centrifuged at 105,000g for 30 min. The supernatant was removed by aspiration, supplemented with unlabeled enzyme carrier (2 units), and mixed with 60 μg of the affinity-purified immune or preimmune IgG. Samples were incubated at 34°C for 30 min followed by overnight incubation at 4°C with gentle rocking. Following these incubations, the immunoprecipitates were collected by centrifugation. The immune/preimmune precipitates were washed five times with Tris-NaCl-SDS buffer by suspension and centrifugation, dissolved in urea-SDS and electrophoresed on 7% polyacrylamide gels (SDS-PAGE) under reducing conditions. Gels were exposed to Biomax MR film, and the radioactive bands were revealed by processing after 1 week (Figure 4). In pulse-chase studies, individual bands were cut from the gels, solubilized in NCS solubilizer, and the radioactivity was measured by liquid scintillation spectroscopy [11,58].

Following incubation of spermatogenic cells (spermatocytes/round spermatids) for 30 min with [^{35}S]methionine (pulse), the cells were pelleted, washed and incubated for 10-h in non-radioactive methionine medium (see above) for chase. The newly synthesized β-D-galactosidase from spermatogenic cells (after pulse or pulse-chase) was immunoprecipitated, and the molecular forms were detected by electrophoresis and autoradiography as above. Data from these studies presented in Figure 4 demonstrate that both spermatocytes and round spermatids synthesize two high molecular weight precursor forms (90-/88-kD) that are converted to lower molecular weight mature forms (64-/62-kD) after 10-h of chase.

The protocol reported here and elsewhere [10,11] can be adopted to study synthesis of other proteins/glycoproteins in cell culture or to assess the effect of specific agents on spermatogenesis. For reproducible results, it is important to use a high titer specific antibody (preferably affinity purified IgG fraction). The antibody should first be titrated against a carrier antigen to establish the amount of antibody needed to quantitatively immunoprecipitate the known quantity of the carrier protein. Since the newly synthesized radio-labeled antigen contributes very little protein/glycoprotein, the addition of a carrier antigen is highly recommended to ensure that the antigen:antibody complex forms a visible pellet during washing procedure.

The time course of conversion of precursor forms (90-/88-kD) in the spermatocytes and round spermatids to mature forms (64- and 62-kD) were quantified in pulse-chase studies. At various time intervals of the chase, individual bands were cut from the gels, solubilized and radioactivity counted as above. Data for the time course of conversion of precursor forms to mature forms are presented in Figure 5. No radioactivity was lost from the round spermatids for up to 2-h after initiation of the chase. In contrast, over 15% of the precursor forms were converted to the 64-kD mature form in spermatocytes within 2-h. Data shown in Figure 5 allowed us to calculate the half-time (t ½) for the turn over of the β-D-galactosidase precursor forms to the mature forms at 6.5 and 9-h in spermatocytes and round spermatids, respectively. Another obvious difference between these germ cells was the apparent molecular weight of the mature forms after 10-h of chase. In spermatocytes, most of the radioactivity was found in the 64-kD mature form, whereas the 62-kD was the predominantly mature form in round spermatids (Figures 4 and 5). To the best of our knowledge, studies from our group are first reports where an exoglycohydrolase has been demonstrated to undergo processing in a cell-specific manner during spermatogenesis. A plausible reason for the longer half-life for round spermatids could be that the acrosome is formed in the early stages of spermiogenesis by the fusion of pro-acrosomal granules [3,15]. Thus, the antigen(s) synthesized in the round spermatids and targeted to the acrosome will be expected to have a longer half-life span. Turnover studies of mature forms (64-/62-kD) in these cells would have given us additional information; however, these studies could not be attempted due to the limited viability of spermatogenic cells in culture [7,34].

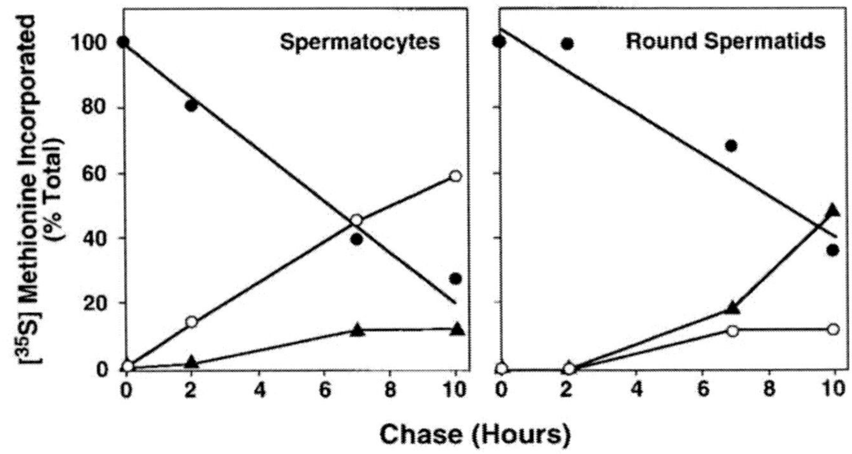

Figure 5. Kinetics of loss of radiolabel from precursor forms of β-D-galactosidase and its simultaneous incorporation into mature forms. Enriched populations of spermatocytes and round spermatids were incubated in 3 ml of MEM supplemented with [^{35}S] methionine (400 μCi/ml) for 30 min (pulse) as described in the legend to Figure 4. The cells were washed and re-incubated in non-radioactive medium for various time intervals (chase). At the indicated time, the cells were washed, extracted and the β-D-galactosidase was immunoprecipitated as described (11). Immunoprecipitates were electrophoresed on 7% SDS-PAGE under reducing conditions. Radioactivity was quantified in the individual bands. In this representative experiment, each point represents the mean of two samples. Apparent molecular weight of 90-/88-kD precursor forms (•) and the 64-kD (○) and 62-kD (▲) mature forms was calculated using [^{14}C]-labeled standard marker proteins. Half-lives of precursor forms were calculated from the linear regression analysis. Reproduced from Skudlarek et al. [11] with permission.

The precursor forms (90-/88-kD) of β-D-galactosidase appear to be related since removal of glycans by the treatment with either endo H (known to remove only the N-linked high mannose/hybrid-type glycans from glycoproteins) or N-glycanase (which removes all N-linked glycans) generated a single polypeptide of an apparent molecular weight 70-kD [10]. This result is consistent with the suggestion that the two forms are differentially glycosylated. Interestingly, the 70-kD polypeptide backbone has been reported in other forms of β-D-galactosidase [58]. Furthermore, the fact that the treatment of precursor forms with Endo H or N-glycanase generated the 70-kD subunit, suggests that the two precursor forms contain different amounts of N-linked high mannose/hybrid-type glycans [10].

It is noteworthy that although spermatocytes and round spermatids synthesize similar precursor forms of β-D-galactosidase, the major mature forms produced after a 10-h chase are different (Figures 4 and 5). In spermatocytes, which have lysosomes but no acrosome, the major mature form had a molecular mass of 64-kD. This form was only a minor constituent in the round spermatids where the acrosome is formed during the Golgi phase by the fusion of small granules to form a large acrosomal granule. These differences in the molecular mass of 64- and 62-kD are apparently due to differential glycosylation of the polypeptide moiety as we have demonstrated for molecular forms of β-D-galactosidase in spermatozoa [38], epididymal luminal fluid [58], and the 90-/88-kD precursor forms synthesized in round spermatids [10]. Since glycan moieties are important in intracellular transport and packaging of glycohydrolases [31,32], it will be of interest to elucidate the structure of oligosaccharide chains of the 64- and 62-kD forms. This information may provide insight into the manner in which β-D-galactosidase and other glycohydrolases are targeted to the acrosome and lysosome.

In human fibroblast and rat epididymal epithelial cells, acid β-D-galactosidase is synthesized in an 84- to 85-kD precursor form which is processed into a mature form of 63-64-kD. Similarly, the enzyme in the mouse macrophages is synthesized in precursor forms of 82-84-kD that are processed to a mature form of 63-kD [61]. In lysosomes, a protective protein has been demonstrated to interact with β-D-galactosidase monomers affecting their multimerization into a high molecular mass aggregate of 600-800-kD [62]. We have obtained no evidence for the presence of high molecular mass aggregates in spermatocytes or round spermatids.

Use of monospecific Antibodies to Acrosomal Glycohydrolases to follow the Successive formation of the Sperm Acrosome

The availability of high titer antibodies against β-D-glucuronidase and β-D-galactosidase has allowed us to use them to follow the successive formation of the acrosome during spermiogenesis in rats using an indirect immunofluorescence assay. In this approach, we examined the binding of preimmune or immune IgG to enriched population of spermatids (round and condensed/elongated) after permeabilization with methanol. Following incubation in diluted FITC-labeled goat anti-rabbit IgG, the cells were washed in PBS, mounted and observed with a confocal microscope using Nomarski differential interference phase-contrast optics [3,11]. Data from these studies presented in figures 6 and 7 demonstrate the distribution of β-D-glucuronidase and β-D-galactosidase, respectively, in the forming/formed acrosome during successive formation of the spermatozoon. To the best of our knowledge, these are the only studies in which antibodies to acid glycohydrolases have been successfully used to examine the formation of the acrosome during spermiogenesis.

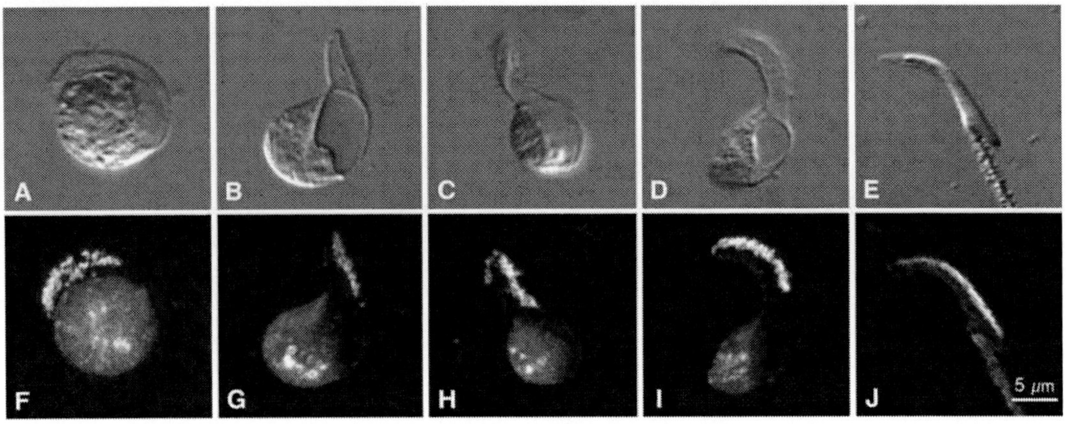

Figure 6. Confocal micrographs showing immunolocalization of β-D-glucuronidase in the forming acrosome during the elongation phase of spermiogenesis in the rat. The mixed testicular germ cells were prepared and immunostained using anti-β-D-glucuronidase IgG (monospecific IgG, 9 μg protein/ml) as primary antibody followed by FITC-labeled anti-rabbit goat IgG as secondary antibody as described [3]. Various phases of acrosome formation were photographed with a confocal microscope using Nomarski differential interference contrast optics (A-E). Note the presence of an intense fluorescence in the forming acrosome during progressive transformation of the elongated spermatids (F-I) into a testicular spermatozoon with fully formed acrosome (J). The elongated spermatids are from the following stages: stage 8, A&F; stage 10, B&G; stage 13, C&H; and stage 16, D&I. The absence of flagellum in the elongated spermatids is due to its loss during preparation of spermatogenic cells by the enzymatic digestion of the testes. Reproduced from Abou-Haila and Tulsiani [3] with permission.

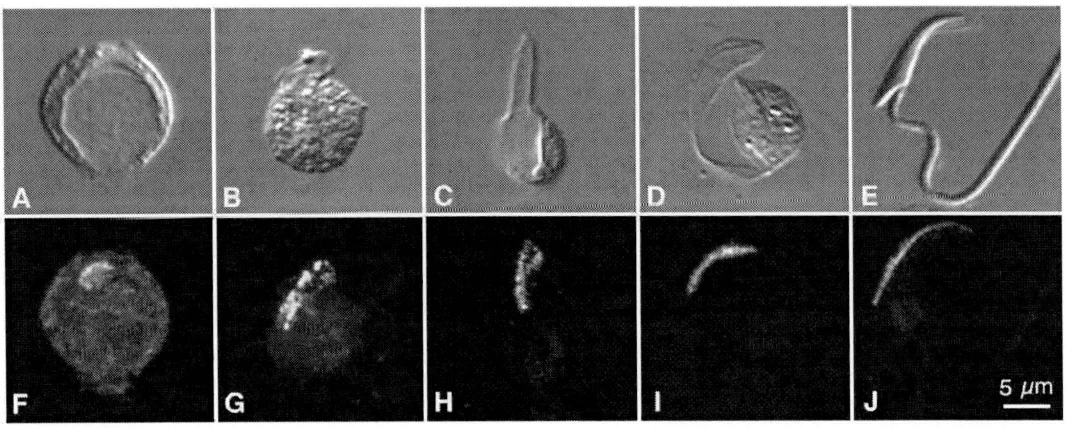

Figure 7. Confocal micrographs showing immunolocalization of β-D-galactosidase using anti-β-D-galactosidase (monospecific IgG, 5 μg protein/ml) as primary antibody and FITC-labeled anti-rabbit goat IgG as the secondary antibody as described [11]. Various phases of acrosome formation were photographed with a confocal microscope using Nomarski differential interference contrast optics (A-E) and immunofluorescence (F-J). Note the presence of intense fluorescence in the Golgi apparatus of stage 6 round spermatids (F) and the forming acrosome during progressive transformation of the elongated spermatids (G-I) into a spermatozoon with fully formed acrosome (J). The elongated spermatids are: stage 8, B & G; stage 10, C & H; and stage 17, D & I; the absence of a flagellum on the elongated spermatids is due to its loss during preparation of the spermatogenic cells by enzymatic treatment of the testes. Reproduced from Skudlarek et al. [11] with permission.

Immunolocalization of Glycohydrolases in Capacitating/Capacitated Spermatozoa

All mammalian spermatozoa studied so far undergo capacitation in the female genital tract (*in vivo* capacitation). Spermatozoa can also be capacitated *in vitro* by incubating in a physiological medium supplemented with BSA [63-65]. The in *vivo/in vitro* capacitation is a prerequisite event before sperm acquire the ability to bind to the zona pellucida, and undergo signal transduction cascade [14,17,63]. The net result is the fusion of the sperm plasma membrane and underlying outer acrosomal membrane at multiple sites and exocytosis of the acrosomal contents. Since spermatozoa undergo priming of the sperm plasma membrane and the outer acrosomal membrane during capacitation [65,66], we used an indirect immunofluorescence (IIF) assay and other staining procedures to assess capacitation-associated membrane priming of mouse spermatozoa. As stated above, the two glycohydrolases are localized in the acrosomal cap region of the cauda epididymal spermatozoa as revealed by the immunopositive staining in the head only when the cells are permeabilized. In order to assess the capacitation-associated membrane priming, we used the two glycohydrolases (β-D-glucuronidase and β-D-galactosidase) as marker enzymes. Briefly, mouse cauda epididymal spermatozoa were incubated for various time periods in an *in vitro* medium that favours capacitation [65]. At a given time, the cells were fixed with paraformaldehyde and stained by IIF to reveal β-D-galactosidase (Figure 8). Data from this approach revealed the presence of a non-specific reaction in the midpiece and postacrosomal region (PAR) when uncapacitated sperm were treated with preimmune/immune IgG without permeabilization. Although a majority of non-permeabilized acrosome-intact spermatozoa (> 90%) had no immunopositive reaction in the sperm acrosome, distinct and progressive immunopositive patterns appear in a time-dependent manner during capacitation. First, an initial reaction was revealed as a thin fluorescent crescent in the anterior acrosome cap (Figure 8B, H) which expanded posteriorly (Figure 8C, I) and laterally (Figure 8D, J) forming an intense diffuse region. As capacitation progressed with the incubation time, fluorescent spots were seen within the acrosome cap of a significant number of spermatozoa (Figure 8E, K); after 60 minutes of incubation, an increased number of spermatozoa had an intense immunofluorescent crescent of patched spots over the head region (Figure 8F, L).

The various immunopositive patterns seen in Figures 3 and 4 of reference 65, when grouped into three stages (stages I, II and III) and plotted as a function of incubation time, revealed a capacitation-dependent progressive membrane priming as evident by the appearance of various immunopositive patterns (see Figure 5 in reference 65). For instance, there were more immunopositive sperm in earlier stage (stage I) than in stages II or III individually or combined after 15 or 30 min of incubation. In contrast, the percentage of immunopositive spermatozoa in stage I showed a dramatic decline at 60 min of incubation: the observed decrease coincided with an increase in the immunopositive patterns in stages II and III, a result implying that spermatozoa in stage I undergo further membrane changes to reveal intense immunopositive staining seen in stages II and III [65]. Combined, these data allowed us to suggest that intra-acrosomal glycohydrolases (and perhaps other acrosomal components) are exposed on the surface of capacitating/capacitated spermatozoa in a time-dependent manner.

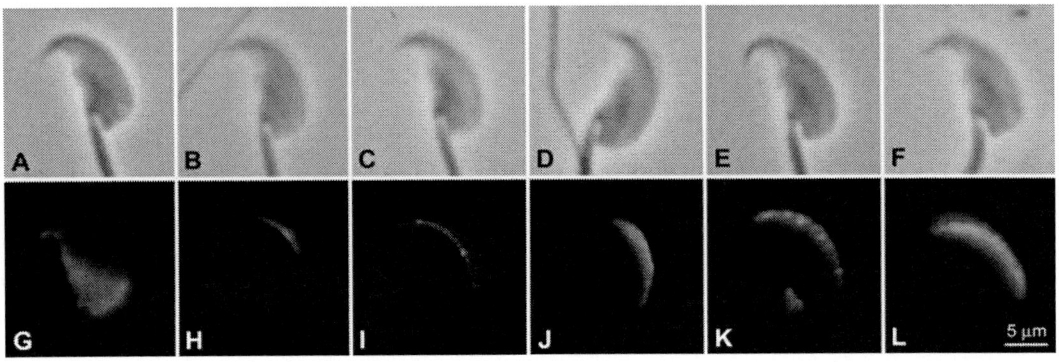

Figure 8. Immunolocalization of β-D-galactosidase in capacitating spermatozoa. Matched phase-contrast (A-F) and indirect immunofluorescence (IIF) patterns (G-L) after the cauda epidydmal spermatozoa were incubated in a medium that favors capacitation [65]. The cells were fixed without permeabilization and immunostained using affinity-purified (monospecific) IgG as primary antibody and FITC-labeled secondary antibody [65]. Being an intra-acrosomal glycohydrolase, the enzyme is not visible on the acrosomal region of uncapacitated spermatozoa (G). However, distinct immunopositive patterns appear (H-L) as capacitation proceeds. For quantification, distinct immunopositive patterns were grouped into three stages: stage I, H, I; Stage II, J; stage III, K, L. Other details are given in reference [65]. Reproduced from Abou-Haila and Tulsiani [65] with permission.

Is the Exposure of Acrosomal Contents on the Surface of Capacitated Spermatozoa Physiologically Significant?

The fact that spermatozoa recognize and bind to the zona pellucida only after capacitation suggests major changes on their cell surface during this process. We have presented *in vitro* evidence suggesting that mouse spermatozoa incubated in a medium that favors capacitation undergo membrane priming in a time-dependent manner [65,66]. The net result of the membrane modifications is the exposure of the two acrosomal glycohydrolases on the surface of capacitating/capacitated (acrosome-intact) spermatozoa (Figure 8). In addition to our studies with acid glycohydrolases, several other intra-acrosomal components have been reported to be present on the surface of capacitated spermatozoa from several species. The growing list of acrosomal molecules suggested to have a role in adhesion to the zona pellucida includes acrosomal matrix component AM67 [67,68], proacrosin/acrosin [69,70], hyaluronidase [71], an acrosomal sperm protein SP-10 [72], and glycohydrolases [65]. The reported presence of several acrosomal components on the surface of capacitated spermatozoa is consistent with our suggestion that the exposure of the intra-acrosomal molecules is physiologically significant.

The irreversible binding of capacitated spermatozoa to the zona pellucida is thought to trigger at least two distinct signalling cascades resulting in the exocytosis of acrosomal contents: a pertussis toxin-sensitive G-protein cascade [73,74] and activation of ion channels [75,76]. These channels are important factors that elevate the intra-sperm pH and the level of intracellular Ca^{2+} ions preceding the acrosome reaction. The typical acrosome reaction involves, swelling of the acrosome followed by the fusion and vesiculation of the outer acrosomal membrane and the overlying sperm plasma membrane at multiple sites [3,14]. Many details of the compounds and potential mechanism that may regulate the membrane fusion before and during acrosomal exocytosis have been discussed in recent reviews [66,77] and will not be repeated here. Suffice it to say that the membrane fusion at multiple sites allows the acrosomal contents to escape, thus exposing the inner acrosomal membrane [64]. The exposed membrane reveals a new set of binding sites specific for ZP2, one of the three glycoproteins that make the zona pellucida [78]. The secondary binding sites hold the acrosome-reacted (hyperactive) sperm and zona-intact egg in contact before the acrosome-reacted sperm penetrates the zona pellucida relying on acrosomal enzymes (glycohydrolases, proteinases etc.) released at the site of sperm-egg binding and enhanced thrust generated by the hyperactivated beat pattern of the bound sperm [21]. Although the potential molecules involved in secondary binding events and sperm-egg fusion will be of interest to many readers of this article, they are beyond the scope of this article. Interested readers are referred to an extended review article by McLeskey et al. [79].

It should be noted that acrosin, a serine-specific proteinase, was considered essential for secondary binding to zona pellucida and/or sperm penetration [79,80]. However, studies published over a decade ago demonstrated that sperm from mice carrying a targeted mutation of the acrosin gene can penetrate the zona pellucida and fertilize the egg, a result consistent with the suggestion that acrosin, at least in mice may not be essential for fertilization [81]. Subsequently it was demonstrated that the absence of acrosin from the mouse sperm does cause a delay in the penetration of the zona pellucida, which may be due to a slower dispersal of sperm acrosomal contents during the acrosome reaction [82].

In summary, this article reviews the biochemical characterization and immunolocalization of two acid glycohydrolases in rodent testicular germ cells and spermatozoa. Germ cell populations were prepared from mouse and rat testes by unit gravity sedimentation using a linear BSA gradient. Fractions rich in spermatocytes, round spermatids (> 95% pure cells) were separately pooled and assayed for several glycohydrolase activities. Most glycohydrolase activities were higher in spermatocytes than in spermatids. Mouse testicular germ cells were found to express the lysosomal form of β-D-glucuronidase with a broad isoelectrofocusing point (pI 5.5-6.0) and a subunit molecular mass of 74-kD. Immunocytochemical studies using high titer antibodies to β-D-glucuronidase and β-D-galactosidase revealed a positive reaction in the Golgi membranes, Golgi-associated vesicles and lysosome-like multivesicular bodies in the late spermatocytes (pachytene spermatocytes) and a stage-specific localization during spermiogenesis. The forming/formed acrosome of the elongated spermatids and cauda epididymal spermatozoa were highly immunopositive for both enzymes. Western blotting approaches on affinity-purified extracts from rat germ cells demonstrated several molecular forms of β-D-galactosidase in spermatocytes and round spermatids; one of these forms (62-kD) was present only in round spermatids. Biosynthetic studies indicate that β-D-galactosidase is synthesized in spermatocytes and round spermatids in high molecular weight precursor forms (90-/88-kD) that undergo processing to lower molecular weight mature forms in a cell-specific manner. The net result is the formation of predominantly 64- and 62-kD forms in spermatocytes and round spermatids, respectively.

ACKNOWLEDGMENTS

The authors acknowledge contributions of several collaborators including Drs. Marjorie D. Skudlarek, Ben Pereira, Marie-Claire Orgebin-Crist, and Catherine A. Chayko during various phases of these studies. We are indebted to Mrs. Loreita Little for preparing the manuscript and to Dr. Benjamin J. Danzo for critically reading it. Data reported in this article were obtained by support in part by grants (HD 25869 and HD 34041) from the National Institute of Child Health and Human Development. Our sincere apologies to many contributors whose relevant articles may not have been cited in this report.

REFERENCES

[1] Clermont, Y; Oko, R; Hermo, L. Biology of mammalian spermatogenesis. In: Desjardins C, Ewing LL editors. *Cell and molecular biology of the testis*. Oxford: University press; 1993; 332-376.

[2] Eddy, EM; O'Brien, DA. The spermatozoon. In: Knobil E, Neill JD editors. *The physiology of reproduction*. New York: Raven press; 1994; 29-77.

[3] Abou-Haila, A; Tulsiani, DRP. Mammalian sperm acrosome: Formation, contents and function. *Arch Biochem Biophys*, 2000, 379, 173-182.

[4] de Kretser, DM; Kerr, JP. The cytology of the testis. In: Knobil E, Neill JD editors. *The physiology of reproduction*. New York: Raven press; 1994; 1177-1290.

[5] Kierszenbaum, AL. Mammalian spermatogenesis *in vivo* and *in vitro*: A partnership of spermatogenic and somatic cell lineages. *Endocr Rev*, 1994, 15, 116-134.

[6] Parvinen, M; Vihko, KK; Toppari, J. Cell interactions during the seminiferous epithelial cycle. *Int Rev Cytol*, 1986, 104, 115-151.

[7] O'Brien, DA. Isolation, separation and short-term culture of spermatogenic cells. *Methods Toxicol*, 1993, 3A, 246-264.

[8] Pereira, BMJ; Abou-Haila, A; Tulsiani, DRP. Rat sperm surface mannosidase is first expressed on the plasma membrane of testicular germ cells. *Biol Reprod*, 1998, 59, 1288-1295.

[9] Abou-Haila, A; Fouquet, JP; Tulsiani, DRP. Characterization and immunolocalization of β-D-glucuronidase in mouse testicular germ cells and spermatozoa. *Exp Cell Res*, 1999, 247, 48-60.

[10] Chayko, CA; Orgebin-Crist, MC; Skudlarek, MD; Tulsiani, DRP. Biosynthesis, processing, and subcellular localization of rat sperm β-D-galactosidase. *Biol Reprod*, 2000, 63, 686-696.

[11] Skudlarek, MD; Abou-Haila, A; Tulsiani, DRP. Rat spermatogenic cell β-D-galactosidase: characterization, biosynthesis, and immunolocalization. *Exp Cell Res*, 2000, 261, 139-149.

[12] Abou-Haila, A; Tulsiani, DRP. Acid glycohydrolases in rat spermatocytes, spermatids and spermatozoa: enzyme activities, biosynthesis and immunolocalization. *Biol Proced Online*, 2001, 3, 35-42.

[13] Benoff, S. Carbohydrates and fertilization: an overview. *Mol Human Reprod*, 1997, 3, 599-637.

[14] Tulsiani, DRP; Abou-Haila, A; Loeser, CR; Pereira, BMJ. The biological and functional significance of the sperm acrosome and acrosomal enzymes in mammalian fertilization. *Exp Cell Res*, 1998, 240, 151-164.

[15] Clermont, Y; Tang, XM. Glycoprotein synthesis in the Golgi apparatus of spermatids during spermiogenesis of the rat. *Anat Rec*, 1985, 213, 33-43.

[16] Hermo, L; Adamali, HL; Mahuran, D; Gravel, RA; Trasler, JM. β-Hexosaminidase immunolocalization and α- and β-subunit gene expression in the rat testis and epididymis. *Mol Reprod Dev*, 1997, 46, 227-242.

[17] Yanagimachi, R. Mammalian fertilization. In: Knobil E, Neill JD editors. *The physiology of reproduction*. New York: Raven press; 1994; 189-317.

[18] Zaneveld, LJD; de Jonge, CR. Mammalian sperm acrosomal enzymes and acrosome reaction. In: Dunbar B, O'Rand M editors. A comparative overview of mammalian fertilization. New York: Plenum; 1991; 63-79.

[19] de Duve, C. The lysosome. *Sci Am*, 1963, 208, 64-72.

[20] Burgess, TL; Kelly, RB. Constitutive and regulated secretion of proteins. *Annu Rev Cell Biol*, 1987, 3, 243-293.

[21] Katz, D; Drobnis, E. Analysis and interpretation of the forces generated by spermatozoa. In: Bavister B, Cummins J, Roldan ERS editors. *Fertilization in mammals*. Norwell: Serono symposium; 1990, 125-137.

[22] Nikolajczyk, BS; O'Rand, MG. Characterization of rabbit testis β-galactosidase and arylsulfatase A: purification and localization in spermatozoa during the acrosome reaction. *Biol Reprod*, 1992, 46, 366-378.

[23] Yang, CH; Srivastava, PN; Williams, WL. Purification and properties of aryl sulfatases from rabbit sperm acrosomes. *Proc Soc Exp Biol Med*, 1974, 145, 721-725.

[24] Tantibhedhyangkul, J; Weerachatyanukul, W; Carmona, E; Xu, H; Anupriwan, A; Michaud, D; Tanphaichitr, N. Role of sperm surface arylsulfatase A in mouse sperm-zona pellucida binding. *Biol Reprod*, 2002, 67, 212-219.

[25] Saling, P. Mammalian sperm interaction with extracellular matrices of the egg. In: Milligan S editor. *Oxford review of reproductive biology*. Oxford: University press, 1989, 339-388.

[26] Tulsiani, DRP; Skudlarek, MD; Orgebin-Crist, MC. Human sperm plasma membranes possess α-D-mannosidase activity but no galactosyl transferase activity. *Biol Reprod*, 1990, 42, 843-858.

[27] Tulsiani, DRP. Glycan-modifying enzymes in luminal fluid of the mammalian epididymis: an overview of their potential role in sperm function. *Mol Cell Endocrinol*, 2006, 250, 58-65.

[28] Conzelmann, E; Sandhoff, K. Glycolipid and glycoprotein degradation. *Adv Enzymol Relat Areas Mol Biol*, 1987, 60, 89-216.

[29] Ichikawa, Y; Look, GC; Wong, CH. Enzyme-catalysed oligosaccharide synthesis. Anal Biochem, 1992, 202, 215-238.

[30] Tulsiani, DRP; Abou-Haila, A. Mammalian sperm molecules that are potentially important in interaction with female genital tract and egg vestment. *Zygote*, 2001, 9, 51-59.

[31] Robbins, A. Endocytosis and compartmentalization of lysosomal enzymes in normal and mutant mammalian cells: mannose 6-phosphate-dependent pathway. In: Das R, Robbins P editors. *Protein transfer and organelle biogenesis*. London: Academic press, 1988, 463-520.

[32] Kornfeld, S; Mellman, I. The biogenesis of lysosomes. *Annu Rev Cell Biol*, 1985, 5, 483-525.

[33] Brown, WJ; Goodhouse, J; Farquhar, MG. Mannose-6-phosphate receptors for lysosomal enzymes recycle between the Golgi complex and endosomes. *J Cell Biol*, 1986, 103, 1235-1247.

[34] O'Brien, DA; Gabel, CA; Rockett, DL; Eddy, EM. Receptor-mediated endocytosis and differential synthesis of mannose 6-phosphate receptors in isolated spermatogenic and Sertoli cells. *Endocrinology*, 1989, 125, 2973-2984.

[35] Hancock, LW; Raab, LS; Aronson, NN Jr. Synthesis and processing of rat sperm-associated α-L-fucosidase. *Biol Reprod*, 1993, 48, 1228-1238.

[36] Tanii, I; Toshimori, K; Araki, S; Oura, C. Appearance of an intra-acrosomal antigen during the terminal step of spermiogenesis in the rat. *Cell Tissue Res*, 1992, 267, 203-208.

[37] Tanii, I; Toshimori, K; Araki, S; Oura, C. Extra-Golgi pathway of an acrosomal antigen during spermiogenesis in the rat. *Cell Tissue Res*, 1992, 270, 451-457.

[38] Skudlarek, MD; Tulsiani, DRP; Nagdas, SK; Orgebin-Crist, MC. β-D-galactosidase of rat spermatozoa: subcellular distribution, substrate specificity and molecular changes during epididymal maturation. *Biol Reprod*, 1993, 49, 204-215.

[39] Nagdas, SK; Skudlarek, MD; Orgebin-Crist, MC; Tulsiani, DRP. Biochemical alterations in the proacrosin-acrosin system during epididymal maturation of the rat spermatozoa. *J Androl*, 1992, 13, 36-43.

[40] Kurth, BE; Wright, RM; Flickinger, CJ; Herr, JC. Stage-specific detection of mRNA for the sperm antigen SP-10 in human testes. *Anat Rec*, 1993, 236, 619-625.

[41] Anakwe, OO; Gerton, GL. Acrosome biogenesis begins during meiosis: evidence from the synthesis and distribution of an acrosomal glycoprotein, acrogranin, during guinea pig spermatogenesis. *Biol Reprod*, 1990, 42, 317-328.

[42] Mukerji, SK; Meizel, S. Rabbit testis proacrosin. Purification, molecular weight estimation, and amino acid and carbohydrate composition of the molecule. *J Biol Chem*, 1979, 254, 11721-11728.

[43] Baba, T; Kashiwabara, S; Watanabe, K; Itoh, H; Michikawa, Y; Kimura, K; Takada, M; Fukamizu, A; Arai Y. Activation and maturation mechanisms of boar acrosin zymogen based on the deduced primary structure. *J Biol Chem*, 1989, 264, 11920-11927.

[44] Baba, T; Niida, Y; Michikawa, Y; Kashiwabara, S; Kodaira, K; Takaneka, M; Kohno, N; Gerton, GL; Arai, Y. An acrosomal protein, sp32, in mammalian sperm is a binding protein specific for two proacrosins and an acrosin intermediate. *J Biol Chem*, 1994, 269, 10133-10140.

[45] Abou-Haila, A; Tulsiani, DRP; Skudlarek, MD; Orgebin-Crist, MC. Androgen regulation of molecular forms of β-D-glucuronidase in the mouse epididymis: comparaison with liver and kidney. *J Androl*, 1996, 17, 194-207.

[46] Tulsiani, DRP; Six, H; Touster, O. Rat liver microsomal and lysosomal β-glucuronidase differ in both carbohydrate and amino acid composition. *Proc Natl Acad Sci USA*, 1975, 7, 3080-3084.

[47] Tomino, S; Paigen, K; Tulsiani, DRP; Touster, O. Purification and chemical properties of mouse liver lysosomal (L form) β-glucuronidase. *J Biol Chem*, 1975, 250, 8503-8509.

[48] Tulsiani, DRP; Keller, KK; Touster, O. The preparation and chemical composition of the multiple forms of β-glucuronidase from the female rat preputial gland. *J Biol Chem*, 1975, 250, 4770-4776.

[49] Zhen, L; Baumann, H; Novak, EK; Swank, RT. The signal for retention of the egasyn-glucuronidase complex within the endoplasmic reticulum. *Arch Biochem Biophys*, 1993, 304, 402-414.

[50] Dott, HM; Dingle, JJ. Distribution of lysosomal enzymes in the spermatozoa and cytoplasmic droplets of bull and ram. *Exp Cell Res*, 1968, 52, 523-540.

[51] Garbers, DL; Wakabayashi, T; Reed, PW. Enzyme profile of the cytoplasmic droplet from bovine epididymal spermatozoa. *Biol Reprod*, 1970, 3, 327-337.

[52] Garner, DL; Reamer, SA; Johnson, LA; Lesley, BA. Failure of immunoperoxidase staining to detect acrosin in the equatorial segment of spermatozoa. J Exp Zool, 1977, 201, 309-315.

[53] Skudlarek, MD; Orgebin-Crist, MC. Effect of swainsonine on rat epididymal glycosidases. *J Reprod Fertil*, 1988, 84, 611-617.

[54] Majumder, GC; Turkington, RW. Acrosomal and lysosomal isoenzymes of β-galactosidase and N-acetyl-β-glucosaminidase in rat testis. *Biochemistry*, 1974, 13, 2857-2864.

[55] Skudlarek, MD; Novak, EK; Swank, RT. Processing of lysosomal enzymes in macrophages and kidney. In: Dingle JJ, Dean RT, Sly W editors. *Lysosomes in biology and pathology*. New York: Elsevier; 1984, 17-43.

[56] von Figura, K; Hasilik, A. Lysosomal enzymes and their receptors. *Annu Rev Biochem*, 1986, 55, 167-193.

[57] Tulsiani, DRP; Buschiazzo, HO; Tolbert, B; Touster, O. Changes in plasma hydrolase activities in hereditary and streptozotocin-induced diabetes. *Arch Biochem Biophys*, 1977, 181, 216-227.

[58] Tulsiani, DRP; Skudlarek, MD; Araki, YA; Orgebin-Crist, MC. Purification and characterization of two forms of β-D-galactosidase from rat epididymal luminal fluid: evidence for their role in the modification of sperm plasma membrane glycoprotein(s). *Biochem J*, 1995, 305, 41-50.

[59] Levine, N; Kelly, H. Measurement of pH in the rat epididymis *in vivo*. *J Reprod Fert*, 1988, 52, 333-335.

[60] Griffiths, G; Warren, G; Stuhlfauth, I; Jockusch, BM. The role of clathrin-coated vesicles in acrosome formation. *Eur J Cell Biol*, 1981, 26, 52-60.

[61] Skudlarek, MD; Swank, RT. Turnover of two lysosomal enzymes in macrophages. *J Biol Chem*, 1981, 256, 10137-10144.

[62] Hoogeveen, AT; Verheijen, FW; Galgaard, H. The relation between human lysosomal β-galactosidase and its protective protein. *J Biol Chem*, 1983, 258, 12143-12146.

[63] Tulsiani, DRP; Yoshida-Komiya, H; Araki, Y. Mammalian fertilization: a carbohydrate-mediated event. *Biol Reprod*, 1997, 57, 487-494.

[64] Huang, TTF; Yanagimachi, R. Inner acrosomal membrane of mammalian spermatozoa: its properties and possible function in fertilization. *Am J Anat*, 1985, 174, 249-268.

[65] Abou-Haila, A; Tulsiani, DRP. Evidence for the capacitation-associated membrane priming of mouse spermatozoa. *Histochem Cell Biol*, 2003, 119, 179-187.

[66] Tulsiani, DRP; Abou-Haila, A. Is sperm capacitation analogous to early phases of Ca^{2+}-triggered membrane fusion in somatic cells and viruses? *Bioessays*, 2004, 26, 281-290.

[67] Foster, JA; Friday, BB; Maulit, MT; Blobel, C; Winfrey, VP; Olson, GE ; Kim, KS ; Gerton, GL. AM67, a secretory component of the guinea pig sperm acrosomal matrix, is related to mouse sperm protein sp56 and the complement component 4-binding proteins. *J Biol Chem*, 1997, 272, 12714-12722.

[68] Kim, KS; Cha, MC; Gerton, GL. Mouse sperm protein sp56 is a component of the acrosomal matrix. *Biol Reprod*, 2001, 64, 36-43.

[69] Jones, R; Brown, CR. Identification of a zona-binding protein from boar sperm as proacrosin. *Exp Cell Res*, 1987, 171, 503-508.

[70] Tesarik, J; Drahorad, J; Testart, J; Mendoza, C. Acrosin activation follows its surface exposure and precedes membrane fusion in human sperm acrosome reaction. *Development*, 1990, 110, 391-400.

[71] Meyers, SA; Rosenberger, AE. A plasma membrane-associated hyaluronidase is localized to the posterior acrosomal region of stallion sperm and is associated with spermatozoal function. *Biol Reprod*, 1999, 61, 444-451.

[72] Coonrod, SA; Herr, JC; Westhusin, ME. Inhibition of bovine fertilization *in vitro* by antibodies to SP-10. *J Reprod Fertil*, 1996, 107, 287-297.

[73] Ward, CR; Storey, BT; Kopf, GS. Activation of a Gi protein in mouse sperm membranes by solubilized proteins of the zona pellucida, the egg's extracellular matrix. *J Biol Chem*, 1992, 267, 14061-14067.

[74] Ward, CR; Storey, BT; Kopf, GS. Selective activation of Gi1 and Gi2 in mouse sperm by the zona pellucida, the egg's extracellular matrix. *J Biol Chem*, 1994, 269, 13254-13258.

[75] Florman, HM; Corron, ME; Kim, TD; Babcock, DF. Activation of voltage-dependent calcium channels of mammalian sperm is required for zona pellucida-induced acrosomal exocytosis. *Dev Biol*, 1992, 152, 304-314.

[76] Arnoult, C; Cardullo, RA; Lemos, JR; Florman, HM. Activation of mouse sperm T-type Ca^{2+} channels by adhesion to the egg zona pellucida. *Proc Natl Acad Sci U S A*, 1996, 93, 13004-13009.

[77] Tulsiani, DRP; Zeng, HT; Abou-Haila, A. Biology of sperm capacitation: evidence for multiple signalling pathways. *Soc Reprod Fertil Suppl*, 2007, 63, 257-272.

[78] Wassarman, PM. Zona pellucida glycoproteins. *Annu Rev Biochem*, 1988, 57, 415-442.

[79] McLeskey, SB ; Dowds, C ; Carballada, R ; White, RR ; Saling, PM. Molecules involved in mammalian sperm-egg interaction. *Int Rev Cytol*, 1998, 177, 57-113.

[80] Saling, P. Mammalian sperm interaction with extracellular matrices of the egg. In: Milligan S editor. *Oxford review of reproductive biology*. Oxford: University press, 1989, 339-388.

[81] Baba, T; Azuma, S; Kashiwabara, S; Toyoda, Y. Sperm from mice carrying a targeted mutation of the acrosin gene can penetrate the oocyte zona pellucida and effect fertilization. *J Biol Chem*, 1994, 269, 31845-31849.

[82] Yamagata, K ; Murayama, K ; Okabe, M ; Toshimori, K ; Nakanishi, T ; Kashiwabara, S ; Baba, T. Acrosin accelerates the dispersal of sperm acrosomal proteins during acrosome reaction. *J Biol Chem*, 1998, 273, 10470-10474.

In: Cell Differentiation Research Developments
Editor: L. B. Ivanova, pp. 97-123
ISBN: 978-1-60021-939-9

Chapter IV

Spermatogenesis in Mammals: A Very Peculiar Cell Differentiation Process

Adriana Geisinger
Unidad Asociada Biología Molecular (IIBCE), Facultad de Ciencias;
Departamento de Biología Molecular, Instituto de Investigaciones Biológicas Clemente Estable, 11600 Montevideo, Uruguay.

Abstract

Spermatogenesis is a very complex terminal cell differentiation process, essential for all the species with sexual reproduction.

Despite of its importance for sexual reproduction and as a main source of variability, mammalian spermatogenesis is still poorly understood at the molecular level, mainly due to some difficulties it presents for its study. One of the main drawbacks is that the testis is a very complex tissue, with the spermatogenic cells of different stages - somatic spermatogonia, meiotic cells and haploid spermatids in different phases of elongation - coexisting with somatic Sertoli cells in the seminiferous tubules of adult animals. Besides, spermatogenic cells do not maintain the differentiation process *in vitro*, thus reducing the possibilities of the use of germ cells culture for experimentation. Therefore, a prerequisite for the analysis of differential gene expression during spermatogenesis is the availability of methods that allow rapid obtainment of highly enriched populations of a certain germ cell type.

In spite of these drawbacks, a few experimental approaches have been recently developed in some laboratories for the analysis of gene expression in the germ line of the male. The studies are revealing testis as a very interesting system for the analysis of differential gene expression because of the astonishingly large number of genes that are differentially expressed and the peculiar patterns of regulation that govern gene expression in spermatogenic cells. Moreover, candidate genes for roles in the regulation of fertility and possible contraceptive targets are starting to be identified.

Here, different approaches for the analysis of gene expression along spermatogenesis employed in various laboratories as well as in our own will be reviewed, and the main results will be discussed.

INTRODUCTION

Infertility affects at least 20-25% of couples that are in reproductive age. In nearly 30% of all infertility cases, the origin is attributed to a male factor and in an additional 30% of cases the cause is attributed to both male and female factors. Statistics vary but it would seem that around 30% of men are sub-fertile and at least 2% are totally infertile. 90% of male infertility is generated by the failure to produce enough sperm that will be healthy, properly formed, and mobile. Some of the most common disorders that may affect sperm production and quality are varicocele, infections, diabetes, environmental factors (pollution, exposure to heavy metals or synthetic estrogens, drugs, and even a nutrient-deficient diet) and congenital abnormalities including genetic disorders. The importance of genetic disorders is such that estimates indicate that in approximately 60% of male infertility cases, genetic causes would be involved (Lilford *et al.*, 1994). Since male fertility has been experiencing an alarming decrease during the last 50 years (*e.g.* Sharpe and Irvine, 2004; Albert *et al.*, 2006), a need for a deeper knowledge on the molecular bases of male germ cell production is evident.

Experimental approaches for the analysis of differential gene expression at the protein level such as the construction and inmunofluorescent screening of monoclonal antibodies libraries (*e.g.* Heyting and Dietrich, 1989; Offenberg *et al.*, 1991; Smith and Benavente, 1992a; 1992b; Vester *et al.*, 1993; Kralewski *et al.*, 1997; Alsheimer *et al.*, 1998) and the comparison of two-dimensional protein patterns (Cossio *et al.*, 1995; 1997) have allowed the identification of important male germ cell-specific products. Moreover, methodologies for the identification of differentially expressed mRNAs such as subtractive hybridization (*e.g.* Wang *et al.*, 2001; Fujii *et al.*, 2002; Beissbarth *et al.*, 2003), mRNA differential display (Geisinger *et al.*, 1996; Catalano *et al.*, 1997; Anway *et al.*, 2003; Luk *et al.*, 2003; Almstrup *et al.*, 2004), SAGE (Yao *et al.*, 2004; Wu *et al.*, 2004; Chan *et al.*, 2006) and mainly microarrays (Yu *et al.*, 2003, Rossi *et al.*, 2004; Maratou *et al.*, 2004; Almstrup *et al.*, 2004; Shima *et al.*, 2004; Ellis *et al.*, 2007; Feig *et al.*, 2007) have rendered high numbers of male germ cell development and differentiation-related genes. These studies have revealed testis as a very interesting system because of the astonishingly large number of genes that are differentially expressed (reviewed in Wrobel and Priming, 2005) and the peculiar patterns that govern the regulation of gene expression in germ cells (Elliott, 2003; Huang *et al.*, 2005; Iguchi *et al.*, 2006). The identification of candidate genes for roles in the regulation of fertility is contributing to the finding of possible male contraceptive targets (Schultz *et al.*, 2003, Schlecht *et al.*, 2004; Chan *et al.*, 2006).

On the other hand, the analysis of the biology of patients with genetic syndromes and the examining of the phenotypes of mouse lines harboring targeted gene mutations are shedding light on genes that are involved in mammalian male germ cells formation and viability. Besides being implicated in pathologies such as hereditary polyposis colorectal cancer (Baker *et al.*, 1995), Bloom Syndrome (Moens *et al.*, 2000), Nijmegen Breakage Syndrome (Resnick *et al.*, 2002), Ataxia-Talangiectasia (Boder, 1975), Fanconi Anemia (Alter, 2003), and many others (*e.g.* Sapiro *et al.*, 2002), some of those genes cause infertility by interrupting an essential event in the process that leads to sperm production.

Unfortunately, however, the molecular processes underlying germ cell differentiation and development are still poorly understood mainly because of some intrinsic difficulties the testicular tissue presents for its study. As a consequence, in most of the infertility cases where genetic causes are presumably involved, the specific cause is still unknown. A deeper comprehension of the molecular events that take place is essential for the understanding of the bases of sexual reproduction, and for the development of therapeutic and biomedical procedures as well.

GENERAL ASPECTS OF SPERMATOGENESIS

Spermatogenesis is the process that leads to the formation of male gamete, the mature sperm. It is a very complex proliferation and terminal cell differentiation process essential for all the species with sexual reproduction that occurs in the testes of adult individuals.

Mammalian testes are ovoid organs that appear wrapped in a dense fibrous membrane, the *tunica albuginea*, and whose inside contains a large number of entangled seminiferous tubules. Seminiferous tubules, which are in charged of sperm production, are covered by a highly specialized stratified epithelium, named germinative or seminiferous epithelium. In adult males, seminiferous epithelium is formed by two different cell populations: a non-proliferative population of large supporting cells (Sertoli cells), that lie on the basement membrane of the tubule, and a proliferative population of germ cells, that constantly migrate from the periphery of the tubule to the lumen, as they differentiate (Greep, 1954).

The germ cell population is composed of different stages of developing cells and comprise, from the periphery to the lumen, spermatogonia, spermatocytes and spermatids (Figure 1). Spermatogonia are stem cells located near the basement membrane of the tubule, which proliferate by mitosis. Some of these progeny cells remain near the basement membrane throughout the entire life of the individual and serve as a self-renewing stem cell population, while others sequentially differentiate into progressively more advanced spermatogonial stages (A1, A2, A3, A4, intermediate and type B spermatogonia) (de Rooij, 2001) and generate primary spermatocytes, that are the cells that undergo meiosis. Meiosis is a special type of cell division exclusive to germ cells that occurs in all eukaryotes, from yeast to men. It involves two successive nuclear divisions with only one round of DNA replication, resulting in haploid cells, the round spermatids. Thus, meiosis is the mechanism by which sexually reproducing organisms compensate for doubling at fertilization. The process of meiosis is mainly related to the behavior of chromosomes during prophase and anaphase of the first meiotic division (meiotic prophase I is very long and has been divided into different stages), with highly significant events such as homologous chromosome synapsis, recombination and segregation taking place (*e.g.* reviews by Kleckner, 2006; Zickler, 2006; Neale and Keeney, 2006; Li and Ma, 2006). Synapsis and recombination appear mediated by a meiosis-specific tripartite protein structure called the synaptonemal complex (reviewed in Wettstein and Sotelo, 1971; von Wettstein *et al.*, 1984; Kleckner, 2006; Zickler, 2006).

Particularly, the homologous recombination that occurs during meiosis (crossing over) has vital importance as a means of genetic variability, and therefore constitutes a main source of biodiversity (Rice and Chippindale, 2001). After the second meiotic division, haploid spermatids enter a phase of differentiation called spermiogenesis, in which they undergo profound progressive nuclear and cytoplasmic changes in order to become mature sperm (Greep, 1954).

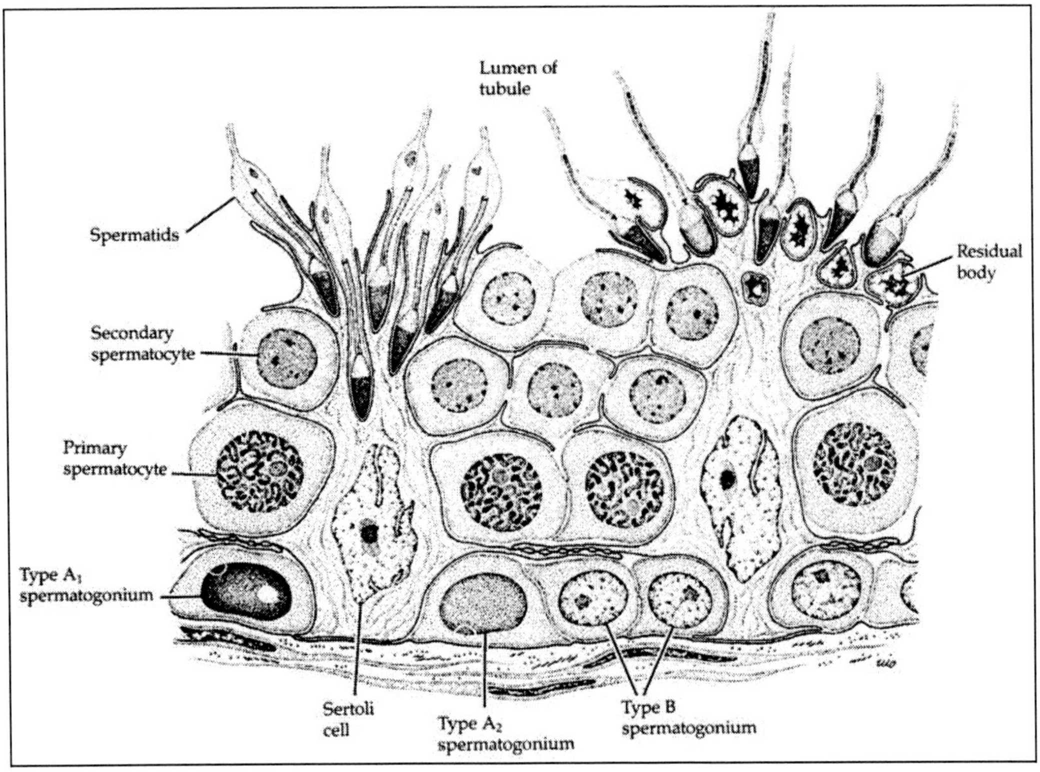

Figure 1. Drawing of a section of a seminiferous tubule (after Dym, 1977).

During the spermatogonial divisions, cytokinesis is not complete. The cells form a syncytium, where each cell is connected to the other via a 1μm diameter cytoplasmic bridge. The successive divisions produce clones of interconnected cells with ions and molecules passing across the bridges (Dym and Fawcett, 1971). In transversal sections of the tubules, all the cells in each concentric layer of the seminiferous epithelium are found at the same differentiation stage. The different cell stages in a transversal section are arranged forming a definite number of cellular associations, which were defined long ago by Clermont for different mammalian species (Leblond and Clermont, 1952; Clermont, 1960; 1963).

The development of germ cells takes place in intimate association with Sertoli cells, which have cytoplasmic prolongations that surround the clonal groups of cells providing a specialized, protected environment (reviewed in Griswold, 1998). Besides, Sertoli cells form tight junctions with each other, thus creating a barrier that protects the inside of the seminiferous tubule; this is known as the blood-testis barrier. Furthermore, they respond to follicle-stimulating hormone (FSH) and testosterone by initiating signaling pathways that contribute to the support of spermatogenesis (reviewed in Walker and Cheng, 2005).

In addition to germ cells at different maturation stages and Sertoli cells, many other cell types exist in the testis (Figure 2). Seminiferous tubules are surrounded by loose areolar connective tissue (interstitial tissue) that consists of numerous reticular fibers, capillary and lymphatic vessels, macrophages, fibrocytes, mastocytes, lymphocytes, and Leydig cells; these latest are responsible for the synthesis and secretion of the steroid hormone testosterone (Greep, 1954). Moreover, peritubular myoid cells are found as part of the basal lamina of the seminiferous tubule.

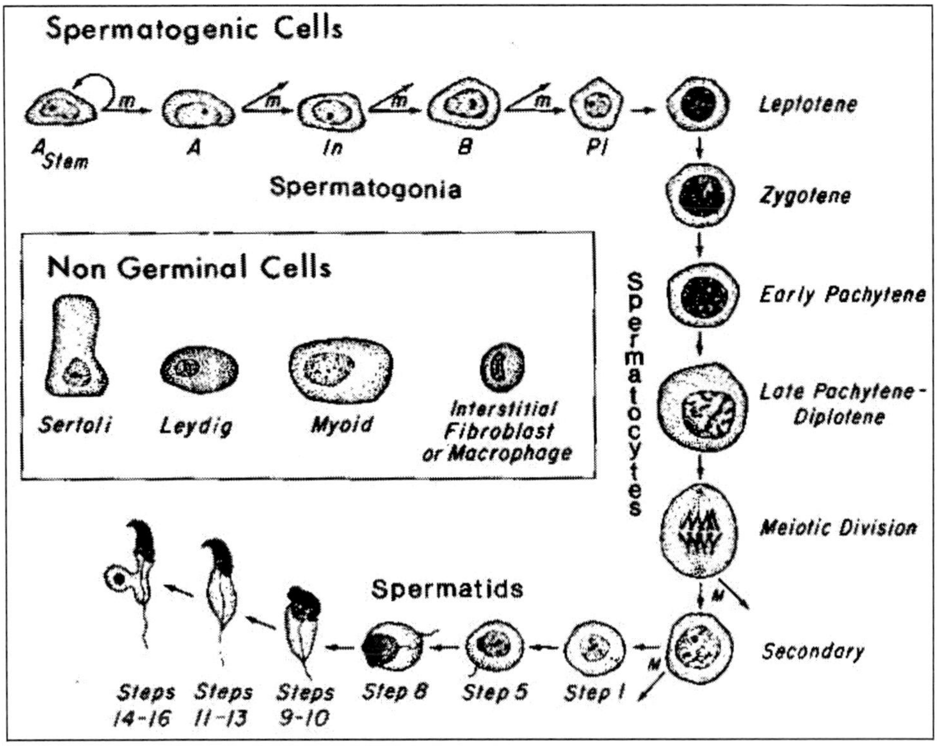

Figure 2. Drawing of the different cell types of the mouse testis (after Meistrich, 1977).

At birth mammalian testes contain somatic Sertoli cells and gonocytes, which are arrested in the G0/G1 phase. The gonocytes start proliferating mitotically, and in mouse and rat by day 6 they have migrated to the basement membrane of the seminiferous tubules and became undifferentiated type A spermatogonia (Bellvé *et al.*, 1977; Malkov *et al.*, 1998). Meiotic prophase begins at approximately postnatal day 9-10 and 13-14 in the mouse and rat respectively. By day 12 in mouse and 17-18 in rat, the predominant cell types are leptotene and zygotene spermatocytes. A few days later the majority of cells reach early pachytene stage (day 14 in mouse and 19-20 in rat), and by day 20 in mouse and 24-25 in rat, round spermatids start to be visible. Elongated spermatids are first detected in the mouse by day 30, and the first cauda spermatozoa appear at the same time (Janca *et al.*, 1986). In rat, the first spermatozoa can be found by day 36 (Malkov *et al.*, 1998).

Difficulties for the Analysis of Spermatogenesis at the Molecular Level

In molecular terms, spermatogenesis can be described as the coordinate execution of three individual programs of gene expression. The first program corresponds to somatic proliferation of spermatogonia, the second provides the primary spermatocyte with the meiosis-specific apparatus for chromosome pairing, recombination and segregation, and the third one is responsible for the profound biochemical and structural changes that haploid spermatids undergo in order to become mature sperm (Erickson, 1990).

In spite of its importance, as already mentioned mammalian spermatogenesis is still poorly understood at the molecular level, mainly due to some difficulties it presents for its study.

One of the main difficulties is the tissue heterogeneity, with all the spermatogenic and somatic cell types coexisting in the testes of adult animals (Bellvé *et al.*, 1977). As a consequence of that heterogeneity, a pre-requisite for the analysis of differential gene expression during spermatogenesis is the availability of methods to allow the assignment of specific transcripts to cells at specific differentiation stages (*e.g.* Meistrich, 1977).

A second drawback is that spermatogenic culture cell lines have not been developed yet. Despite the many attempts to achieve complete spermatogenesis *in vitro*, the results up to now have been very poor. This can be partly due to the difficulties concerning the purification of defined subpopulations of germ cells, but also to the inability to reproduce *in vitro* the intricate relationship between germ cells and Sertoli cells, and the lack of knowledge about the optimal conditions that promote the survival and differentiation of germ cells in long-term cultures (reviewed in Sofikitis *et al.*, 2005). Short-term cultures are feasible, but the cell viability declines very fast (Dirami *et al.*, 1999), cells do not proliferate or the proliferation rate is minimal, and their biochemical and morphological characteristics become quickly altered during the culture. Therefore, the use of germ cell culture for experimentation is not possible yet.

Most of the molecular studies concerning the main events that take place during meiosis have been undertaken in yeast, a simple unicellular organism where culture and mutant analyses are easy to conduct. An important number of genes involved in meiosis have been identified in yeast, and their roles have been unveiled (*e.g.* Chu *et al.*, 1998; Primig *et al.*, 2000; Mata *et al.*, 2002; Gregan *et al.*, 2006, and reviews by Vershon and Pierce, 2000; Roeder and Bailis, 2000; Davis and Smith, 2001; Krogh and Symington, 2001; Whitby, 2005; Hochwagen and Amon, 2006). Nevertheless, the preservation of the morphological aspects of meiosis along evolution is not maintained at the molecular level and, as a consequence, we cannot extrapolate most of the results from yeast to higher eukaryotes. In fact, the sequencing of higher eukaryotic genomes has shown that most of the meiosis-specific genes identified in yeast do not have homologues in other genomes; it seems that function analogy is far more frequent than genetic sequence homology [an example of this are the analogous synaptonemal complex proteins ZIP1 and SCP1, from yeast and higher eukaryotes respectively (Sym *et al.*, 1993; Dong and Roeder, 2000; Meuwissen *et al.*, 1992; Liu *et al.*, 1996; Schmekel *et al.*, 1996)]. Moreover, an important proportion of the meiosis-specific genes from the budding yeast *Saccharomyces cerevisiae* are not even shared by the distantly related fission yeast *Schizosaccharomyces pombe*, and viceversa (*e.g.* Watanabe *et al.*, 2001; Davis and Smith, 2001; Mata *et al.*, 2001; Whitby, 2005), suggesting that the genes involved in meiosis are highly species-specific (Watanabe *et al.*, 2001). On the other hand, some studies have started to show that not all spermatogenesis-related genes in rodents have a human ortholog (Lin *et al.*, 2006; our laboratory, unpublished results). This fact could be interpreted in terms of the rapid evolution of reproductive proteins (Swanson and Vacquier, 2002; Clark *et al.*, 2006).

Due to that variability, the need to conduct molecular studies of spermatogenesis in mammalian models is evident when it comes to learn about the bases of our own reproduction. Although a few studies have been carried out in humans using either testicular biopsies (*e.g.* Lin *et al.*, 2006; Ellis *et al.*, 2007; Feig *et al.*, 2007) or testes from deceased adults and aborted fetuses (*e.g.* Sha *et al.*, 2002), rodents have been by far the mostly chosen models because of obvious practical and ethical reasons.

Methods for the Enrichment in Stage-Specific Cell Populations from Rodent Testicles

Basically two different strategies have been employed in order to allow the comparison of gene expression in different testicular cell populations:

1. Use of whole Testes from Prepuberal Animals

One of the approaches has been the use of young individuals that have not completed the first spermatogenic wave. By comparing gene expression in juvenile animals of different ages (for instance, whose seminiferous tubules have only Sertoli cells and spermatogonia but lack meiotic cell types, are enriched in meiotic cells but lack spermatids, have reached the round spermatid stage but lack spermatozoa) and with adult individuals, changes in mRNA levels can be correlated to expression in a specific cell type. This approach has been widely used (*e.g.* Vilardell *et al.*, 1989; Thomas *et al.*, 1989; Almstup *et al.*, 2004; Marathou *et al.*, 2004; Shima *et al.*, 2004; Iguchi *et al.*, 2006), and has proved to be adequate since, in general terms, microarray analyses have shown significant associations between the expression profiles of coclustered genes (pools of genes that become more or less abundant on the same days) and cell type specific UniGene libraries (Maratou *et al.*, 2004).

This strategy has the main advantage of little handling and sample preparation time, thus reducing the impact of RNA degradation on the reliability and reproducibility of the expression data. However, the approach has several limitations. First, gene-expression analysis on total RNA prepared from such a complex tissue with so many different cell types cannot precisely distinguish which cell types express a specific mRNA. As a consequence, results have to be validated by *in-situ* hybridization or some other techniques that can address the cell-type composition of a tissue. Second, since the cell-type composition of the testis changes during development as a result of proliferation and differentiation, we cannot assure whether an mRNA that is more abundant in a given RNA sample is either upregulated or expressed in a cell type that has become more abundant or larger. Third, since normal spermatogenesis depends on interactions between somatic and germ cells, we think that it is not possible to directly attribute a certain transcript detectable in one RNA sample and not in another to a specific, recently appeared cell type. This new RNA could be synthesized *de novo* in a previously present cell type (*e.g.* Sertoli cells), in response to the new environmental or regulatory conditions. Moreover, we do not know if gene expression in a specific testicular cell type is identical during the first spermatogenic wave to that of the same cell type in adult individuals that have already undergone complete spermatogenesis.

2. Use of Cell Separation Techniques for the Obtainment of Highly Enriched Stage-Specific Cell Populations

A second approach to allow the analysis of gene expression along different stages of spermatogenesis has been the use of cell separation techniques. These techniques yield large quantities of highly enriched populations of spermatogenic cells at specific stages of differentiation. The strategy precisely reveals which transcripts are present in a certain cell type. However, it has the disadvantage of involving laborious cell preparation procedures and therefore in some cases expression levels may change to a certain extent during the purification process (Wrobel and Priming, 2005), since complicated handling can selectively damage certain cell types. Moreover, the duration of the process is especially critical for the representation of some short life RNAs and proteins.

The first step for any cell separation method is the preparation of a cellular suspension. A good testicular cell suspension will be one that maximizes the number of viable cells while preventing cell clumps, avoids selective damage of specific cell types, and minimizes the formation of multinucleates (Meistrich, 1977), which tend to form because of the syncitial nature of the seminiferous epithelium. Protocols using mechanical dissociation for the preparation of testicular cell suspensions have been described (Lam *et al.*, 1970), but the suspensions mechanically prepared aggregate very fast, the yield of viable cells is at most 80% (generally less), and results are not highly reproducible since, to a certain degree, they are operator-dependant; more important, some cell types are selectively damaged (Meistrich, 1977). Methods involving a combination of gentle mechanical action with enzymatic treatment using trypsin, DNase (Meistrich *et al.*, 1973) and collagenase (Romrell *et al.*, 1976) have rendered better results in terms of yield and cell type representation, while minimizing cell aggregation. However, they are time-consuming, involve a lot of handling, and the use of enzymes could affect cell integrity and/or preservation of some macromolecules of interest.

We have recently developed a protocol to rapidly obtain cellular suspensions from testis material using a Medimachine system (DakoCytomation, Carpinteria, CA). The procedure is very fast (the whole procedure takes 15 min including animal dissection), reproducible, and although a mechanical method, guarantees good viability and does not selectively damage any specific cell type. Since the method involves very little manipulation and avoids enzymes and other harmful reagents, we expect that it will become an ideal choice for delicate downstream applications such as gene expression studies (Rodríguez *et al.*, submitted).

Once the cell suspension has been prepared it is submitted to a cell separation technique, in order to obtain large quantities of homogeneus populations of spermatogenic cells at specific differentiation stages. The most commonly used approaches have been based on the fact that the different cell types in the spermatogenic cell suspension differ greatly in size and, as a consequence, can be separated because of their different sedimentation velocities.

The first devised method for cell separation consists on the gravimetric decantation of cells in an albumin gradient and is called Staput (Miller and Phillips, 1969). It is composed of a cylindrical chamber with a conical base (Staput chamber) where the cells are sedimented for 4 hours at 4^0C, a gradient former located higher than the chamber in order to generate a liquid flow by gravity, and a fraction collector (Lam *et al.*, 1970; Meistrich, 1977). The Staput approach has been used for the isolation of spermatogenic cell populations from adult mice (Lam *et al.*, 1970; Romrell *et al.*, 1976; Yu *et al.*, 2003) and rats (Go *et al.*, 1971), allowing an important enrichment in pachytene spermatocytes and round spermatids (75% and 81% respectively for mice; Meistrich, 1977); an enrichment in spermatogonia and early meiotic stages has been attained by the use of prepuberal specimens (Bellvé *et al.*, 1977). We used to employ a Staput chamber for spermatogenic cell separations in our laboratory in the past (Figure 3). In our experience, the method is extremely laborious and takes too much time (to the 4-hour sedimentation period, the fractions collection time must be added), thus turning it inappropriate for gene expression studies.

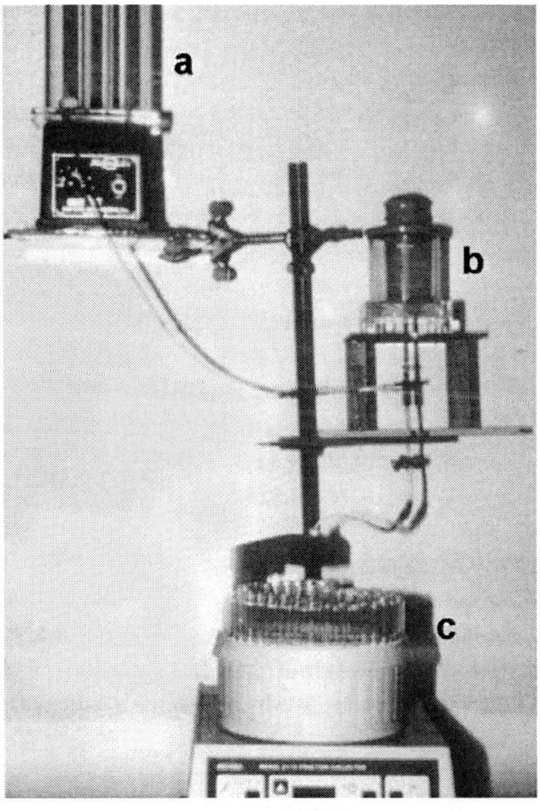

Figure 3. Staput. a: gradient former; b: sedimentation chamber; c: fraction collector. This chamber is loaded and unloaded from the bottom.

The most widely used method for testicular cell separation aiming at differential gene expression analysis has been centrifugal elutriation (*e.g.* Geisinger *et al.*, 1996; 2002; 2005; Alsheimer and Benavente, 1996; Alsheimer *et al.*, 1997; Cossio *et al.*, 1997; Weiss *et al.*, 1977; Schlecht *et al.*, 2004, and many others). This method is based on the use of a special centrifuge rotor with only one separation chamber (Figure 4), and combines liquid flow velocity with intensity of centrifugal forces. A gradient of flow rates is established inside the elutriator chamber, and the different cell populations are collected according to their different sedimentation rates (for a detailed description of the method see Meistrich, 1977). The procedure is much faster than Staput (20-90 minutes), and the number of cells that can be loaded is higher (more than 3 x 10^9, compared to 1 x 10^9 which is the highest number of cells that could be loaded in a Staput chamber). Fractions enriched in pachytene spermatocytes and round spermatids can be obtained from adult animals, with a purity of over 80% each (Meistrich, 1977, and our own experience). A third fraction highly enriched in elongated spermatids and residual bodies can be collected. It is possible to obtain an enrichment in other cell populations, but the purity of the collected fractions is notoriously lower. Some researchers have improved the purity of the spermatocyte populations by using prepuberal animals whose testes are devoid of round spermatids - a common contaminant of pachytene fractions when using adult testis - for centrifugal elutriation (Rossi *et al.*, 2004).

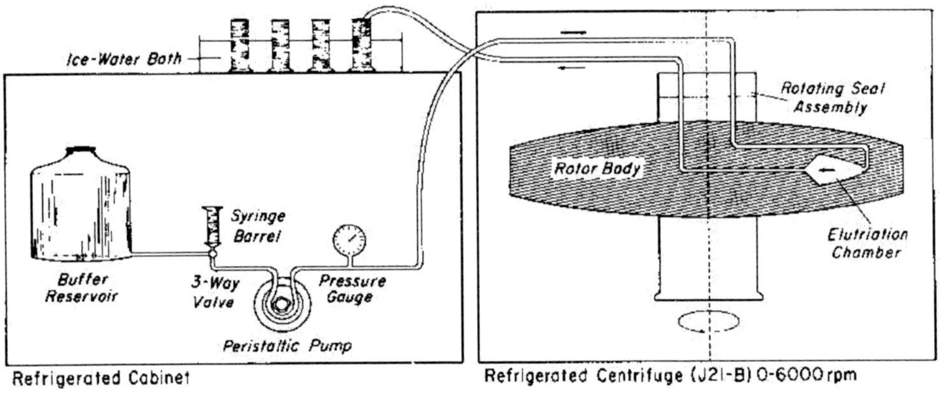

Figure 4. Representation of an elutriator (after Meistrich, 1977).

The sequential use of two different separation techniques such as centrifugal elutriation and equilibrium density centrifugation on Percoll gradients has been described (Meistrich *et al.*, 1981; Bucci *et al.*, 1986). This highly increases the purity of the collected cell populations but, again, the additional time and handling involved make it unsuitable for delicate gene expression studies.

A different method that can be used for the obtainment of spermatogenic stage-specific cell populations is cell sorting. Several laboratories have used flow citometry of testicular cell suspensions for the identification of cell subpopulations either for basic research or for clinical diagnostic purposes. DNA (or other components) has been stained with a fluorescent dye, and the cells have been classified based on the differences in DNA content in the C, 2C and 4C populations (*e.g.* Clausen *et al.*, 1982; Janca *et al.*, 1986; Aravindan *et al.*, 1990;

Hittmair *et al.*, 1992; Kostakopoulos *et al.*, 1997; Aslam *et al.*, 2002; Wistuba *et al.*, 2003; Neubauer *et al.*, 2004). Others used DNA content and cell size simultaneously for classification [dual parameter flow cytometry (Vigodner *et al.*, 2003)]. In some cases, the cell subpopulations were sorted afterwards (*e.g.* Grogan *et al.*, 1981; Van Kroonenburgh *et al.*, 1985; Petit *et al.*, 1995; Lasalle *et al.*, 1999). Malkov *et al.* (1998) have described the use of four-parameter flow cytometry (DNA content, cell size, nuclear size, and cell complexity) for the characterization of the testicular developmental schedule of the rat and its comparison to that of the mouse, using a FACS machine (fluorescence-activated cell sorter).

In our laboratory we have recently started to separate early meiotic prophase cells (leptotene/zygotene) from *Cavia porcellus* (guinea pig) by FACS, in order to conduct gene expression studies in those early stages. Although important genes whose products are involved in synapsis and recombination are presumably expressed during early meiotic prophase, very little knowledge about those stages is available up to now due to their short duration in the studied mammalian models and, as a consequence, the difficulty to obtain highly enriched cell populations (As previously mentioned, only pachytene spermatocytes, round spermatids and elongated spermatids can be obtained with relatively high purity levels by elutriation). It has been shown in our laboratory that early meiotic prophase stages are unusually long in guinea pig; thus, the proportion of cells that can be found at a certain moment in those stages is much higher than in any other studied Mammal (Rodríguez and Wettstein, 2004). In a work to be published elsewhere, we have proved that early meiotic prophase cells in guinea pig can be resolved as a discrete peak and sorted with a very high purity level by FACS (Rodríguez *et al.*, manuscript in preparation).

In our experience, the use of high performance FACS machines allows the obtainment of high amounts of testicular purified cell populations in a short time. The obtained cell populations are almost totally pure (our laboratory, unpublished results), and the purity of the collected samples can be assessed by re-sorting of an aliquot if desired (Purity of the cell populations has become a especially important issue nowadays, since the most commonly used methods for studying gene expression - mainly RT-PCR-based - are very sensitive and, therefore, contaminant cells could cause misleading results). Moreover, any testicular cell population can be purified as long as it can be individualized in the dot plots (or a cell type-specific antibody is available). Another advantage of FACS is that since it is multi-purpose equipment, it is more frequently available at research institutions than an elutriator. We have found that the precise identification of the testicular cell subpopulations together with the advantage of efficiently separating them by FACS represents a powerful tool for studying mammalian spermatogenesis at the molecular level.

Gene Expression During Mammalian Spermatogenesis: Particular Traits

Recent progress in molecular biology techniques has encouraged investigations of complex biological processes such as mammalian spermatogenesis. Although this process is still poorly understood at the molecular level, an increasing number of genes related to the developmental stages of the testis are being identified. The advances are mainly a result of the combination of the different strategies for studying specific testicular cell populations (use of total testis samples from animals at different stages of sexual maturity or purified germ cell populations) with techniques for differential gene expression analysis such as subtractive hybridization, differential display, and especially microarray studies. Furthermore, targeted mutagenesis in mice provides a good way of addressing the contribution of individual genes to spermatogenesis. The amount of genes identified up to now as differentially or specifically expressed during meiosis and spermiogenesis is very large and cannot be listed here (lists of testis-specific and/or differentially expressed transcripts along spermatogenesis can be found in Rossi *et al.*, 2003; Schultz *et al.*, 2003; Yu *et al.*, 2003; Almstrup *et al.*, 2004; Maratou *et al.*, 2004; Schlecht *et al.*, 2004; Shima *et al.*, 2004, and others). Besides the identification of a high number of genes, large-scale transcriptional profiling experiments have started to yield insight into the overall gene expression patterns associated with normal germ cell differentiation.

What follows only seeks to be an enumeration of some of the particular characteristics – many of them quite peculiar - of mammalian spermatogenesis in relation to gene expression patterns, arisen from those studies.

1. Testis is a very complex organ in terms of gene expression, based on the high number of genes that would be involved in testis development and/or spermatogenesis (50% of the murine genome would be expressed during testis development according to Shima *et al.*, 2004; 58% according to Schultz *et al.*, 2003).

2. Around 30% of the murine genome appears as differentially expressed during testicular development (Shima *et al.*, 2004), meaning that the testis would express a higher number of genes than most other tissues in a specific or differential way; this again, is indicative of the testicular complexity. Some examples of testis-specific genes are those coding for synaptonemal complex components (Meuwissen *et al.*, 1992; Lammers *et al.*, 1994; Offenberg *et al.*, 1998; Costa *et al.*, 2005), and for transition proteins (reviewed in Meistrich *et al.*, 2003) and protamines (reviewed in Oliva, 2006), the two groups of basic proteins that sequentially replace histones during chromatin condensation in elongating and elongated spermatids.

3. Pachytene spermatocytes and round spermatids exhibit the highest number of specific transcripts in the testis (Schultz *et al.* 2003; Shima *et al.* 2004). The fact that so many genes are exclusively expressed in the male germ line provides a vast number of potential targets for germ cell-directed contraception, since inhibitory drugs would be expected to have few side effects (Schultz *et al.* 2003).

4. Among testicular somatic cells, spermatogonia would be the cell type expressing the smallest number of specific transcripts (Schultz *et al.* 2003; Shima *et al.* 2004). A large portion of Sertoli cell-specific transcripts are expressed coincidentally with meiosis, with very few transcripts showing elevated expression earlier or later in testis development; this suggests that many of the Sertoli cell transcripts would be involved in the adaptation of these cells to the newly forming germ cells (Shima *et al.* 2004). Leydig cells exhibit a number of specific transcripts that correspond to genes involved in steroidogenesis. Finally, myoid cells would have the greatest number of enriched and specific transcripts of the somatic testicular cells, especially during the early postnatal stages. This indicates that these cells would have a

more important role in relation to seminiferous tubules formation and spermatogenesis than originally thought (Shima *et al.* 2004).

5. Testis and brain exhibit the highest similarity of gene expression patterns among 17 studied tissues (Guo *et al.* 2003). This would support the notion that testis and brain share some particular correlation (and perhaps some regulatory mechanisms), with important possible implications for human speciation and sexual selection (Wilda *et al.*, 2000).

6. Testis has an unusually high number of tissue-specific alternative transcripts, which would be generated by the use of alternative promoters (Huang *et al.*, 2005), alternative polyadenilation sites (Wallace *et al.*, 1999), and mainly alternative splicing (reviewed in Kleene, 2001, Venables, 2002 and Elliott and Grellscheid, 2006). In fact, genome-wide analyses of expressed sequence tags (ESTs) have shown that testis would be one of the tissues with the highest levels of alternative splicing (Xu *et al.*, 2002; Yeo *et al.*, 2004). In our experience, when searching for testicular differentially expressed genes, we very often isolate cDNAs corresponding to previously unidentified (presumably testis-specific) splice variants of already known genes; frequently even no ESTs for those alternative transcripts are present in databases (unpublished results). An important number of the alternative transcripts encode spermatogenic-specific protein isoforms. Among these are isoforms for metabolic enzymes (*e.g.* Rossi *et al.*, 2004), histones (reviewed by Govin *et al.*, 2004), chaperones (Eddy, 1999), lamins (Furukawa and Hotta, 1993; Furukawa *et al.*, 1994; Nakajima and Abe, 1995) and transcription factors such as the spermiogenesis-specific factor CREB (reviewed in Kimmins *et al.* 2004). On the other hand, a compared mapping of alternative splice sites in mouse and human showed that species-specific alternative splicing is more frequent in the testis than in other tissues; if conservation is used as a surrogate for functional significance, then testis could be the tissue with the highest non-functional splicing (Kan *et al.*, 2005).

Concerning the use of alternative promoters, it is important to point out that germ line cells exhibit a marked preference for the use of TATA-independent promoters (Kleene, 2001).

7. A striking trait of spermatogenic cells is the unusually high levels of translational repression. One of the mechanisms employed for translational repression in meiotic and post-meiotic cells is mRNA sequestration as free ribonucleoprotein particles (RNPs), and their shift to polysomes at the moment of translation (Iguchi *et al.*, 2006). Some spermatogenic transcripts exhibit upstream open reading frames (uORFs), a strategy usually considered to reduce the rate of translation (reviewed by Lovett and Rogers, 1996 and Vilela and McCarthy, 2003), while some others repress translation by regulating the length of their poly(A) tails (translational repression in spermatogenic cells has been vastly reviewed by Kleene, 2001, Kleene, 2003 and Elliott, 2003). A good example of translational repression is represented by the mRNAs for protamines 1 and 2; those mRNAs are translationally inert in round and early elongating spermatids, and translated seven days after transcription, in elongated spermatids (Kleene, 1989). Due to translational repression most of the mRNAs in meiotic and haploid cells appear to be over-expressed, since the levels of protein are low relative to mRNA levels (Kleene, 2001).

8. Testis expresses an extremely high number of tissue-specific RNA-binding proteins. These include specific translation factors, proteins involved in translational repression and activation (Kwon and Hecht 1993; Giorgini *et al.*, 2001; Giorgini *et al.*, 2002; Kleene, 2003;

Elliott and Grellscheid, 2006), and specific components of the splicing machinery (Venables and Eperon, 1999; Elliot and Grellscheid, 2006). These latter would account for the high levels of testis-specific splicing, and at least some of them would possibly bind RNA through testis-specific mechanisms (Skrisovska *et al.*, 2007).

```
tBC2-9          GGGGTTGGGGATTTAGCTCA-GTGGTAGAGCGCTT--GCCTAGGAAGCGCAAGGCCCTGG
tBC2-22         GGGGTTGGGGATTTAGCTCA-GTGGTAGAGCGCTT--GCCTAGCAAGCGCAAGGCCCTGG
tBC2-15         GGGGTTGGGGATTTAGCTCA-GTGGTAGAGTGCTT--GCCTAGGAAGCGCAAGGCCCTGG
tBC2-13         GGGGTTGGGGATTTAGCTCA-GTGGTAGAGCGCTT--ACCTAGGAAGCGCAAGGCCCTGG
tBC2-2          GGGGTTGGGGATTTAGCTCA-GTGGTAGAGCGCTT--GCCTAGCAAGCGCAAGGCCCTGG
tT3-3           GGGGTTGGGGATTTAGCTCA-GTGGTAGAGCGCTT--GCCTAGGAAGCGCAAGGCCCTGG
tT3-6           GGGGTTGGGGATTTAGCTCA-GTGGTAGAGCGCTT--GCCTAGGAAGCGCAAGGCCCTGG
tT3-13          GGGGTTGGGGATTTAGCTCA-GTGGTAGAGCGCTT--GCCTAGGAAGCGCAAGGCCCTGG
tT3-1           GGGGTTGGGGATTTAGCTCA-GTGGTAGAGCGCTT--GCCTAGCAAGCGCAAGGCCCTGG
tT3-17          GGGGTTGGGGATTTAGCTCAGTTGGTAGAGCGCTT--GCCTAGGAAGCGCAAGGCCCTGG
BC1             GGGGTTGGGGATTTAGCTCA-GTGGTAGAGCGCTT--GCCTAGCAAGCGCAAGGCCCTGG
                ********************  ******** ** *   ***** ****************

probe           GTTCGGTCCCCAGCTCCG------------------------------------------
tBC2-9          GTTCGGTCCCCAGCTCTGAAAAAAAGAACCAAAAAAAAAAAAAAACAAAAAAAAATTT--
tBC2-22         GTTCGGTCCCCAGCTCCGAAAAAAAAAAAAAAAAAAAAGAAAAAAAAAAAAA--------
tBC2-15         GTTCGGTCCCCAGCTCCGAAAAAAAGAACCAAAAAAAAAAAAAAAAAGAATAACAAAGCA
tBC2-13         GTTCGGTCCCCAGCTCCGAAAAAAAGAACCAAAAAAAAAAA-------------------
tBC2-2          GTTCGGTCCCCAGCTCCGAAGGAAAAAAAAAAAAAGAAAAATGTGTTCAGCAAGTTTT--
tT3-3           GTTCGGTCCCCAGCTCCG------------------------------------------
tT3-6           GTTCGGTCCCCAGCTCCG------------------------------------------
tT3-13          GTTCGATCCCCAGCTCCA------------------------------------------
tT3-1           GTTCGGTCCTCAGCTTCG------------------------------------------
tT3-17          GTTCGGTCCCCAGCTCC-------------------------------------------
BC1             GTTCGGTCCTCAGCTCCGAAAAAAAAAAAAAAAAAAAAAAGACAAAATAACAAAAAGACC
                ***** *** *****

probe           -----------------------------------
tBC2-9          -----------------------------------
tBC2-22         -----------------------------------
tBC2-15         GGACTT-----------------------------
tBC2-13         -----------------------------------
tBC2-2          -----------------------------------
tT3-3           -----------------------------------
tT3-6           -----------------------------------
tT3-13          -----------------------------------
tT3-1           -----------------------------------
tT3-17          -----------------------------------
BC1             AAAAAAAAACAAGGTAACTGGCACACACAACCTTT

Sequence 1: probe          78 bp
Sequence 2: tBC2-2        115 bp Accession number: U25028
Sequence 3: tBC2-9        115 bp Accession number: U25467
Sequence 4: tBC2-13        98 bp Accession number: U25470
Sequence 5: tBC2-15       123 bp Accession number: U25468
Sequence 6: tBC2-22       109 bp Accession number: U25031
Sequence 7: tT3-1          75 bp Accession number: U25496
Sequence 8: tT3-3          75 bp Accession number: U25507
Sequence 9: tT3-6          75 bp Accession number: U25506
```

Figure 5A. Analysis of ID expression pattern in the rat by Northern-blot. A 5′ consensus 78 bp sequence which is common to BC1, BC2 and T3 RNAs was used as a probe. A. A sequence alignment of the probe with BC1 and any five BC2 and T3 clones from Genbank made with ClustalW is shown. The BC1-specific probe used by Muslimov *et al.* (2002) was complementary to the 60 3′-most nucleotides of BC1 RNA.

9. Testicular cells transcribe high levels of non-coding polyadenylated RNAs (Kleene, 2001). A group of such RNAs in rodents is composed of ID ("Identifier") elements (Sutcliffe *et al.*, 1984). These constitute a family of short interspersed repetitive elements (SINEs) that are transcribed by RNA polymerase III in a tissue-specific way mainly in brain, where they have been proposed to be related to the regulation of neuronal-specific gene expression (McKinnon 1986). However, ID expression has been also detected in some other tissues such as testis (Kim *et al.*, 1995). Muslimov *et al.* (2002) have analyzed the expression pattern of BC1 – proposed as a master gene for ID repetitive elements, although this role has been recently controverted (Johnson and Brookfield, 2006) – in the murine testis by *in situ* hybridization with a BC1-specific probe. They have demonstrated that BC1 was expressed in the murine testis, especially in spermatogonia, and hypothesized that it could be participating in the mediation of translational control in the syncytial ensembles of cells; BC1 RNA expression levels decreased during meiosis, with negligible amounts in round spermatids. Using a consensus DNA sequence for the ID family members as a probe (Figure 5A), we have studied the transcription pattern of ID elements in different tissues of the rat by Northern-blot. Our results show that ID transcripts are present at the highest levels in the brain, followed by testis (Figure 5B). A lower signal was detected in ovary. Expression was not detectable in the other analyzed tissues after overnight exposures, although a low signal was visible in almost every tissue after a five-day over-expossure (not shown). Within the testis, ID RNA was detectable in purified pachytene spermatocytes, round spermatids and in whole testis as well (Figure 5C). The different testicular expression patterns of BC1 RNA alone and of total ID elements suggest that maybe different roles could exist for the different ID family members in the testis.

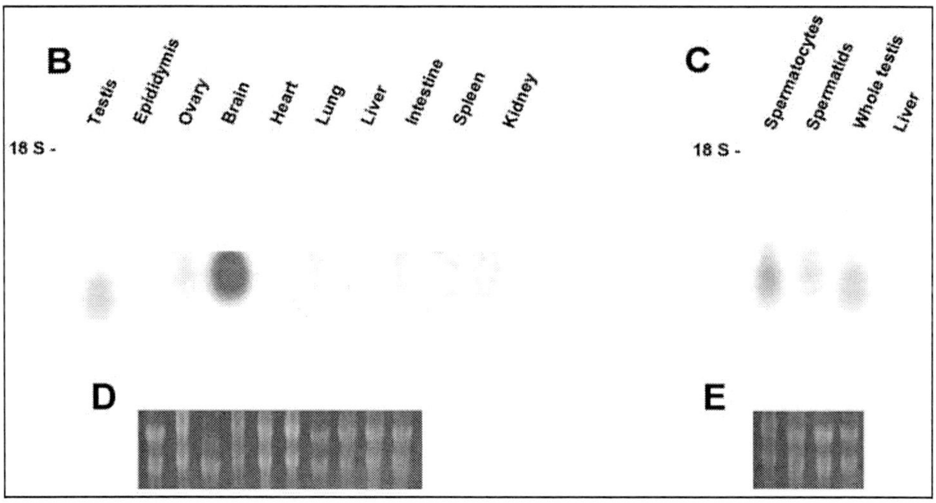

Figure 5(B,C,D,E). Analysis of ID expression pattern in the rat by Northern-blot. A 5′ consensus 78 bp sequence which is common to BC1, BC2 and T3 RNAs was used as a probe. B. Analysis of the expression pattern in different tissues. C. Analysis in whole testis and highly enriched testicular cell populations separated by elutriation. The hybridization pattern, visualized as spots instead of bands, reflects the size heterogeneity of the different family members. The position of 18S rRNA in B and C is shown. D and E show the ethidium bromide stained agarose gels to document that all lanes were equally loaded.

Although the reasons why the patterns of gene expression during testicular cell differentiation differ so greatly from those in most tissues are poorly understood, these peculiar patterns are usually attributed to important functions in spermatogenesis. Kleene (2005) has suggested that the atypical expression patterns could be explained in light of the powerful selective pressures that influence male reproductive success, known as sexual selection. Sexual selection would be characterized by intense competition of males for females, an enormous variety of the strategies to maximize male reproductive success, exaggerated male traits at all levels of biological organization, co-evolution of sexual traits in males and females, etc.

The sequential replacement of histones with transition proteins and protamines allows DNA compaction of the round spermatid nucleus to a volume of about 5% of that of a somatic cell nucleus (Sassone-Corsi, 2002). That amazing compaction - together with nucleus reshaping and cytoplasm elimination - produces a decrease in hydrodynamic resistance that enhances motility, thus contributing to facilitate the sperm journey (Allen *et al.*, 1996). The large number of spermatogenic-specific metabolic enzymes (*e.g.* Rossi *et al.*, 2004) could be related to supplying ATP to the sperm flagellum (Besides, since the testes are descended in Mammals, testis-specific enzyme isoforms could accommodate better to testicular temperature).

X-chromosome is transcriptionally inactivated in male meiosis as a mechanism to avoid recombination between X and Y-chromosomes (reviewed in Turner, 2007). The transcriptional inactivation of the X-chromosome would lead to the selection of autosomal genes to replace the functions of X-linked genes (*e.g.* McCarrey and Thomas, 1987; Wallace *et al.*, 1999; Elliot *et al.*, 2000).

The atypical importance of post-transcriptional regulation could be related to the design of a strategy to regulate protein synthesis in cells that do not transcribe (elongating and elongated spermatids, after histone replacement). The high levels of alternative splicing would account for the huge amounts of novel tissue/cell type-specific proteins. Specially, post-transcriptional regulation would select for novel, testis-specific RNA-binding proteins (Kleene, 2001). According to Elliott and Grellscheid (2006), an increase in alternative splicing might be one of the mechanisms that contribute to the development of sophisticated and rapidly evolving metazoan tissues and, due to sexual selection, testis evolves very fast (also brain is a rapidly evolving tissue). Another advantage of post-transcriptional regulation mechanisms is that they are acutely sensitive and can respond fast to environmental changes, as required in reproductive systems (Elliott, 2003).

CONCLUSION

Testis is a relatively unique tissue because it is the site of an extensive adult developmental process. Male germ cell development continues during the entire life, and the pathway involves the coordinated division and differentiation of huge numbers of cells. Germ line cells have some exclusive traits: they are the only cells in the body that undergo meiosis, the cells that result from meiosis have a haploid and individually reorganized genome and, most importantly, are able to give rise to totipotent diploid zygotes. Three individual programs of gene expression take place during mammalian spermatogenesis in an overlapping and coordinated way, and changes in gene expression occur in intimate association with testicular somatic cells.

The developmental and differentiation programs that lead to the production of mammalian male germ cells are still poorly understood at the molecular level due to some difficulties they present for their study, mainly tissue heterogeneity. As a consequence of that heterogeneity, the availability of methods to allow the assignment of certain transcripts to specific cell types is required. Basically two approaches have been used: the comparison of expression patterns using whole testes from animals at different stages of sexual maturity, and the use of highly enriched cell populations. Unfortunately, none of these strategies for studying gene expression in such a complex system meets all the requirements, since both approaches have advantages and drawbacks. A good approach would be one to allow the obtainment of big amounts of highly purified cell populations in the shortest possible time, to ensure good preservation of delicate macromolecules. In that sense, we consider that the use of high-performance cell sorting combined with a fast and reliable method for the preparation of testicular cell suspensions can be the most suitable strategy developed up to now as a first step for differential gene expression studies in male stage-specific germ line cells.

A number of laboratories have recently approached the study of mammalian male germ cell differentiation using the modern available techniques for gene expression analyses. The studies are revealing testis as a very interesting system because of the peculiarities it presents in comparison to somatic tissues. However, much remains to be learned about the function of the investigated genes. Particularly, the similarities in expression patterns between testis and brain are intriguing and should deserve further study. A deeper understanding of the molecular processes that regulate gene expression during spermatogenesis is essential for the comprehension of the normal functioning of the male reproductive system in Mammals, as well as for the treatment of testicular pathologies.

Acknowledgements

The author thanks Rodolfo Wettstein and Rosana Rodríguez for critical reading of the manuscript. This work was supported by CSIC (I+D grant C19) and DINACYT (FCE grant 10171).

References

Allen, M. J.; Lee 4th, J. D.; Lee, C. & Balhorn, R. (1996). Extent of sperm chromatin hydration determined by atomic force microscopy. *Mol. Reprod. Dev., 45*, 87-92.

Almstrup, K.; Nielsen, J. E.; Hansen, M. A.; Tanaka, M.; Skakkebæk, N. E. & Leffers, H. (2004). Analysis of Cell-Type-Specific Gene Expression During Mouse Spermatogenesis. *Biol. Reprod., 70*, 1751–1761.

Alsheimer, M. & Benavente, R. (1996). Change of karyoskeleton during mammalian spermatogenesis: expression pattern of nuclear lamin C2 and its regulation. *Exp. Cell Res., 228*, 181-188.

Alsheimer, M.; Fecher, E. & Benavente, R. (1998). Nuclear envelope remodelling during rat spermiogenesis: distribution and expression pattern of LAP2/thymopoietins. *J. Cell Sci., 111*, 2227-2234.

Alsheimer, M.; Imamichi, Y.; Heid, H. & Benavente, R. (1997). Molecular characterization and expression pattern of XY body-associated protein XY40 of the rat. *Cromosoma, 106*, 308-314.

Alter, B. P. (2003). Cancer in Fanconi anemia, 1927-2001. *Cancer, 97*, 425-440.

Anway, M. D.; Li, Y.; Ravindranath, N.; Dym, M. & Griswold, M. D. (2003). Expression of testicular germ cell genes identified by differential display analysis. *J. Androl., 24*, 173-184.

Aravindan, G. R.; Ravindranath, N.; Gopalakrishnan, K. & Moudgal, N. R. (1990). DNA flow-cytometric analysis of testicular germ cell populations of the bonnet monkey (*Macaca radiata*) as a function of sexual maturity. *J. Reprod. Fertil., 89*, 397-406.

Aslam, H.; Schneiders, A.; Perret, M.; Weinbauer, G. F. & Hodges, J. K. (2002). Quantitative assessment of testicular germ cell production and kinematic and morphometric parameters of ejaculated spermatozoa in the grey lemur, *Microebus murinus*. *Reproduction, 123*, 323-332.

Baker, S. M.; Bronner; C. E.; Zhang, L.; Plug, A. W.; Robatzek, M.; Warren, G.; Elliott, E. A.; Yu, J.; Ashley, Y.; Arnheim, N.; Flavell, R. A. & Liskay, R. M. (1995). Male mice defective in the mismatch repair gene *PMS2* exhibit abnormal chromosome synapsis in meiosis. *Cell, 82*, 309-319.

Beissbarth, T.; Borisevich, I.; Horlein, A.; Kenzelmann, M.; Hergenhahn, M.; Klewe-Nebenius, A.; Klaren, R.; Korn, B.; Schmid, W.; Vingron, M. & Schutz, G. (2003). Analysis of CREM-dependent gene expression during mouse spermatogenesis. *Mol. Cell Endocrinol., 212*, 29-39.

Bellvé, A. R.; Cavicchia, J. C.; Milletfe, C. F., O´Brien, D. A.; Bhatnagar, Y. M. & Dym, M. (1977). Spermatogenic cells of the prepubral mouse. Isolation and morphological characterization. *J. Cell Biol., 74*, 68-85.

Boder, E. (1975). Ataxia-telangiectasia: some historic, clinical and pathologic observations. *Birth Defects Orig Artic Ser., 11*, 255-270.

Bucci, L. R.; Brock, W. A.; Johnson, T. S. & Meistrich, M. L. (1986). Isolation and biochemical studies of enriched populations of spermatogonia and early primary spermatocytes from rat testes. *Biol. Reprod., 34*, 195-206.

Catalano, R. D.; Vlad, M. & Kennedy, R. C. (1997). Differential display to identify and isolate novel genes expressed during spermatogenesis. *Mol. Hum. Reprod., 3*, 215-221.

Chan, W. Y.; Lee, T. L.; Wu, S. M.; Ruszczyk, L.; Alba, D.; Baxendale, V. & Rennert, O. M. (2006). Transcriptome analyses of male germ cells with serial analysis of gene expression (SAGE). *Mol. Cell Endocrinol., 16*, 8-19.

Chu, S.; DeRisi, J.; Eisen, M.; Mulholland, J.; Botstein, D.; Brown, P.O. & Herskowitz, I. (1998). The transcriptional program of sporulation in budding yeast. *Science, 282*, 699-705. *Erratum in Science, 282*, 1421.

Clark, N. L.; Aagaard, J. E. & Swanson, W. J. (2006). Evolution of reproductive proteins from animals and plants. *Reproduction, 131*, 11-22.

Clausen, O. P.; Parvinen, M. & Kirkhus, B. (1982). Stage-related variations in DNA fluorescence distribution during rat spermatogenic cycle measured by flow cytometry. *Cytometry, 2*, 421-425.

Clermont, Y. (1960). Cycle of the seminiferous epithelium of the guinea pig. A method for identification of the stages. *Fertil. Steril., 11*, 563-573.

Clermont, Y. (1963). The cycle of the seminiferous epithelium in man. *Am. J. Anat., 112*, 35-51.

Cohen, P. E. & Pollard, J. W. (2001). Regulation of meiotic recombination and prophase I progression in mammals. *Bioessays, 23*, 996-1009.

Cossio, G.; Sanchez, J.C.; Golaz, O.; Wettstein, R. & Hochstrasser, D.F. (1995). Spermatocytes and round spematids of rat testis: protein patterns. *Electrophoresis, 16*, 1225-1230.

Cossio, G.; Sanchez, J. C.; Wettstein, R. & Hochstrasser, D. F. (1997). Spermatocytes and round spermatids of rat testis: the difference between in vivo and in vitro protein patterns. *Electrophoresis, 18*, 548-552.

Costa, Y., Speed, R.; Ollinger, R.; Alsheimer, M.; Semple, C. A.; Gautier, P.; Maratou, K.; Novak, I.; Hoog, C.; Benavente, R. & Cooke, H. J. (2005). Two novel proteins recruited by synaptonemal complex protein 1 (SYCP1) are at the centre of meiosis. *J. Cell Sci., 118*, 2755–2762.

Davis, L. & Smith, G. R. (2001). Meiotic recombination and chromosome segregation in *Schizosaccharomyces pombe*. *Proc. Natl. Acad. Sci. U.S.A., 98*, 8395-8402.

de Rooij, D. G. (2001). Proliferation and differentiation of spermatogonial stem cells. *Reproduction, 121*, 347–354.

Dirami, G.; Ravindranath, N; Pursel, V. & Dym, M. (1999). Effects of stem cell factor and granulocyte macrophage-colony stimulating factor on survival of porcine type A spermatogonia cultured in KSOM. *Biol. Reprod., 61*, 225–230.

Dong, H. & Roeder, G. S. (2000). Organization of the yeast Zip1 protein within the central region of the synaptonemal complex. *J. Cell Biol., 148*, 417-426.

Dym, 1977. The male reproductive system. In L. Weiss & R. O. Greep (Eds.), *Histology* (4th Ed., pp. 979-1038). New York: McGraw-Hill.

Dym, M. & Fawcett, D. W. (1971). Further observations on the number of spermatogonia, spermatocytes, and spermatids connected by intercellular bridges in the mammalian testis. *Biol. Reprod., 4*, 195-215.

Eddy, E. M. (1999). Role of heat shock protein HSP70-2 in spermatogenesis. *Rev. Reprod., 4*, 23-30.

Elliott, D. J.; Venables, J. P.; Newton, C. S.; Lawson, D.; Boyle, S.; Eperon, I. C. & Cooke, H.J. (2000). An evolutionarily conserved germ cell-specific hnRNP is encoded by a retrotransposed gene. *Hum. Mol. Genet., 9*; 2117-2124.

Elliott, D. (2003). Pathways of post-transcriptional gene regulation in mammalian germ cell development. *Cytogenet. Genome Res., 103*, 210-216.

Elliott, D. J. & Grellscheid, S. N. (2006). Alternative RNA splicing regulation in the testis. *Reproduction, 132*, 811–819.

Ellis, P.J.; Furlong, R. A.; Conner, S.J.; Kirkman-Brown, J.; Afnan, M.; Barrat, C.; Griffin, D. K. & Affara, N. A. (2007). Coordinated transcriptional regulation patterns associated with infertility phenotypes in men. *J. Med. Genet.* (in press).

Erickson, R. P. (1990). Post-meiotic gene expression, *Trends Genet., 6*, 264-268.

Feig, C.; Kirchhoff, C.; Ivell, R.; Naether, O.; Schulze, W. & Spiess, A. N. (2007). A new paradigm for profiling testicular gene expression during normal and disturbed human spermatogenesis. *Hum. Mol. Reprod., 13*, 33-43.

Fujii, T.; Tamura. K.; Masai, K.; Tanaka, H.; Nishimune, Y. & Nojima, H. (2002). Use of stepwise subtraction to comprehensively isolate mouse genes whose transcription is up-regulated during spermiogenesis. *EMBO Rep., 3*, 367–372.

Furukawa, K.; Inagaki, H. & Hotta, Y. (1994). Identification and cloning of an mRNA coding for a germ cell-specific A-type lamin in mice. *Exp. Cell Res., 212*, 426-430.

Furukawa, K. & Hotta, Y. (1993). cDNA cloning of a germ cell specific lamin B3 from mouse spermatocytes and analysis of its function by ectopic expression in somatic cells. *EMBO J., 12*, 97-106.

Geisinger, A.; Alsheimer, M.; Wettstein, R. & Benavente, R. (2005). The mammalian gene pecanex-1 is differentially expressed during spermatogenesis. *Biochim. Biophys. Acta, 1728*, 34-43.

Geisinger, A.; Dos Santos, A.; Benavente, R. & Wettstein, R. (2002). Identification and characterization of Srsp1, a rat gene differentially expressed during spermatogenesis and coding for a serine stretch-containing protein. *Cytogenet. Genome Res., 98*, 249-254.

Geisinger, A.; Wettstein, R. & Benavente, R. (1996). Stage-specific gene expression during rat spermatogenesis: application of the mRNA differential display method. *Int. J. Dev. Biol., 40*, 385-388.

Giorgini, F.; Davies, H. G. & Braun, R. E. (2001). MSY2 and MSY4 bind a conserved sequence in the 3′ untranslated region of protamine mRNA in vitro and in vivo. *Mol. Cell Biol., 21*, 7010–7019.

Giorgini, F.; Davies, H. G. & Braun, R. E. (2002). Translational repression by MSY4 inhibits spermatid differentiation in mice. *Development, 129*, 3669–3679.

Go, V. L. W.; Vernon, R. G. & Fritz, I. B. (1971). Studies on spermatogenesis in rats. I. Application of the sedimentation velocity technique to an investigation of spermatogenesis. *Can. J. Biochem., 49*, 753-760.

Govin, J.; Caron, C.; Lestrat, C.; Rousseaux, S. & Khochbin, S. (2004). The role of histones in chromatin remodelling during mammalian spermiogenesis. *Eur. J. Biochem., 271*, 3459–3469.

Greep, R. O. (1954). Aparato genital masculino. In R. O. Greep (Ed.), *Histología* (1st edition, pp. 751-787). Buenos Aires: El Ateneo.

Gregan, J.; Rabitsch, P. K.; Rumpf, C., Novatchkova, M.; Schleiffer, A. & Nasmyth, K. (2006). High-throughput knockout screen in fission yeast. *Nat. Protoc., 1*, 2457-2464.

Griswold, M.D. (1998). The central role of Sertoli cells in spermatogenesis. Semin. Cell Dev. Biol., 9, 411-416.Grogan, W.M.; Farnham, W.F. & Sabau, J.M. (1981). DNA analysis and sorting of viable mouse testis cells. *J. Histochem. Cytochem., 29*, 738–746.

Guo, J.; Zhu, P.; Wu, C.; Yu, L.; Zhao, S. & Gu, X. (2003). In silico analysis indicates a similar gene expression pattern between human brain and testis. *Cytogenet. Genome Res., 103*, 58-62.

Heyting, C. & Dietrich, A. J. J. (1989). Synaptonemal complex proteins. Genome, 31, 81-87.

Hittmair, A.; Rogatsch, H.; Offner, F.; Feichtinger, H.; Ofner, D. & Mikuz, G. (1992). Deoxyribonucleic acid flow cytometry and semiquantitative histology of spermatogenesis: a comparative study. *Fertil. Steril., 58*, 1040-1045.

Hochwagen, A. & Amon, A. (2006). Checking your breaks: surveillance mechanisms of meiotic recombination. *Curr. Biol. 16*, R217-228.

Huang, X.; Li, J.; Lu, L.; Xu, M.; Xiao, J.; Yin, L.; Zhu, H.; Zhou, Z. & Sha, J. (2005). Novel development-related alternative splices in human testis identified by cDNA microarrays. *J. Androl., 26*, 189-196.

Iguchi, N.; Tobias, J. W. & Hecht, N. B. (2006). Expression profiling reveals meiotic male germ cell mRNAs that are translationally up- and down-regulated. *Proc. Natl. Acad. Sci. U.S.A., 103*, 7712-7717.

Janca, F. C.; Jost, L. K. & Evenson, D. P. (1986). Mouse testicular and sperm cell development characterized from birth to adulthood by dual parameter flow cytometry. *Biol. Reprod., 34*, 613-623.

Johnson, L. J. & Brookfield, J. F. (2006). A test of the master gene hypothesis for interspersed repetitive DNA sequences. *Mol. Bio. Evol., 23*, 235-239.

Kan, Z.; Garrett-Engele, P. W.; Johnson, J. M. & Castle, J. C. (2005). Evolutionarily conserved and diverged alternativesplicing events show different expression and functional profiles. *Nucleic Acids Res., 33*, 5659–5666.

Kim, J.; Kass, D. H. & Deininger, P. L. (1995). Transcription and processing of the rodent ID repeat family in germline and somatic cells. *Nucleic Acids Res., 23*, 2245-2251.

Kimmins, S.; Kotaja, N.; Davidson, I. & Sassone-Corsi, P. (2004). Testis-specific transcription mechanisms promoting male germ-cell differentiation. *Reproduction, 128*, 5–12.

Kleckner, N. (2006). Chiasma formation: chromatin/axis interplay and the role(s) of the synaptonemal complex. *Chromosoma 115*, 175-194.

Kleene, K.C. (1989). Poly(A) shortening accompanies the activation of translation of five mRNAs during spermiogenesis in the mouse. *Development, 106*, 367–373.

Kleene, K.C. (2001). A possible meiotic function of the peculiar patterns of gene expression in mammalian spermatogenic cells. *Mech. Dev., 106*, 3-23.

Kleene, K.C. (2003). Patterns, mechanisms, and functions of translation regulation in mammalian spermatogenic cells. *Cytogenet. Genome Res., 103*, 217–224.

Kleene, K.C. (2005). Sexual selection, genetic conflict, selfish genes, and the atypical patterns of gene expression in spermatogenic cells. *Dev. Biol., 277*, 16–26.

Kostakopoulos, A.; Georgoulakis, J.; Deliveliotis, C.; Spanakis, G.; Filippidou, A. & Tamvakis, N. (1997). Deoxyribonucleic acid flow cytometry in the assessment of spermatogenesis. *J. Urol., 158*, 79-81.

Kralewski, M.; Novello, A. & Benavente, R. (1997). A novel Mr 77,000 protein of the XY body of mammalian spermatocytes: its localization in normal animals and in Searle's translocation carriers. *Chromosoma, 106*, 160-167.

Krogh, B. O. & Symington, L. S. (2004). Recombination proteins in yeast. *Annu. Rev. Genet. 38*, 233-271.

Kwon, Y. K. & Hecht, N. B. (1993). Binding of a phosphoprotein to the 3′ untranslated region of the mouse protamine 2 mRNA temporally represses its translation. *Mol. Cell Biol., 13*, 6547–6557.

Lam, D. M. K.; Furrer, R. & Bruce, W. R. (1970). The separation, physical characterization, and differentiation kinetics of spermatogonial cells of the mouse. *Proc Natl. Acad. Sci. U.S.A., 65*, 192-199.

Lammers, J. H. M.; Offenberg, H. H.; van Aalderen, M.; Vink, A. C. G.; Dietrich, A. J. J. & Heyting, C. (1994). The gene encoding a major component of the lateral elements of synaptonemal complexes of the rat is related to X-linked lymphocyte-regulated genes. *Mol. Cell Biol., 14*, 1137-1146.

Lasalle, B.; Ziyyat, A.; Testart, J.; Finaz, C. & Lefevre, A. (1999). Flow cytometric method to isolate round spermatids from mouse testis. *Hum Reprod., 14*, 388-394.

Leblond, C.P. & Clermont, Y. (1952). Definition of the stages of the cycle of the seminiferous epithelium in the rat. *Ann. N. Y. Acad. Sci., 55*, 548-573.

Li, W. & Ma, H. (2006). Double-stranded DNA breaks and gene functions in recombination and meiosis. *Cell Res., 16*, 402-412.

Liang, P.; Averboukh, L. & Pardee, A. B. (1993). Distribution and cloning of eukaryotic mRNAs by means of differential display: refinements and optimization. *Nucleic Acids Res., 21*, 3269-3275.

Lilford, R.; Jones, A. M.; Bishop, D. T.; Thornton, J. & Mueller, R. (1994). Case-control study of whether subfertility in men is familial. *BMJ, 309*, 570–573.

Lin, Y. H.; Lin, Y. M.; Teng, Y. N.; Hsieh, T. Y. T.; Lin, Y. S & Kuo, P. L. (2006). Identification of ten novel genes involved in human spermatogenesis by microarray analysis of testicular tissue. *Fertility and Sterility, 86*, 1650-1658.

Liu, J. G.; Yuan, L.; Brundell, E.; Bjorkroth, B.; Daneholt, B. & Hoog, C. (1996). Localization of the N-terminus of SCP1 to the central element of the synaptonemal complex and evidence for direct interactions between the N-termini of SCP1 molecules organized head-to-head. *Exp. Cell Res., 226*, 11-19.

Lovett, P. S. & Rogers, E. J. (1996). Ribosome regulation by the nascent peptide. *Microbiol. Rev., 60*, 366-385.

Luk, J. M.; Mok, B. W.; Shum, C. K.; Yeung, W. S.; Tam, P. C.; Tse, J. Y.; Chow, J. F.; Woo, J.; Kam, K. & Lee, K. F. (2003). Identification of novel genes expressed during spermatogenesis in stage-synchronized rat testes by differential display. Biochem. *Biophys. Res. Commun., 307*, 782-390.

Malkov, M.; Fisher, Y. & Don, J. (1998). Developmental schedule of the postnatal rat testis determined by flow cytometry. *Biol. Rep., 59*, 84-92.

Maratou, K.; Forster, T.; Costa, Y., Taggart, M.; Speed, R. M.; Ireland, J.; Teague, P.; Roy, D. & Cooke, H. J. (2004). Expression profiling of the developing testis in wild-type and Dazl knockout mice. *Molec. Reprod. Dev., 67*, 26–54.

Mata, J.; Lyne, R.; Burns, G. & Bahler, J.(2002). The transcriptional program of meiosis and sporulation in fission yeast. *Nat Genet., 32*, 143-147.

Mauduit, C.; Florin, A.; Amara, S.; Bozec, A.; Siddeek, B.; Cunha, S.; Meunier, Selva, L. J.; Albert, M.; Vialard, F.; Bailly, M. & Benahmed, M. (2006). Effets à long terme des perturbateurs endocriniens environnementaux sur la fertilité masculine. *Gynecol. Obstet. Fertil., 34*, 978-84.

McCarrey, J. R. & Thomas, K. (1987). Human testis-specific human PGK gene lacks introns and possesses characteristics of a processed gene. *Nature, 326*, 501-505.

McKinnon, R. D.; Shinnick, T. M. & Sutcliffe, J. G. (1986). The neuronal identifier element is a *cis*-acting positive regulator of gene expression. *Proc. Natl. Acad. Sci. U.S.A., 83*, 3751-3755.

Meistrich, M. L. (1977). Separation of spermatogenic cells and nuclei from rodent testes. *Methods Cell. Biol., 15*, 15–54.

Meistrich, M. L.; Longtin, J.; Brock, W. A.; Grimes, S. R. & Mace, M. L. (1981). Purification of rat spermatogenic cells and preliminary biochemical analysis of these cells. *Biol. Reprod., 25*, 1065-1077.

Meistrich, M. L.; Mohapatra, B.; Shirley, C. R. & Zhao, M. (2003). Roles of transition nuclear proteins in spermiogenesis. *Chromosoma, 111*, 483-488.

Meistrich, M. L; Bruce, W. R & Clermont, Y. (1973). Cellular composition of fractions of mouse testis cells following velocity sedimentation separation. *Exp. Cell Res., 79*, 213-227.

Meuwissen, R. L.; Offenberg, H. H.; Dietrich, A. J.; Riesewijk, A.; van Iersel, M. & Heyting, C. (1992). A coiled-coil related protein specific for synapsed regions of meiotic prophase chromosomes. *EMBO J., 11*, 5091–5100.

Miller, R. G. & Phillips, R. A. (1969). Separation of cells by velocity sedimentation. *J. Cell Physiol., 73*, 197-201.

Moens, P. B.; Freire, R.; Tarsounas, M.; Spyropoulos, B. & Jackson, S. P. (2000). Expression and nuclear localization of BLM, a chromosome stability protein mutated in Bloom's syndrome, suggest a role in recombination during meiotic prophase. *J. Cell Sci., 113*, 663-672.

Muslimov, I. A.; Lin, Y.; Heller, M.; Brosius, J.; Zakeri, Z. & Tiedge, H. (2002). A small RNA in testis and brain: implications for male germ cell development. *J. Cell Sci., 115*, 1243-1250.

Nakajima, N. & Abe, K. (1995). Genomic structure of the mouse A-type lamin gene locus encoding somatic and germ cell-specific lamins. *FEBS Lett., 365*, 108-114.

Neale, M. J. & Keeney, S. (2006). Clarifying the mechanics of DNA strand exchange in meiotic recombination. *Nature, 13*, 153-158.

Neubauer, K.; Jewgenow, K.; Blottner, S.; Wildt, D. E. & Pukazhenthi, B. S. (2004). Quantity rather than quality in teratospermic males: a histomorphometric and flow cytometric evaluation of spermatogenesis in the domestic cat (*Felis catus*). *Biol. Reprod., 71*, 1517-1524.

Offenberg, H. H.; Dietrich, A. J. J. & Heyting, C. (1991). Tissue distribution of two major components of synaptonemal complexes of the rat. *Chromosoma, 101*, 83-91.

Offenberg, H. H.; Schalk, J. A. C.; Meuwissen, R. L. J.; van Aalderen, M.; Kester, H. A.; Dietrich, A. J. J. & Heyting, C. (1998). SCP2: a major protein component of the axial elements of synaptonemal complexes of the rat. *Nucleic Acids Res., 26*, 2572-2579.

Oliva, R. (2006). Protamines and male infertility. *Hum. Reprod. Update, 12*, 417-435.

Petit, M. J.; Ratinaud, M. H.; Cordelli, E.; Spano, M. & Julien, R. (1995). Mouse testis cell sorting according to DNA and mitochondrial changes during spermatogenesis. *Cytometry, 19*, 304–312.

Primig, M.; Williams, R. M.; Winzeler, E. A.; Tevzadze, G. G.; Conway, A. R.; Hwang, S. Y.; Davis, R. W. & Esposito, R. W. (2000). The core meiotic transcriptome in budding yeasts. *Nat. Genet., 26*, 415-423.

Resnick, I. B.; Kondratenko, I.; Togoev, O.; Vasserman, N.; Shagina, I.; Evgrafov, O.; Tverskaya, S.; Cerosaletti, K. M.; Gatti, R. A. & Concannon, P. (2002). Nijmegen breakage syndrome: clinical characteristics and mutation analysis in eight unrelated Russian families. *J. Pediatr., 140*, 355-61.

Rice, W. R. & Chippindale, A. K. (2001). Sexual recombination and the power of natural selection. *Science, 294*, 555-559.

Rodríguez, R. & Wettstein, R. (2004). Quantitative study on guinea pig spermatogenesis shows a relative high percentage of early meiotic prophase stages. *Anat. Rec., 278*, 493-504.

Roeder, G. S. & Bailis, J. M. (2000). The pachytene checkpoint. Trends. Genet., 16, 395-403.

Romrell, L. J.; Bellvé, A. R. & Fawcet, D. W. (1976). Separation of mouse spermatogenic cells by sedimentation velocity. *Dev. Biol., 19*, 119-131.

Rossi, P.; Dolcia, S.; Settea, C.; Capolunghia, F.; Pellegrinia, M.; Loiarroa, M.; Di Agostinoa, S.; Paronettoa, M.P.; Grimaldia, P.; Mericob, D.; Marteganib, E. & Geremia, R. (2004). Analysis of the gene expression profile of mouse male meiotic germ cells. *Gene Expr. Patterns, 4*, 267–281.

Sapiro, R.; Kostetskii, I.; Olds-Clarke, P.; Gerton, G. L.; Radice, G. L. & Strauss III, J. F. (2002). Male infertility, impaired sperm motility, and hydrocephalus in mice deficient in sperm-associated antigen 6. *Mol. Cell Biol., 22*, 6298-305.

Sassone-Corsi, P. (2002). Unique chromatin remodeling and transcriptional regulation in spermatogenesis. *Science, 296*, 2176-2178.

Schlecht, U.; Demougin, P.; Koch, R.; Hermida, L.; Wiederkehr, C.; Descombes, P.; Pineau, B. J. & and Primig, M. (2004). Expression profiling of mammalian male meiosis and gametogenesis identifies novel candidate genes for roles in the regulation of fertility. *Molec. Biol. Cell, 15*, 1031–1043.

Schmekel, K.; Meuwissen, R. L.; Dietrich, A. J.; Vink, A. C.; van Marle, J., van Veen, H. & Heyting, C. (1996). Organization of SCP1 protein molecules within synaptonemal complexes of the rat. *Exp. Cell Res., 226*, 20-30.

Schultz, N.; Hamra, F. K. & Garbers, D. L. (2003). A multitude of genes expressed solely in meiotic or postmeiotic spermatogenic cells offers a myriad of contraceptive targets. *Proc. Natl. Acad. Sci. U.S.A., 100*, 12201–12206.

Schwarzacher, T. (2003). Meiosis, recombination and chromosomes: a review of gene isolation and fluorescent in situ hybridization data in plants. *J. Exp. Bot., 54*, 11-23.

Sha, J.; Zhou, Z.; Li, J.; Yin, L.; Yang, H.; Hu, G.; Luo, M.; Chan, H. C. & Zou, K. (2002). Identification of testis development and spermatogenesis-related genes in human and mouse testes using cDNA arrays. *Molec. Hum. Reprod., 8*, 511-517.

Sharpe, R. M. & Irvine D. S. (2004). How strong is the evidence of a link between environmental chemicals and adverse effects on human reproductive health? *BMJ, 328*, 447-451.

Shima, J. E.; McLean, D. J.; McCarrey, J. R. & Griswold, M. D. (2004). The murine testicular transcriptome: characterizing gene expression in the testis during the progression of spermatogenesis. *Biol. Reprod., 71*, 319–330.

Skrisovska, L.; Bourgeois, C. F.; Stefl, R.; Grellscheid, S. N.; Kister, L.; Wenter, P.; Elliot, D. J.; Stevenin, J. & Allain, F. H. (2007). The testis-specific human protein RBMY recognizes RNA through a novel mode of interaction. *EMBO Rep., 8*, 372-379.

Smith, A. & Benavente, R. (1992a). Identification of a structural component of rat synaptonemal complexes. *Exp. Cell Res., 198*, 291-297.

Smith, A. & Benavente, R. (1992b). Meiosis-specific protein selectively associated with sex chromosomes of rat pachytene spermatocytes. *Proc. Natl. Acad. Sci. U.S.A., 89*, 6938-6942.

Sofikitis, N.; Pappas, E.; Kawatani, A.; Baltogiannis, D.; Loutradis, D.; Kanakas, N.; Giannakis, D.; Dimitriadis, F.; Tsoukanelis, K.; Georgiou, I.; Makrydimas, G.; Mio, Y.; Tarlatzis, V.; Melekos, M. & Miyagawa, I. (2005). Efforts to create an artificial testis: culture systems of male germ cells under biochemical conditions resembling the seminiferous tubular biochemical environment. *Hum. Reprod. Update, 11*, 229–259.

Sutcliffe, J. G.; Milner, R. J.; Gottesfeld, J. M. & Reynolds, W. (1984). Control of neuronal gene expression. *Science, 225*, 1308-1315.

Swanson, W. J. & Vacquier, V. D. (2002). The rapid evolution of reproductive proteins. *Nat. Genet., 3*, 137– 44.

Sym, M., Engebrecht, J. A. & Roeder, G. S. (1993). ZIP1 is a synaptonemal complex protein required for meiotic chromosome synapsis. *Cell, 72*, 365-378.

Thomas, K. H.; Wilkie, T. M.; Tomashefsky, P.; Bellvé, A. R. & Simon, M. I. (1989). Differential gene expression during spermatogenesis. *Biol. Reprod., 41*, 729-739.

Turner, J. M. (2007). Meiotic sex chromosome inactivation. *Development, 134*, 1823-1831.

Van Kroonenburgh, M. J.; Beck, J. L.; Scholtz, J. W.; Hacker-Klom, U. & Herman, C. J. (1985). DNA analysis and sorting of rat testis cells using two parameter flow cytometry. *Cytometry, 6*, 321–326.

Venables, J. P. & Eperon, I. C. (1999). The roles of RNA-binding proteins in spermatogenesis and male infertility. *Curr. Opin. Genet. Dev., 9*, 346-354.

Venables, J. P. (2002). Alternative splicing in the testes. Curr. Opin. Genet. Dev., 12, 615-619.

Vershon, A. K. & Pierce, M. (2000). Transcriptional regulation of meiosis in yeast. *Curr. Opin. Cell Biol., 12*, 334-339.

Vester, B.; Smith, A.; Krohne, G. & Benavente, R. (1993). Presence of a nuclear lamina in pachytene spermatocytes of the rat. *J. Cell Sci., 1*04, 557-563.

Vigodner, M.; Lewin, L. M.; Shochat, L.; Mittelman, L, & Golan, R. (2003). Meiosis in the golden hamster: a confocal microscopy and flow cytometric analysis. *Molec. Reprod. Dev., 64*, 86-95.

Vilardell, J.; Coll, M. D.; Querol, E. & Egozcue, J. (1989). Histone electrophoretic pattern in the characterization of synaptonemal complexes. *Cell. Mol. Biol., 35*, 201-214.

Vilela, C. & McCarthy, J. E. (2003). Regulation of fungal gene expression via short open reading frames in the mRNA 5' untranslated region. *Mol. Microbiol., 49*, 859-867.

Von Wettstein, D.; Rasmussen, S. W. & Holm, P. B. (1984). The synaptonemal complex in genetic segregation. *Ann. Rev. Genet.,* 18, 31-413.

Walker, W. H. & Cheng, J. (2005). FSH and testosterone signaling in Sertoli cells. *Reproduction, 130*, 15–28.

Wallace, A. M.; Dass, B.; Ravnik, S. E.; Tonk, V.; Jenkins, N. A.; Gilbert, D. J.; Copeland, N. G. & MacDonald, C. C. (1999). Two distinct forms of the 64,000 Mr protein of the cleavage stimulation factor are expressed in mouse male germ cells. *Proc. Natl. Acad. Sci. U.S.A., 96*, 6763-6768.

Wang, P. J.; McCarrey, J. R.; Yang, F. & Page, D. C. (2001). An abundance of X-linked genes expressed in spermatogonia. *Nat. Genet., 27*, 422–426.

Watanabe, T.; Miyashita, K.; Saito, T. T.; Yoneki, T.; Kahihara, Y.; Nabeshima, K.; Kishi, Y. A.; Shimoda, C. & Nojima, H. (2001). Comprehensive isolation of meiosis-specific genes identifies novel proteins and unusual non-coding transcripts in *Schizosaccharomyces pombe. Nucleic Acids Res., 29*, 2327-2337.

Weiss, M.; Vigier, M.; Hue, D.; Perrard-Sapori, M. H.; Marret, C.; Avallet, O. & Durand, P. (1997). Pre- and postmeiotic expression of male germ cell-specific genes throughout 2-week cocultures of rat germinal and Sertoli cells. *Biol Reprod., 57*, 68-76.

Wettstein, R. & Sotelo, J. R. (1971). The molecular architecture of synaptonemal complexes. In E. J. DuPraw (Ed.), *Advances in Cell and Molecular Biology* (vol. 1, pp.109-152). New York: Academic Press.

Whitby, M. C. (2005). Making crossovers during meiosis. *Biochem. Soc. Trans., 33*, 1451-1455.

Wilda, M.; Bachner, D.; Zechner, U.; Kehrer-Sawatzki, H.; Vogel, W. & Hameister, H. (2000). Do the constraints of human speciation cause expression of the same set of genes in brain, testis and placenta? *Cytogenet. Cell Genet., 91*, 300-302.

Wistuba, J; Schrod, A.; Greve, B.; Hodges, J. K.; Aslam, H.; Weinbauer, G. F. & Luetjens, M. (2003). Organization of seminiferous epithelium in primates: relationship to spermatogenic efficiency, phylogeny, and mating system. *Biol. Reprod., 69*, 582-591.

Wrobel, G. & Primig, M. (2005). Mammalian male germ cells are fertile ground for expression profiling of sexual reproduction. *Reproduction, 129*, 1–7.

Wu, S. M.; Baxendale, V.; Chen, Y.; Pang, A. L.; Stitely, T.; Munson, P. J.; Leung, M. Y.; Ravindranath, N.; Dym, M.; Rennert, O. M. & Chan, W. Y. (2004). Analysis of mouse germ-cell transcriptome at different stages of spermatogenesis by SAGE: biological significance. *Genomics, 84*, 971-981.

Xu, Q.; Modrek, B. & Lee, C. (2002). Genome-wide detection of tissue-specific alternative splicing in the human transcriptome. *Nucleic Acids Res., 30*, 3754-3766.

Yao, J.; Chiba, T.; Sakai, J.; Hirose, K.; Yamamoto, M.; Hada, A.; Kuramoto, K.; Higuchi, K. & Mori, M. (2004). Mouse testis transcriptome revealed using serial analysis of gene expression. *Mamm. Genome, 15*, 433-451.

Yeo, G.; Holste, D. Kreiman, G. & Burge, C. B. (2004). Variation in alternative splicing across human tissues. *Genome Biol., 5*, R74.

Yu, Z.; Guo, R.; Ge, Y.; Ma, J.; Guan, J.; Li, S.; Sun, X.; Xue, S. & Han, D. (2003). Gene expression profiles in different stages of mouse spermatogenic cells. *Biol. Reprod., 69*, 37-47.

Zickler, D. (2006). From early homologue recognition to synaptonemal complex formation. *Chromosoma, 115*, 158-174.

In: Cell Differentiation Research Developments
Editor: L. B. Ivanova, pp. 125-150
ISBN: 978-1-60021-939-9

Chapter V

THE STATE OF SPERMATOGENESIS IN WHITE MALE RATS AFTER ACUTE HUPOXIC AND ISHEMIC INFLUENCES

Olga N. Shevantaeva, Yury I. Kosuyga and Igor V. Chelyshev
Nizhny Novgorod State Medical Academy, Nizhny Novgorod,
"Medical diagnostics", Ltd., Nizhny Novgorod, Russia.

ABSTRACT

So far there are only few works concerning changes in male reproductive system after extremal hypoxic and ischemic influences. In the current study we have investigated the state of spermatogenesis of white male rats after modeling of clinical death and acute hypobaric hypoxia. It has been demonstrated that the observed impairment and death of the considerable part of germ cells as well as Sertoli cells and Leydig cells was due to increased production of free radicals. The level of free-radical oxidation was higher in the early reperfusion period than in the early post-hypoxic period. It has been found that lactate intratesticular level decreased during first hours after clinical death modeling whereas it elevated after acute hypobaric hypoxia modeling; it provided the more expressed preservation of germ cells in the second case. By the 21st day of the both experiments the tendency towards the increase in number of germ epithelial cells as well as Leydig and Sertoli cells has been revealed yet by the 60th day of the observation the cellular picture of the testicles has not been fully recuperated (compared to the intact animals). The low intensity of the processes of cell division and maturation in the late post-reanimation and post-hypoxic periods is obliged to the small number of both Sertoli and Leydig cells, as these cells provide metabolic and hormonal support of gametogenesis. . The found disorders of spermatogenesis lead to qualitative and quantitative changes of ejaculate indices.

INTRODUCTION

Spermatogenesis is the complex cyclic process of cells transformation that [process] takes place in the coiled seminiferous tubules of testes of mammalian males. During spermatogenesis the germ cells proliferate and mature being sequentially spermatogones, spermatocytes, early and late spermatids and spermatozoids [17,18,29]. The full series of morphological changes observed between an appearance of two identical cells associations in a same part of a seminiferous tubule is called "spermatogenous epithelium cycle"; the cells associations are called "stages of the cycle" [44]. As the rate of development of a germ cell is constant throughout spermatogenesis, the different generations of spermatogenous epithelium mature consecutively, with a constant rate for each stage. As a result, the length of spermatogenous epithelium cycle equals invariably to 12,9 days in rats [15,63].

According to Huckins C. [38], spermatogenous epithelium cycle may be represented from periodical formation of spermatogones A_I and the subsequent development. The differentiation of generations of spermatogones A_I takes place every 13 days; this differentiation results in formation of spermatogones B. The B cells undergo division forming preleptotenic spermatocytes which continue to evolve through the second cycle; simultaneously, the second generation of spermatogones A_I emerges from the non-differentiated generation of spermatogones A. After the next 13 days the group of preleptotenic spermatocytes matures into pachytenic spermatocytes; the second generation of cells A_I completes the spermatogonial differentiation forming spermatogones B; simultaneously, the third generation of spermatogones A_I is formed and so on. The fully developed spermatogenous epithelium four identical cycles.

It was established, that the time-length of development of spermatogenous epithelium is stable and species-specific [58]. The total length of the whole process of spermatogenesis is 48 days in rats [33].

It has been shown lately that the preservation of amount of germ cells is achieved via the balance between processes of cells proliferation and elimination by means of apoptosis [18,19,62]. Sertoli cells play the important role in governing these processes: they produce pro- and anti-apoptotic agents and germ cells with energy and material substances [29,48,55,71,26,42,56,60]. Lactate is one of the main substances which are produced by Sertoli cells: it is used, first of all, by the cells of spermatogenous epithelium for ATP synthesis [60,30,52,6]. K. Erkkilä et al (2002) demonstrated that lactate plays the important role in regulation of apoptosis in testicular tissue, supporting cells survival.

The researches of recent years demonstrated that hypoxia, ischemia and subsequent reperfusion cause the cascade of metabolic and structural changes which lead to impairment and death of cells of essential organs and systems. So far, there are just few works concerning changes in male reproductive system after the influence of acute hypobaric hypoxia and clinical death with total ischemia [4,23,47]. The results of these works are quite contradictory. There are no works where the estimation of spermatogenesis state would be represented by a collection of several indices relevant for the reproductive system.

The objective of the current study is to investigate the state of spermatogenesis and possible mechanisms of its impairment in extremal hypoxic and ischemic conditions.

MATERIALS AND METHODS

The study was performed on 336 white male outbred rats with body weight 180-230 grams. The experimental group was comprised of 306 animals which were studied for spermatogram indices and some biochemical indices of testicular tissue after modeling of acute hypobaric hypoxia and clinical death at 40 min and days 1, 3, 7, 14, 21, 30, 45, 60.

The control group was composed of intact animals (30), that were studied similarly except for acute hypobaric hypoxia and clinical death modeling. It was found that indices studied did not differ at different times of observation.

Modeling of Acute hypobaric Hypoxia

Modeling of acute hypobaric hypoxia was performed in a pressure chamber. The rats were placed into the microenvironment corresponding to atmospheric "elevation" up to 11500-12000 meters. The rats stayed at this "area of death" until the appearance of agonal breathing [46].

Modeling of Clinical Death

The clinical death was modeled in narcotized animals (intraperitoneal injection of Aethaminalum-natrium, 25 mg/kg) by means of complete cross-clamping of cardiac vascular fascicle; after that the closed-chest cardiac massage and artificial pulmonary ventilation had been performed until the natural breathing and cardiac activity were restored [43].

Determining amount of Spermatozoids in Testicular Tissue

A suspensions of testicular cells was prepared in a glass homogenizer from 100 mg of wet testicular tissue, with an addition of 1 ml of phosphate buffer (pH=7,4). The suspension obtained was diluted 1:10. Amount of spermatozoids was then counted in Goryaev's chamber as follows:

$$X = (A / 2) \times 10^7$$

where

X – amount of spermatozoids in 1 g of testicular tissue;

A- amount of spermatozoids in 5 big squares of Goryaev's chamber.

Cytological Examination

The quantitative cytological technique [39] was used for estimation of a state of spermatogenesis: smears of suspension of testicular tissue were prepared and stained according to Romanovsky. Identification and counting (up to the total number of 500) of different types of spermatogenous epithelium and Sertoli and Leydig cells were done in light microscopy in order to get a spermatogram – percentages of different types of testicular cells. The amount of spermatogenous epithelium cells in 1 g of testicular tissue was calculated via proportions, using data of spermatozoids count in Goryaev's chamber.

Gaining Ejaculate

Ejaculate was obtained by the technique of electric stimulation of seminal hillock via rectal mucosa [65]. Parameters of the electric stimulation: amplitude 2,5 – 6 V, frequency 0,8 – 1 Hz, an impulse length 0,1 - 0,5 ms, stimulation length up to 120 s, an impulse shape – rectangular.

The stimulation was performed in morning hours in order to exclude the influence of circadian fluctuations of sex hormones.

Determining Quantitative and Qualitative Indices of Spermatogram

Ejaculate collected was placed in 2 ml of medium #199 at 37^{0}C and then diluted tenfold. An amount of spermatozoids was counted in 5 big squares of Goryaev's chamber. The following spermatozoids was counted separately: with progressive movements (normokinesis), with vibratory movements (hypokinesis) and motionless spermatozoids (akinesis).

The calculations were made according to the following formula:

$$X = A \times 10^6$$

where

A - amount of spermatozoids in 5 big squares of Goryaev's chamber;

X – total amount of spermatozoids in ejaculate.

Determining Lactate and Pyruvate Levels in Testicular Tissue

The levels of lactate and pyruvate in testicular tissue were detected by enzymatic technique employing lactate dehydrogenase [5]. The amount of lactate was calculated from the level of synthesized NADH, the amount of pyruvate was calculated from the decrease of NADH registered on a spectrophotometer at 340 nm.

Lactate and pyruvate amounts were expressed as μmol per 1 g of tissue.

The Chemiluminescence Technique of Detecting free Radical and Antioxidative Activity

The presence of free radicals in testicular tissue was detected by the chemiluminescence technique [16]. The technique is based on the fact that the catalytic decomposition of H_2O_2 takes place in the mixture of H_2O_2 and ferric sulfate due to the presence of Fe^{2+} ions (Fenton's reaction). The formed free radicals participate in the activation of free radical oxidation in a biological substrate which leads to formation of the instable tetroxide; the latter decays and emit photons that are registered on the apparatus BCL-06. The informative indices are: I max (mV) – the maximal intensity of luminescence of the experimental sample, which reflects the free radical activity of the sample; summary chemiluminescence during the certain time-span (S), which reflects the general activity of anti-oxidative defensive systems.

The data obtained was processed using Mann-Whitney U test. The differences were believed to be significant at $p < 0{,}05$.

RESULTS

The Influence of Acute Hypobaric Hypoxia on white Male Rats Spermatogenesis

Cytological Study of Testicles of white Male Rats at Different Times after Hypoxia

At the 40th minute after acute hypobaric hypoxia modeling the amount of all types of germ epithelium cells did not differ significantly compared to the levels in intact animals (Figure 1).

Beginning the 3rd day, the decrease of all types of germ epithelium cells was observed. Starting the 21st day, a some elevation of number of all germ cells was registered, yet it did not reach those levels in intact animals even by the 60th day of the experiment.

Spermatogones. Amount of spermatogones at the 1st day of the experiment did not differ from that in intact animals. By the 3rd day their amount decreased more than ten-fold and were minimal at the 14th day.

Spermatocytes. By the 24 hour after modeling of acute hypobaric hypoxia amount of spermatocytes was twice of that in intact animals, yet it decreased substantially at the 3rd day. There were 5,4 ± 0,4 millions of spermatocytes in experimental animals by the 14th day (132,2 ± 7,5 millions in intact animals).

Early and late spermatids. Decrease of amount of early and late spermatids was noted by the 1st day of the experiment. There were no early spermatids at the 14th day of the experiment and the amount of late spermatids was diminished more than 90-fold.

Spermatozoids. The average amount of spermatozoids in intact animals was 81,0 ± 1,6 millions per 1 gram of wet tissue. By the 3rd day of the experiment the amount of spermatozoids decreased more then twice: down to 32,0 ± 1,7 millions. Afterwards the amount of spermatozoids continue to decrease and became minimal by the 14th day of the experiment - 12,5 ± 0,8 millions, that was 6-fold lower than in intact animals. Beginning the

21st day of the experiment the amount of spermatozoids starts to raise. However, the amount of spermatozoids was 39,0 ± 1,2 millions by the 60th day, i.e. two-fold lower than in intact animals.

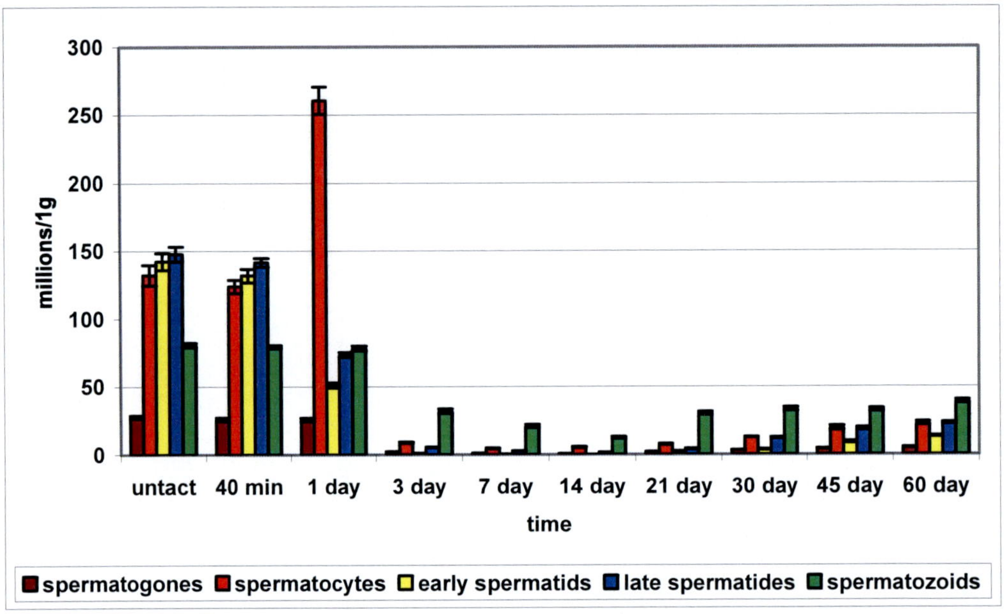

Figure 1. Amounts of cells of spermatogenous epithelium at different times after acute hypobaric hypoxia modeling.

In addition to spermatogenous epithelium cells, we identified and counted Sertoli cells (nuclei) and Leydig cells (Figure 2).

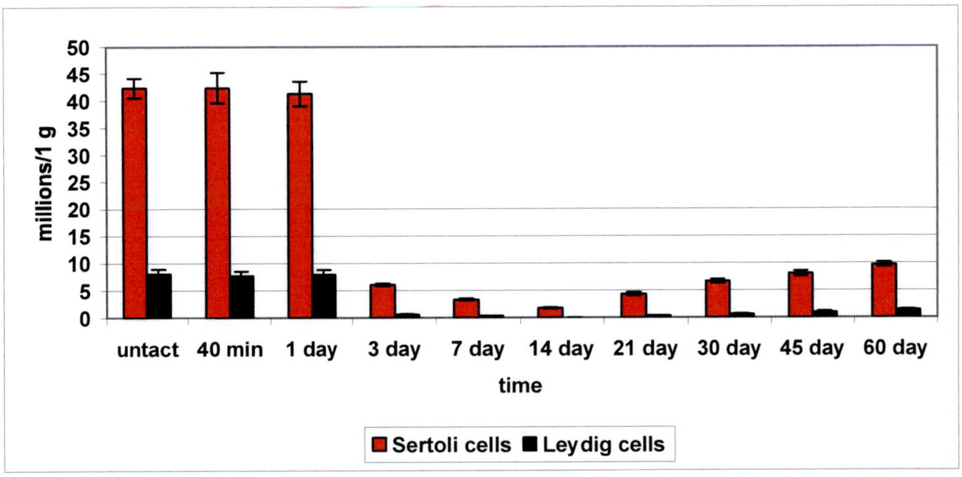

Figure 2. Amounts of Sertoli cells and Leydig cells in testicles of white male rats at different times after acute hypobaric hypoxia modeling.

Sertoli cells. The amount of Sertoli cells per 1 gram of testicular tissue was 42,4 ± 1,8 millions in intact animals. Beginning the 3rd day of the experiment, the amount of Sertoli cells in testicular tissue decreased significantly and remained low till the end of observation period.

Leydig cells. The amount of Leydig cells, which are the main producers of testosterone in testicles, decreased significantly beginning the 3rd day of post-hypoxic period. By the 14th day of the experiment Leydig cells was found in testicles of just 20% of experimental animals. Afterwards the gradual elevation of number of Leydig cells was noted. However, their amount was 5,8-fold lower than in intact animals even by the 60th day of observation.

Concentrations of Glycolysis Substrates in Testicular Tissue after Modeling of Acute Hypobaric Hypoxia

Concentrations of pyruvate and lactate in testicular tissue of intact animals were 4,20 ± 0,2 and 0,26 ± 0,02 μmol per 1 gram of tissue, respectively. By the 40th minute after modeling of acute hypobaric hypoxia the lactate and pyruvate concentrations raised by 19% and 40%, respectively (Figure 3.).

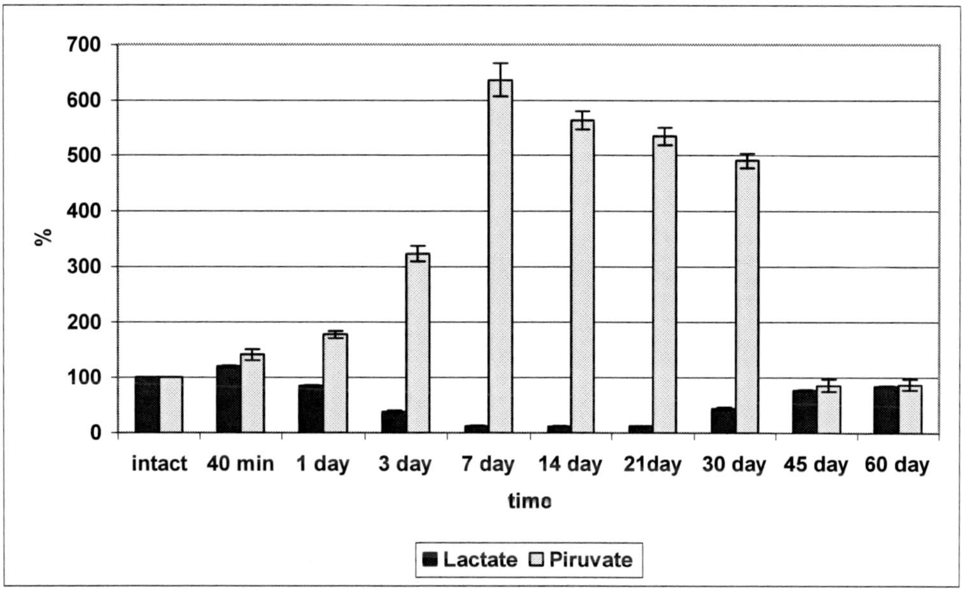

Figure 3. Concentrations of glycolysis substrates (% from the concentrations of the same substrates in intact animals) in testicular tissue of white male rats at different times of post-hypoxic period.

By the 1st day of observation the lactate level was decreased whereas the pyruvate level continued to grow. At the 7th day of post-hypoxic period the level of lactate decreased by almost 90% whereas the pyruvate concentration elevated more than 5-fold compared with control data. Beginning the 14th day, the gradual decrease of the pyruvate level was noted. By the 45th day of the experiment its concentration was close to the initial value. An increase of the lactate level was found at the 30th day of the experiment. However, at the 45th and 60th days of the experiment the lactate concentrations were significantly lower than in intact animals.

Activity of free Radical Oxidation and Anti-oxidative System in Testicular Tissue after Modeling of Acute Hypobaric Hypoxia

Our researches demonstrated that the activity of free radical oxidation was high at 40th minute and at 1st and 3rd days of post-hypoxic period (Figure 4). At the 7th and 14th days the levels of free radical oxidation were not different from those in intact animals. At the 21st and 30th days a some increase of chemiluminescence was noted.

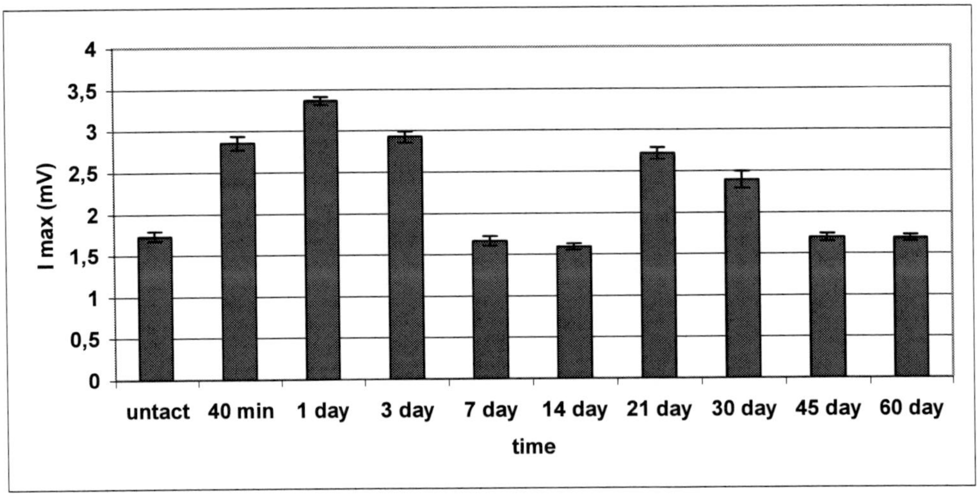

Figure 4. Activity of free radical oxidation in testicular tissue of white male rats at different times after modeling of acute hypobaric hypoxia. (Imax – the peak value of chemiluminescence intensity).

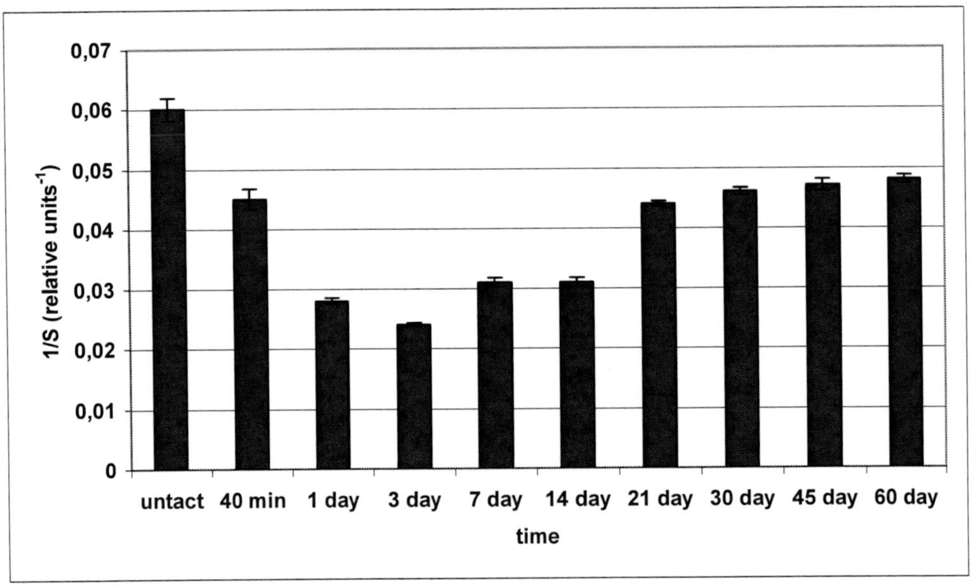

Figure 5. Activity of anti-oxidative system in testicular tissue of white male rats at different times after modeling of acute hypobaric hypoxia (S – area under the curve of chemiluminescence against time; 1/S – the reverse index, reflecting activity of anti-oxidative system).

From the 40th minute till the 3rd day of post-hypoxic period the index of light sum of chemiluminescence was high which fact indicated the low activity of anti-oxidative system in testicular tissue. Beginning the 7th day of the experiment the activity of anti-oxidative system increased somewhat, yet it did not reach the level of intact animals by the 60th day of the experiment (Figure 5).

Quantitative and Qualitative Changes of Spermogram at Different Times after Modeling of Acute Hypobaric Hypoxia

The research qualitative characteristics of spermatozoids in ejaculate and their motility began 24 hours after modeling of acute hypobaric hypoxia.

The quantitative and qualitative changes of spermogram in post-hypoxic period were compared with the same indices of ejaculate of the group of intact (control) animals. The sperm of intact male rats is characterized by high concentration of spermatozoids in a limited volume of ejaculate and their high motility. The amount of spermatozoids in ejaculate of intact rats was 19,3 ± 0,37 millions, the percentage of mobile gametes was 70,8 ± 1,8%, progressive movements were registered in 42,9 ± 1,6% of cells.

The analysis of qualitative changes of spermograms in post-hypoxic period revealed the significant decrease of total amount of spermatozoids in ejaculate of male rats: at the 1st day after modeling of acute hypobaric hypoxia the amount of sex cells decreased 3-fold and was 6,9 ± 0,5 millions. Afterwards, the amount of spermatozoids in ejaculate continued to diminish (Figure 6).

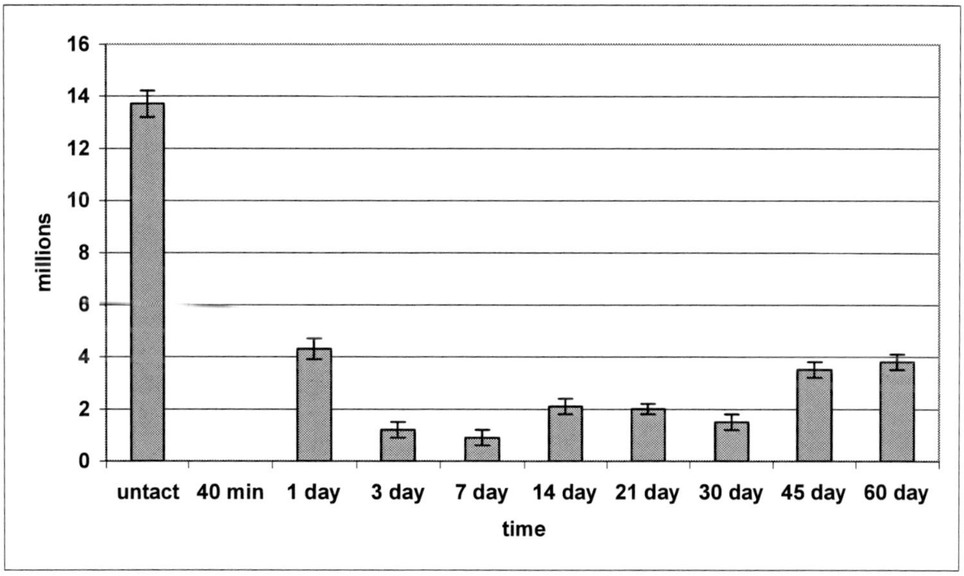

Figure 6. Total amount of spermatozoids in ejaculate of white male rats at different times after modeling acute hypobaric hypoxia.

The minimal amount of sex cells in ejaculate was registered at 30th day of the experiment (2,9 ± 0,3 millions).

The amount of spermatozoids in ejaculate began to raise at the 45th day after hypobaric hypoxia influence and reached 7,0 ± 0,3 millions by the 60th day, that was approximately 3-fold lower, than in intact animals.

The total amount of spermatozoids and of their mobile pool were decreased (Figure 7).

By the end of the observation period the amount of mobile spermatozoids was just 3,8 ± 0,3 millions, i.e. 3,6-fold lower than in intact animals.

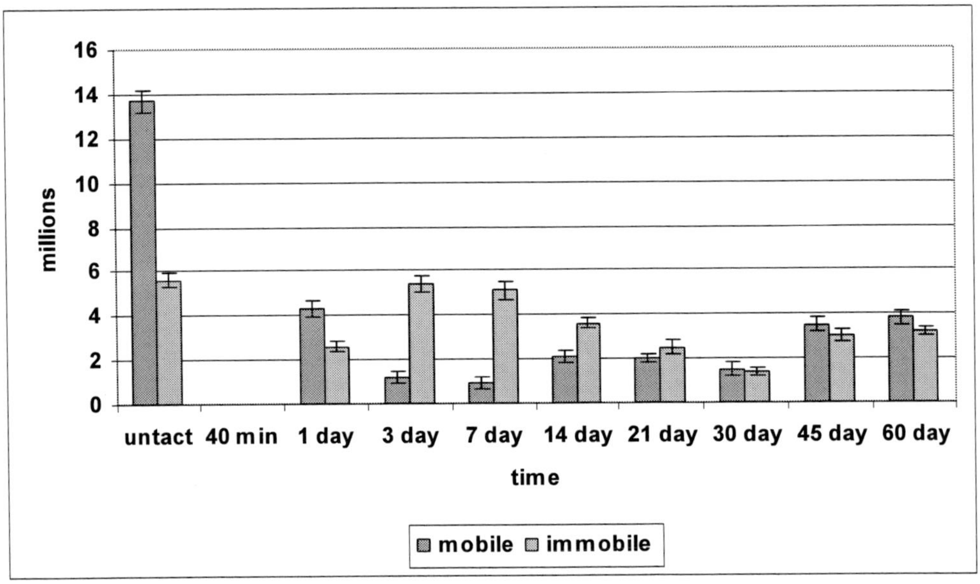

Figure 7. The total amounts of mobile and immobile spermatozoids in ejaculate of white male rats at different times after modeling of acute hypobaric hypoxia.

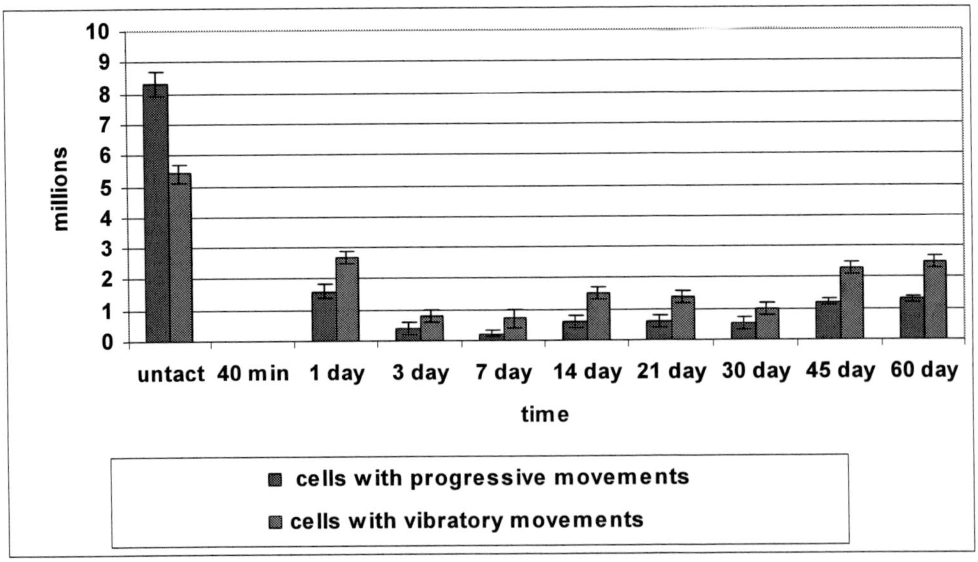

Figure 8. Absolute amounts of spermatozoids with progressive and vibratory movements in ejaculate of white male rats at different times after modeling of acute hypobaric hypoxia.

By the end of the observation period the proportion between mobile and immobile sex cells had not completely restored.

We analyzed the dynamics of change of absolute amounts of spermatozoids with progressive and vibratory movements (Figure 8).

The figure 8 demonstrates that during the whole experiment the total number of cells with both progressive and vibratory movements was significantly decreased. It is important to note, that in contrast to intact animals, at all times of the observation after modeling of acute hypobaric hypoxia the spermatozoids with vibratory movements prevailed in ejaculate of white male rats.

State of Spermatogenesis of white Male Rats after Modeling of Clinical Death

Cytological Study of Testicles of white Male Rats at Different Times of Post-reanimation Period

The experiments demonstrated that at 40th minute after modeling of clinical death the spermatograms of experimental and intact animals were the same (Figure 9).

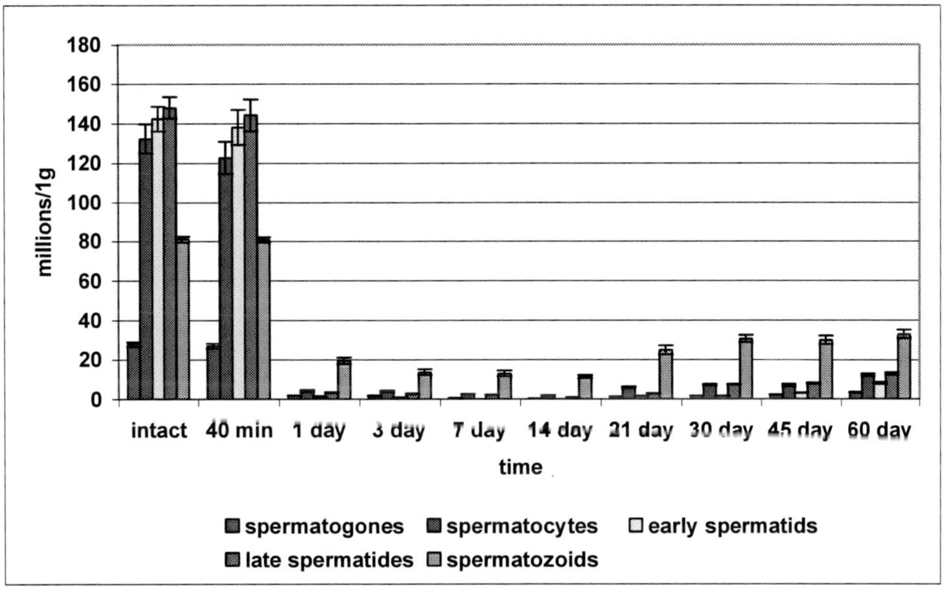

Figure 9. Amounts of cells of spermatogenous epithelium at different times after clinical death modeling.

However, beginning the 1st day of the experiment, the absolute numbers of all types of testicular cells decreased significantly.

Spermatogones. At the 1st day the amount of spermatogones was diminished almost 15-fold compared to the control data. The minimal amount of spermatogones was noted at the 14th day of the experiment (0,3 ± 0,04 millions). Beginning the 21st day of the observation the amount of spermatogones began to raise, yet it remained significantly reduced up to the end of the experiment.

Spermatocytes. At the 1st day of reperfusion the absolute number of spermatocytes was reduced more than 30-fold. The minimal amounts of spermatocytes were noted by the 7th and 14th days: 2,6 ± 0,1 millions and 2,0 ± 0,1 millions, respectively. Beginning the 21st day of the experiment we observed somewhat increase of number of spermatocytes, yet even by the 60th day of the observation their amount was almost 10-fold lower than in intact animals.

Early and late spermatids. The decrease of absolute amount of these cells was found at the 1st day of the experiment. There were no early spermatids in smears of testicular tissue suspension at 7th and 14th days. Afterwards a some raise of amount of spermatids was observed. However, the absolute amounts of these cells did not reach the intact levels by the end of observation period.

Spermatozoids. The sharp diminution of the amount of spermatozoids was noted at the 1st day of the experiment. Afterwards the absolute amount of spermatozoids in testicular tissue continued to diminish and became 11,4 ± 0,9 millions at the 14th day. Beginning the 21st day of the post-reanimation period the gradual elevation of the amount of spermatozoids was registered. By the 60th day of the experiment the amount of spermatozoids was 32,9 ± 2,1 millions, i.e. significantly lower than in intact animals.

Sertoli cells. The significant decrease of number of Sertoli cells in testicular tissue was observed beginning the 1st day of the experiment (Figure 10); the absolute amount of Sertoli cells diminished almost 40-fold by the 14th day. A somewhat increase of the amount of Sertoli cells was noted between 21st and 60th days of the experiment, yet the amount of Sertoli cells was significantly lower than in intact animals.

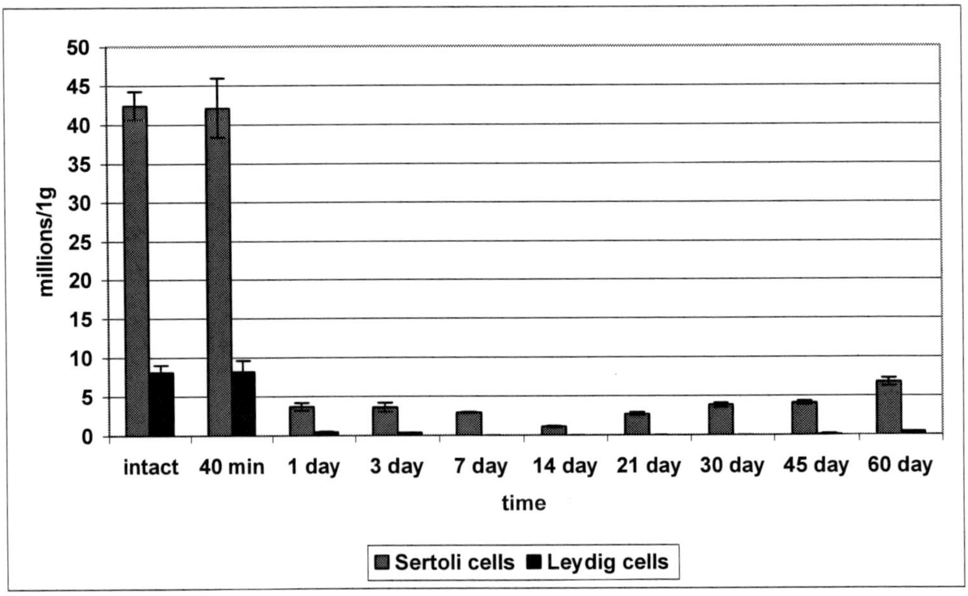

Figure 10. Amounts of Sertoli cells and Leydig cells in testicular tissue of white male rats at different times of post-reanimation period after modeling of clinical death.

Leydig cells. The twenty-fold diminution of the amount of Leydig cells (down to 0,4 ± 0,1 millions {8,1 ± 0,9 millions in intact animals}) was observed by the 1st day of the reperfusion period. There were no Leydig cells in the smears of testicular tissue at 7th , 14th, 21st and 30th days of the experiment. Leydig cells reappeared in the spermatograms by the 45th day, yet their amount was only 0,4 ± 0,1 by the 60th day (Figure 10).

Concentrations of Glycolysis Substrates in Testicular Tissue after Modeling of Clinical Death

The significant decrease of pyruvate and lactate levels in testicular tissue of white male rats was noted at the 40th minute of reperfusion period (Figure 11). The lactate level increased somewhat at the 1st day of reperfusion period, wet it was by 38% below the initial level. The pyruvate level in testicular tissue increased and was significantly higher than in intact animals.

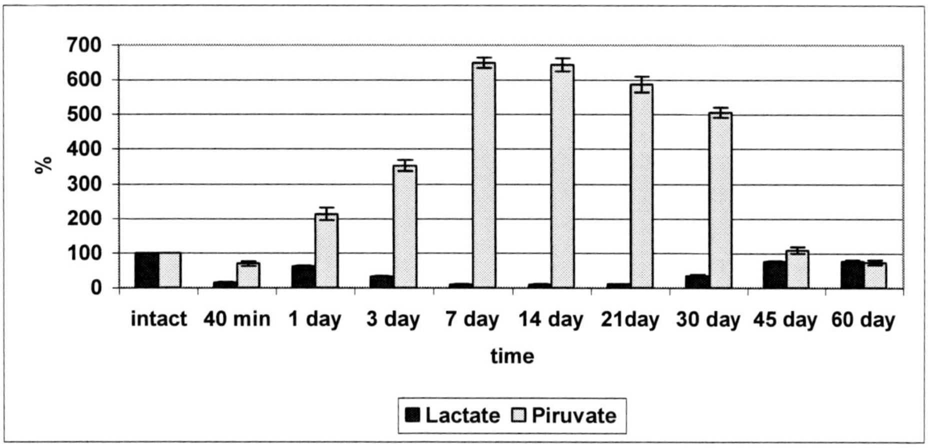

Figure 11. Concentrations of substrates of glycolysis (as % from intact levels) in testicular tissue of white male rats at different times of post-reanimation period.

The concentration of lactate was minimal between 3rd and 21st days of post-reanimation period. The pyruvate concentration increased sharply; at the 14th day its concentration was more than 5-fold higher than the initial level.

Beginning the 21st day the gradual decrease of pyruvate concentration was noted. At the 60th day of the observation the pyruvate concentration was 27% lower than in intact animals.

The elevation of lactate concentration was registered at the 30th day of post-reanimation period; however, this elevation did not reach the level of intact animals by the end of the observation period.

Activity of free Radical Oxidation and Anti-oxidative System in Testicular TIssue after Modeling of Clinical Death

Our researches indicated that at the 40th minute of reperfusion period and at the 1st and 3rd days of post-reanimation period the intensity of chemiluminescence, reflecting the activity of free-radical oxidation, was twice as high as in intact animals (Figure 12). At the 7th and 14th days the indices of chemiluminescence intensity did not differ from the intact levels.

As it is shown at the Figure 13, from the 40th minute of reperfusion period till the 3rd day the reverse index of the light sum of chemiluminescence was low, which fact indicates the low activity of anti-oxidative system in testicular tissue of rats subjected to 10 min of clinical death. Beginning the 7th day of the experiment the activity of anti-oxidative increased somewhat. However, even at 60th day of post-reanimation period the activity of anti-oxidative system was significantly lower than in intact animals.

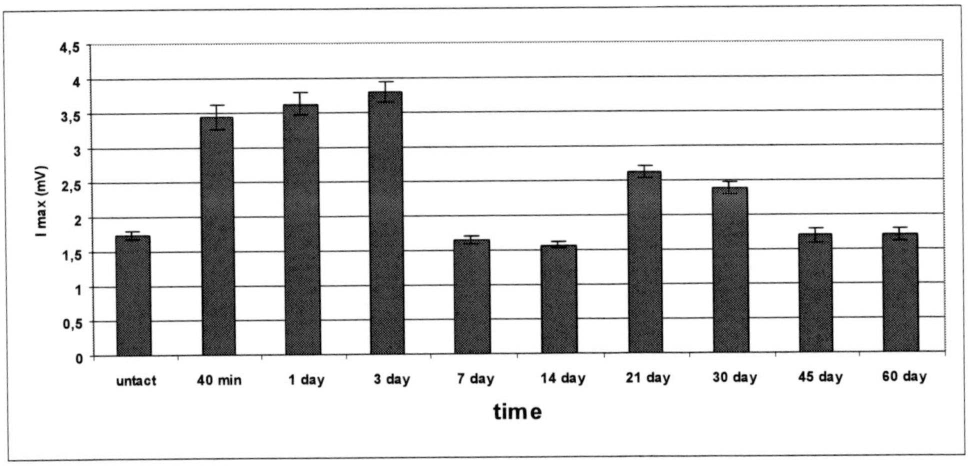

Figure 12. Activity of free radical oxidation in testicular tissue of white male rats at different times after modeling of clinical death. (Imax – the peak value of chemiluminescence, mV).

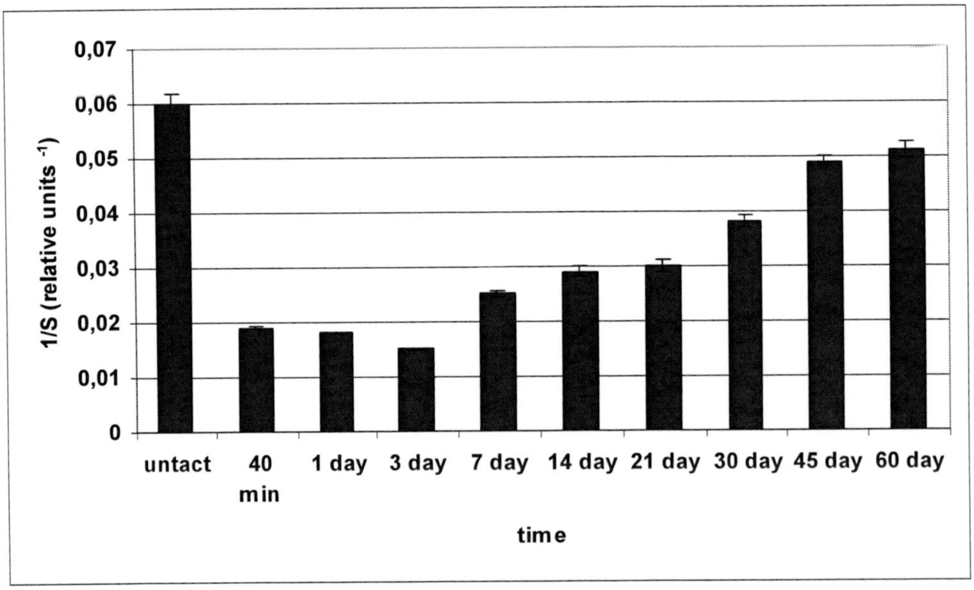

Figure 13. Activity of anti-oxidative system in testicular tissue of white male rats at different times after modeling of clinical death. (S – area under the curve of chemiluminescence against time; 1/S – the reverse index).

Quantitative and Qualitative Changes of Spermogram at Different Times of Post-reanimation Period

The rats subjected to clinical death demonstrated more severe recuperation period than the rats subjected to acute hypobaric hypoxia; besides, the rats subjected to clinical death showed aggressive behavior afterwards.

The white male rats produced no ejaculate at the electric stimulation up to the 14th day of the experiment. The analysis of spermograms demonstrated that the total amount of spermatozoids in ejaculate at 14th day of post-reanimation period was significantly lower than in intact animals: 4,3 ± 0,4 millions and 19,3 ± 0,4 millions, respectively (Figure 14).

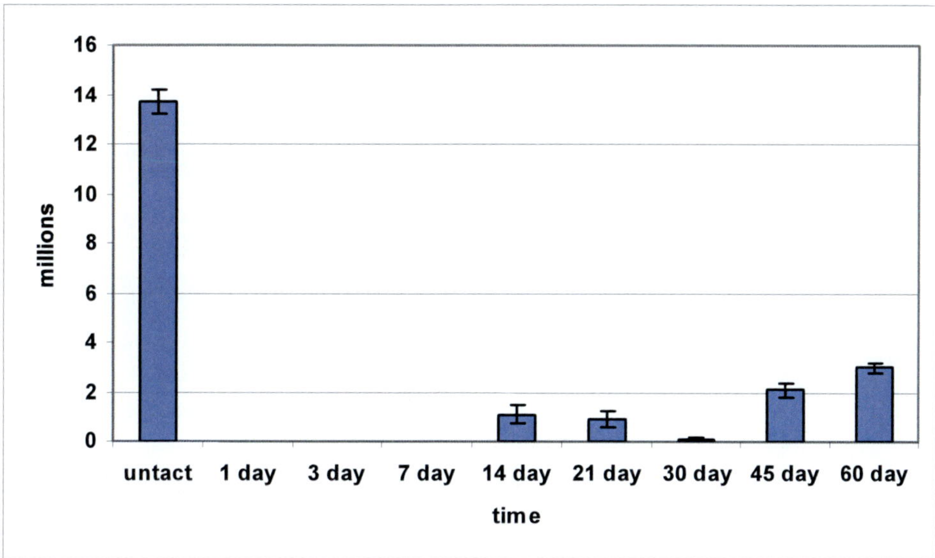

Figure 14. Total amount of spermatozoids in ejaculate of white male rats at different times of post-reanimation period after modeling of clinical death.

At the 21st day the amount of spermatozoids in semen decreased compared to the amount at the 14th day, and by the 30th day was 1,4 + 0,2 millions. At the 45th day the amount of spermatozoids raised up to the 4,9 ± 0,3 millions and remained so till the 60th day of the observation (Figure 14).

It should be noted that the absolute amounts of mobile and immobile spermatozoids in ejaculate of white male rats during post-reanimation period were drastically reduced (Figure 15).

By the end of the observation period the amount of mobile spermatozoids was 4,5-fold lower than in intact animals. At the 14th and 21st days of the experiment just one out of 7 experimental animals had spermatozoids with progressive movements in semen.

At the 30th day of the experiment there were no animals having spermatozoids with progressive movements and there was a just one animal out of 7 having spermatozoids with vibratory movements (Figure 16.).

Beginning the 45th day of the experiment a somewhat increase of number of mobile spermatozoids was noted. Cells with vibratory movements prevailed in spermogram. There was no complete recovery of all spermogram indices by the 60th day of the experiment.

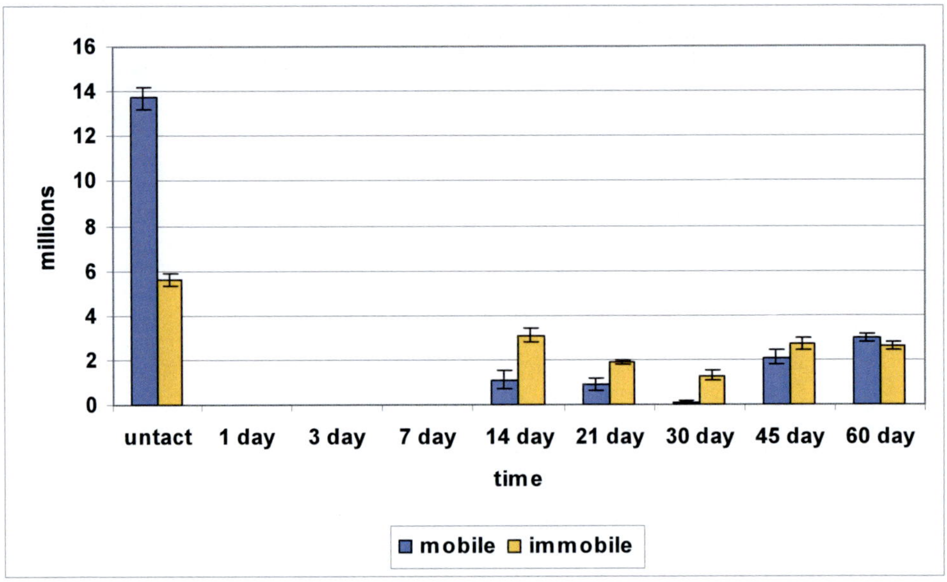

Figure 15. Absolute amounts of mobile and immobile spermatozoids in ejaculate of white male rats at different times of post-reanimation period.

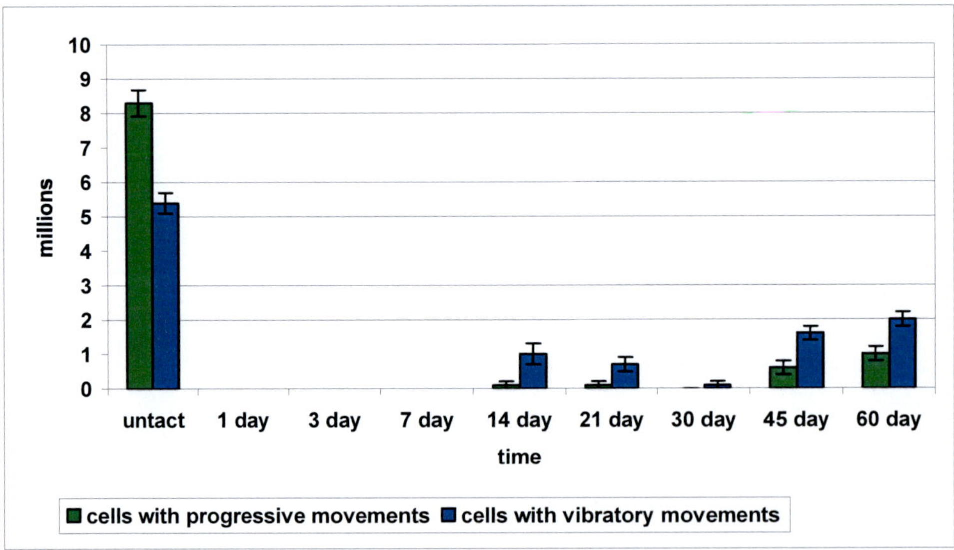

Figure 16. Absolute amounts of spermatozoids with progressive and vibratory movements in ejaculate of white male rats at different times of post-reanimation period.

Discussion

Our researches revealed the definitive connection between extremal hypoxic and ischaemic influences and disorders of spermatogenesis.

The most substantial changes in testicular tissue after modeling of acute hypobaric hypoxia and clinical death were noted between 7th and 14th day, when there were no early spermatids and Leydig cells in rats spermatograms.

At the 21st day of the experiment the tendency towards activation of spermatogenesis appeared as somewhat elevation of numbers of cells of spermatogenous epithelium, that was related to the beginning of a new stage of the cycle of development of cells of spermatogenous epithelium [25]. It may be hypothesized the low rates of division and maturation of sex cells between 21st and 60th day of the experiment were connected with small numbers of Sertoli and Leydig cells in testicular tissue. Sertoli cells influence formation of gametes by means of metabolic supply of cells of spermatogenous epithelium; they participate in supporting integrity of hematotesticular barrier and maintain preservation of the medium, enriched with potassium and bicarbonate, in testicular tubules, as such medium is necessary for performing meiosis and completion of spermatozoids development [57,69]. Leydig cells are main producers of testosterone; a shortage of the latter leads to suppression of meiotic division of spermatocytes and upsets their transformation into spermatids [59].

It is important to note, that more substantial changes of cells amounts in testicular tissue were observed after modeling of clinical death. At all times of the observation the amounts of spermatozoids predecessors (spermatogones, spermatocytes and spermatids) and Sertoli and Leydig cells were significantly lower than after modeling of acute hypobaric hypoxia; that fact is, probably, due to the decreased delivery of not just oxygen but also oxidation substrates in ischemic conditions. In addition, it is known that development of pathological changes in cells of ischemized organ is not restricted by the period of ischemia. Structural and functional disorders progress even after restoration of blood flow. Reperfusion, being the necessary condition of restoration of an organ viability, entails negative consequences as well [27], leading to development of secondary hypoxia [64].

Notably, amount of spermatozoids in testicular tissue exceeded amount of late spermatids during recovery period (beginning the 21st day), regardless of experimental model. Normally, the amount of spermatozoids in testicular tissue is always less than the amount of late spermatids due to elimination of defective spermatozoids and movement of mature spermatozoids to epididymis [72]. It is known that spermatozoids formed are unable to move themselves from testicle to epididymis and transferred with flow of liquid produced by Sertoli cells [57,72]; the amount of the latter was significantly reduced in our experiments up to the end of the observation period.

We demonstrated that the amount of Sertoli cells was decreased in post-reanimation period to the greater extent than in post-hypoxic period; therefore, the processes of spermatozoids movement from testicle to epididymis were slowed. It probably explains why we did not find significant differences between amounts of spermatozoids in testicles after modeling of clinical death and acute hypobaric hypoxia, beginning the 14th day.

In order to study mechanisms of changes of cellular composition of testicular tissue we performed biochemical estimation of glycolysis state and activity of free radical oxidation system and anti-oxidative system in testicular tissue after modeling of acute hypobaric hypoxia and clinical death.

The significant increase of concentrations of lactate and pyruvate in testicular tissue was registered as early as at 40th minute after modeling of acute hypobaric hypoxia; this fact may indicate glycolysis activation in testicular tissue due to suffered acute hypoxia.

It is interesting that at the 1st day of post-hypoxic period the amounts of spermatogones, spermatozoids and Leydig cells and Sertoli cells in testicular tissue remained normal and the amount of spermatocytes even increased. It is probably related to the membrane-stabilizing effect of lactate (which concentration increased during first hours of post-hypoxic period) and its ability to suppress processes of cells death, irrespective of levels of adenylic nucleotides (ATP, ADP and AMP) in testicles [21]. It is also known that lactate is the main metabolic substrate for spermatogones and, partially, spermatocytes; it activates division of the cells [6]. The drop of lactate concentration at 1st day after modeling of acute hypobaric hypoxia is probably related to its high consumption by dividing cells of spermatogenous epithelium (spermatocytes).

In addition, the substantial decrease of amounts of early and late spermatids was noted at the 1st day of the observation. It was demonstrated [50], that spermatids, in contrast to other cells of spermatogenous epithelium, were characterized by high activity of tricarbonic acid cycle enzymes; that is why spermatids are most sensitive to oxygen deficiency.

The decrease of lactate at the 3rd day of the experiment may be explained by diminution of amount of Sertoli cells. It is known that Sertoli cells consume up to 95% of intraluminal glucose in seminiferous tubules and are main lactate suppliers for germ cells [67,22,41,66]. As amount of Sertoli cells raised, the level of testicular lactate increased.

The increase of pyruvate level in testicular tissue was, possibly, related to: 1) disturbance of pyruvate transformation into acetyl-CoA and lactate in the conditions of disordered tissue respiration during post-hypoxic period [13]; 2) decrease of amounts of spermatocytes and spermatids that use pyruvate as an energy substrate [31,35,52,61].

The noted at the 3rd day of post-hypoxic period decrease of amounts of all germ cells types and Sertoli and Leydig cells may be explained by their death due to activation of free radical oxidation and exhaustion of anti-oxidative system. It is known, that the deficiency of oxygen and macroergic phosphates (first of all, ATP) in hypoxia disintegrates activity of Ca-transporting ATP-ases; it entails accumulation of Ca ions in cytosol, and , in turn, activation of lipolysis. So, the permeability of mitochondrial membranes increases and depletion of tissue respiration is observed. Mitochondria become swollen; it leads to disruption of their external membrane cytochrome C release into cytosol. The free cytochrome C may activate lipid peroxidation and raise level of free radicals [14,37].

Our researches revealed the significant increase of free radical oxidation in testicular tissue at 40th minute and 1st and 3rd days of post-hypoxic period.

It should be noted that the some intensification of free radical oxidation was found at 21st and 30th days of the experiment, which might be due to activation of cells division in testicular tissue [10].

Our researches demonstrated that free radicals levels at 45th and 60th days of post-hypoxic period did not differ significantly from that level in intact animals, yet the activity of anti-oxidative system remained at lower level; we believe, that it was related to diminished amounts of spermatogenous epithelium cells and Sertoli and Leydig cells. It is known that testicular cells contain various anti-oxidants [32,53]. For example, the spermatocytes are characterized by the high activity of superoxide dismutase and high level of glutathione; spermatozoids demonstrate the high activity of superoxide dismutase [1]; Sertoli cells are characterized by the high content of glutathione and high activity of superoxide dismutase and glutathione-dependent enzymes [8]. Testosterone, that is produced by Leydig cells, possesses anti-oxidative properties as well; however, the mechanism of its anti-oxidative action has not yet been studied [3,11,20].

The study of glycolysis state after clinical death modeling revealed that as early as at 40th minute of reperfusion period the levels of lactate and pyruvate dropped; it indicated, probably, decrease of glycolysis activity.

The activity of free radical oxidation at the 40th minute of reperfusion period was higher than in intact animals and in animals subjected to modeling of acute hypobaric hypoxia at the same time of the observation. The level of anti-oxidative defense was correspondingly lower.

It may be hypothesized that low concentrations of metabolic substrates at the 40th minute of reperfusion period and high level of free radical oxidation initiated death of testicular cells during the 1st day of reperfusion period.

The sharp decrease of amounts of all cell types of spermatogenous epithelium and Sertoli and Leydig cells was observed at 1st and 3rd days of the experiment. The amounts of Sertoli and Leydig cells were reduced more than 10-fold and almost 20-fold, respectively, compared to corresponding levels in intact animals.

The levels of lactate and pyruvate were significantly raised at 1st day after clinical death modeling, if compared with the 40th minute of the experiment; it indicated, obviously, the blockade of tricarbonic acid cycle (Figure 11). The level of lactate had dropped by the 3rd day of the experiment and remained low up to the end of the observation period; it was, probably, due to the small amount of Sertoli cells.

The most pronounced changes in testicular tissue after clinical death modeling were registered at 7-14th days, when there were no early spermatids and Leydig cells in rats spermatograms.

The tendency towards activation of spermatogenesis began to appear at the 21st day of the post-reanimation period: a somewhat elevation of spermatogenous epithelium cells amounts and a raise of free radical oxidation were noted.

By the 60th day of the experiment the amount of spermatogenous epithelium cells remained significantly lower than in intact animals; it was due to the decreased amounts of Sertoli and Leydig cells.

A complex cytological examination of testicular tissue in combination with a study of ejaculate characteristics allow to gain the full information about spermatogenesis state in experimental modeling of various pathologic conditions. However, we have found no complex studies of spermatogenesis in models of acute hypobaric hypoxia and clinical death.

In all series of experiments on modeling acute hypobaric hypoxia and clinical death we studied the generally accepted indices of ejaculate, specifically, general amount of spermatozoids and their mobility.

Our researches demonstrated that during the whole observation period after acute hypobaric hypoxia modeling the immobile spermatozoids prevailed in rat spermogram, and the cells with vibratory movements were predominant among mobile spermatozoids. The depression of spermatozoids mobility is probably related to the activation of free radical oxidation leading to membranes impairment and a disorder of intracellular metabolism of sex cells [7].

We believe that the low mobility of sex gametes in late post-hypoxic period is related to an altered androgenic saturation of testicular tissue. Our experiments registered the significant decrease of Leydig cells (the main producers of testosterone) amount in the post-hypoxic period. Some authors demonstrated that in the case of androgenic deficiency the concentration of fructose (the main energy source for gametes) decreased in epididymis and seminal vesicles. That, in turn, may lead to decrease of rectilinear mobility of spermatozoids [36].

The following fact was noted after acute hypobaric hypoxia modeling: the total amount of spermatozoids in ejaculate decreased, beginning the 1st day of post-hypoxic period, whereas the total amount of spermatozoids in testicular tissue remained at the level of intact animals. The main reason of decrease of spermatozoids amount in ejaculate may be regarded, according to studies published, their impairment and death in epididymis as a result of free radical activation in hypoxia [12,28]. According to N.I. Boyko [12], the free radicals are the very factor deteriorating spermogram indices.

It has been established that mature spermatozoids are more susceptible to impairing action of free radicals comparing to other cells. It has the following reasons.

First, spermatozoid membranes contain a lot of polyunsaturated fatty acids which are extremely susceptible to peroxidation. J.G. Alvarez, B.T. Storey (1989) [2] demonstrated that epididymal spermatozoids are more susceptible to action of free radicals than testicular ones, and among the former - the caudal pool. It may be related to an elevation of concentrations of polyunsaturated fatty acids and cholesterol in spermatozoids cytoplasmic membranes with their movement from *caput epididymidis* to *cauda epididymidis* [70,51,68,34].

Second, unlike to other cells, spermatozoids have limited potential to amend impaired structures, as their chromatine is inactive and amount of cytoplasm is small.

Third, spermatozoids' defense system against free radicals is weak: catalase is absent [9,54], amounts of glutathione-peroxidase and superoxide dismutase are relatively small [8,40,49].

Taking into account the aforementioned data, it may be supposed that decrease of total amount of spermatozoids in ejaculate at the 1st day of the experiment is related to their death in epididymis, as a result of activation of free radical oxidation during early post-hypoxic period.

The maximal decrease of spermatozoids concentration in testicular tissue was noted at the 14th day, whereas the minimal concentration of spermatozoids in ejaculate was registered at the 30th day of the experiment. This apparent discrepancy may be explained by the fact that spermatozoids of ejaculate was formed at least two weeks prior to sperm obtaining. It is exactly the period of spermatozoids maturation in epididymis.

We had been failing to gain rat sperm for 14 days of post-reanimation period after clinical death modeling. It was, probably, due to disorders of ejaculation processes. According to published works, the correlation exists between activity of ejaculation-related nervous centers and levels of catecholamines (e.g., dopamine) and indolamines (e.g., serotonin) in central nervous system. Catecholamines stimulate these centers, whereas indolamines suppress them. It was demonstrated that serotonergic system at the central level acts as a suppressor of the ejaculation reflex [24]. V.V. Lobov [45] studied serotonin concentration in canine cerebrum after 5 minutes and 10 minutes long clinical death. It happened that the serotonin concentration in canine cerebrum increased in post-reanimation period.

Having in mind these data, we may suppose that in our case the lack of ejaculation in rats during two weeks of post-reanimation period is related to high serotonin concentration in rat cerebrum blocking ejaculation centers. Even the prolonged electric stimulation of rats was not effective during this period.

The study of ejaculate, collected in late post-reanimation period, revealed lower amount of spermatozoids than after modeling of acute hypobaric hypoxia. This fact supports our opinion, that post-reanimation period is characterized by the more severe disorder of spermatozoids transition from testicles to epididymis and further to ejaculate, due to a small amount of Sertoli cells.

CONCLUSION

Experimental acute hypobaric hypoxia and clinical death lead to pronounced disorders of spermatogenesis in white male outbred rats. The more substantial changes of cellular composition of testicles was observed after clinical death modeling than after acute hypobaric hypoxia modeling. Impairment and death of a substantial part of spermatogenous epithelium cells and Sertoli and Leydig cells during early post-hypoxic and post-reanimation periods were related to activation of free radical oxidation. The increase of testicular lactate level during first hours after acute hypobaric hypoxia supports relative stability of cellular composition of testicles and initiates spermatocytes division, whereas the decrease of testicular lactate level during first hours after clinical death modeling reduces resistance of the cells against injuring action of free radicals leading to death of a substantial part of germ cells. The low intensities of processes of division and maturation of spermatogenous epithelium cells in late post-reanimation and post-hypoxic periods are related to small amounts of Sertoli and Leydig cells, which provide metabolic and hormonal support for gametogenesis. The found disorders of spermatogenesis lead to qualitative and quantitative changes of ejaculate indices.

REFERENCES

[1] Alvarez, J.G., Touchstone, J.C., Blasco, L., Storey, Â.Ò. (1987) Spontaneous lipid peroxidation and production of hydrogen peroxide and superoxide in human spermatozoa. Superoxide dismutase as major enzyme protectant against oxygen toxicity. *J. Androl., No 8.*, 338-348.

[2] Alvarez, J.G., Storey, Â.Ò. (1989) Role of glutathione peroxidase in protecting mammalian spermatozoa from loss of motility caused by spontaneous lipid peroxidation. *Gamete Res., Vol. 23*, 77-90.

[3] Amsterdam, À., Dantes, À., Selvaraj, N., Aharoni, D. (1997) Apoptosis in steroidogenic cells: structure-function analysis. *Steroids, Vol. 62*, 207-211.

[4] Apsalyamov, V. Ch., Bazhanov, A.N., Savenko, V.A. (1993) Morphofunctional condition of epitheliospermatogenic layer of the rat seminiferous tubules in postreanimation period. *Morphologia, No 11-12*, 106-113.

[5] Asantin, V.S. (1965) *The new methods of biochemical photometry*. Moscow: Nauka.

[6] Bajpai, M., Gupta, G. and Setty, B.S. (1998) Changes in carbohydrate metabolism of testicular germ cells during meiosis in the rat. *Eur. J. Endocrinol., 138*, 322-327.

[7] Baker, H.W.G., Brindle, J., Irvine, D.S., Aitken R.J. (1996) Protective effect of antioxidants on the impairments of sperm motility by activated polymorphonuclear leukocytes. *Fertil. Steril., Vol. 65*, 411-419.

[8] Bauche, F., Fouchard, Ì.Í., Jegou Â. (1994) Antioxidant systems in rat testicular cells. *FEBS Lett., Vol. 349*, 392-396.

[9] Bize, I., Santander, G, Cabello, Ð. et al. (1991) Hydrogen peroxide is involved in hamster sperm capacitation in vitro. *Biol. Reprod., Vol. 44*, 398-403.

[10] Bogatch, P.G., Kurskiy, M.D., Kucherenko, N.E., Rybalchenko, V.K. (1981) *Structure and function of biological membranes*. Kiev: Visha Shkola.

[11] Boulikas, T. (1997) Gene therapy of prostate cancer: p53, suicidal genes, and other targets *Anticancer Res., Vol. 17*, 1471-1505.

[12] Boyko, N.N. (2004) Disorders of reproduction in prostatitis/syndrome of chronic pelvic pain. *Medunarodniy Medicinskiy Zhurnal, No 3*, 141.

[13] Bylenko, M.V. (1989) *Ischemic and reperfusion damages of organs (molecular mechanisms, ways of prevention, treatment)*. Moscow: Medicine.

[14] Campbell, A. K. (1983) *Intracellular Calcium: Its Universal Role as Regulator. Chichester*: Wiley.

[15] Clermont, Y. (1972) Kinetics of spermatogenesis in mammals: Seminiforous epithelium cycle and spermatogonial renewal. *Physiol. Rev., Vol. 52*, 198-236.

[16] Cuzmina, E.I., Nelyubin, A.S., Schennikova, M.K. (1983). Using of induced chemiluminescence for estimation of free radical reactions in biological substrates. In: *Biochemistry and biophysics of microorganisms*. Collected works. Gorky, 179-183.

[17] de Rooij, D.G. (1998) Stem cells in the testis. *Int. J. Exp. Pathol., 79*, 67-80.

[18] de Rooij, D.G. (2001) Proliferation and differentiation of spermatogonial stem cells. *Reproduction, 121*, 347-354.

[19] Dunkel, L., Hirvonen, V. and Erkkila, K. (1997) Clinical aspects of male germ cell apoptosis during testis development and spermatogenesis. *Cell Death Differ., 4*, 171-179

[20] Erkkila, Ê., Henriksen, Ê., Hirvonen, V. et. al. (1997) Testosterone regulates apoptosis in adult human seminiferous tubules in vitro. *J. Clin. Endocrinol. Metab., Vol. 82*, 314-2321.

[21] Erkkila K., Aito H., Aalto K., Pentikainen V., Dunkel L. (2002) Lactate inhibits germ cell apoptosis in the human testis. *Molecular Human Reproduction, Vol. 8*, No. 2, 109-117.

[22] Evan, R.W., Setchell, Â.Ð. (1978) The effect of rete testis fluid on the metabolism of testicular spermatozoa. *Journal of Reproduction and Fertility, Vol. 52*, 15-19.

[23] Fahim, M.S., Messiha F.S., Girgis, S.M. (1980) Effect of acute and chronic simulated high altitude on male reproduction and testosterone level. *Arch. Androl., Vol. 4*, No 3, 217-219.

[24] Foreman, M.M., Hall, J.L., Love, R.L. (1989) The role of the 5-HT2 receptor in the regulation of sexual performance of male rats. *Life Sci., Vol. 45*, No 14, 1263-1270.

[25] Franca, L.R., Ogawab, Ò., Avarbockb, M.R. et al. (1998) Germ cell genotype controls cell cycle during spermatogenesis in the rat. *Biology of Reproduction, Vol. 59*, 1371-1377.

[26] Francavilla, S., D'Abrizio, P., Rucci, N., Silvano, G., Properzi, G., Straface, E., Cordeschi, G., Necozione, S., Gnessi, L., Arizzi, M. and Ulisse, S. (2000) Fas and Fas ligand expression in fetal and adult human testis with normal or deranged spermatogenesis. *J. Clin. Endocrinol. Metab., 85*, 2692-2700.

[27] Freitas, M. Ischemic syndrome: Reperfusion and myocardial cyto-protection / Ì. Freitas // *Rev. Port. Cardiol.* - 1997. - Vol. 16, ¹ 11. - Ð. 925-931

[28] Gagnon, Ñ., Iwasaki, À., de Lamirande, Å., Kowalski, N. (1991) Reactive oxygen species and human spermatozoa. *Ann. NY Acad. Sci., Vol. 637*, 436-444.

[29] Griswold, M.D. (1998) The central role of Sertoli cells in spermatogenesis. *Semin. Cell Dev. Biol., 9*, 411-416.

[30] Grootegoed, J.A., Jansen, R. and Van der Molen, H.J. (1984) The role of glucose, pyruvate and lactate in ATP production by rat spermatocytes and spermatids. *Biochim. Biophys. Acta, 767*, 248-256.

[31] Grootegoed, J.A., den Boer P.J. (1990) Energy metabolism of spermatids: a review. In: Hamilton, D.W., Waites, G.M.H., Eds. *Scientific Basis of Fertility Regulation Cellular and Molecular Events in Spermatogenesis*. Cambridge: Cambridge University Press, 193-216.

[32] Gu, W., Hecht, N.B. (1996) Developmental expression of glutathione peroxidase, catalase and manganese superoxide dismutase mRNAs during spermatogenesis in the mouse. *J. Androl., Vol. 17*, 256-262.

[33] Heller, C.G., Clermont, Y. (1964) Kinetics of the geminal epithelium in men. *Recent. Prog. Hormone Res., Vol. 20*, 545-575.

[34] Hinton, Â.Ò., Palladino, Ì.À., Rudolph, D., Labus, J.C. (1995) The epididymis as protector of maturating spermatozoa. *Reprod. Fertil. Dev., No 7*, 731-745.

[35] Homm, R.E., Rusticus, Ñ., Hahn, D.W. (1977) The antispermatogenic effect of 5-thio-D-glucose in male rat. *Biol. Reprod., Vol.17*, 698-700.

[36] Hoskins, D.D., Brandt, Í., Acott, T.S. (1978) Initiation of sperm motility in the mammalian epididymis. *Fed. Proc., Vol. 37*, No 11, 2534-2542.

[37] Huang, H.M., Gilson, G.E. (1989) Phosphatidylinositol metabolism during in vitro hypoxia *J. Neurochem, No 3*, 830-835.

[38] Huckins, C. (1978) The morphology and kinetics of spermatogonial degeneration in normal adult rats: an analysis using a simplified classification of the germinal epithelium. *Anat. Record., Vol.190*, 905-926.

[39] Ivanov, Yu. V. (1986) Cytological criteria of spermatogenesis state in toxic-hygienic studies. *Gigiena I Sanitaria, No. 4*, 52-55.

[40] Jeulin, Ñ., Soufir, J.C., Weber Ð. et al. (1989) Catalase activity in human spermatozoa and seminal plasma. *Gamete Res., Vol. 24*, 185-196.

[41] Jutte, N.H.P.M., Jansen, R., Grootegoed, J.A. et al. (1982) Regulation and survival of rat pachytene spermatocyte by lactate supply from Sertoli cells. *Journal of Reproduction and Fertility, Vol. 65*, 431-438.

[42] Koji, T., Hishikawa, Y., Ando, H., Nakanishi, Y. and Kobayashi, N. (2001) Expression of Fas and Fas ligand in normal and ischemia-reperfusion testes: Involvement of the Fas system in the induction of germ cell apoptosis in the damaged mouse testis. *Biol Reprod, 64*, 946-954.

[43] Korpachev, V.G., Lysenkov, S.P, Tell, L.Z. (1982) Modeling of clinical death and post-reanimation disease in rats. *Patologicheskaya phisiologia I experimentalnaya terapia, No. 3*, 78-80.

[44] Leblond, C.P., Clermont, Y. (1952) Spermatogenesis of rat, mouse, hamster and guinea pig as revealed by the "periodic acid fuchsin sulfurous acid" technique. *Amer. J. Androl., Vol. 90*, 167-215.

[45] Lobov V.V., Bychovtsev, A.N. (1991) Monoaminergic and cholinergic mechanisms of regulation of cardiovascular reactions after stopping of blood flow of different time-length. *Krovoobrashchenie, No 1*, 3-9.

[46] Lukyanova L.D. (Eds.) (1990) *The methodical guide for experimental study of drugs designed for clinical trials as anti-hypoxic pharmaceuticals*. Moscow: Pharmacology Committee of USSR Ministry of Health.

[47] Lysenko, A.I., Kirpatovskii, I.D., Glushakov, A.S., Perekrest P.V. (2003) Ultrastructural and enzyme chemical changes in dog testicles as affected by temporary interruption of blood circulation and use of cardiac massage. *Endocrinologia, Vol. 144*, No 7, 2882.

[48] Mauduit, C., Hamamah, S. and Benahmed, M. (1999) Stem cell factor/c-kit system in spermatogenesis. *Hum. Reprod. Update, 5*, 535-545.

[49] Mennella, M.R., Jones R. (1980) Properties of spermatozoa superoxide dismutase and lack of involvement of superoxides in metal-ion-catalyzed lipid-peroxidation reactions in semen. *Biochem. J., Vol. 191*, 289-297.

[50] Mita, M., Hall, P.F. (1982) Metabolism of round spermatids from rats: lactate as the preferred substrate. *Biology of Reproduction, Vol. 26*, 445-455.

[51] Morton, B.E., Sagadraca, R., Fraser, Ñ. (1978) Sperm motility within the mammalian epididymis: species variations and correlations with free calcium levels in epididymal plasma. *J. Fertil. And Steril., Vol. 29*, 695-698.

[52] Nakamura, M., Okinaga, S. and Arai, K. (1984) Metabolism of pachytene primary spermatocytes from rat testes: pyruvate maintenance of adenosine triphosphate level. *Biol. Reprod., 30*, 1187-1197.

[53] Oberley, T.D., Oberley, L.W., Slattery, A.F. et. al. (1990) Immunohistochemical localization of antioxidant enzymes in adult Syrian hamster tissues and during kidney development. *Am. J Physiol., Vol. 137*, 199-214.

[54] Pavlova, S.K., Kanchev, L.N., Alexandrov Ì.Ò. (1994) Peroxidase activities in bull spermatozoa. *Mol. Reprod. Dev., Vol. 37*, 204-209.

[55] Pentikainen, V., Erkkila, K. and Dunkel, L. (1999) Fas regulates germ cell apoptosis in the human testis in vitro. *Am. J. Physiol. Endocrinol. Metab., 276*, E310-316.

[56] Pentikainen, V., Erkkila, K., Suomalainen, L., Otala, M., Pentikainen, M.O., Parvinen, M., Dunkel, L. et al. (2001) TNF alpha downregulates the Fas ligand and inhibits apoptosis in the human testis. *J. Clin. Endocrinol. Metab., 86*, 4480-4488.

[57] Pshenichnikova, T.Ya. (1991) *Infertility in married couples*. Moscow: Medicine.

[58] Raytsina, S.S. (1982) *The modern problems of spermatogenesis*. Moscow: Nauka.

[59] Resnikov, A.G., Varaga, S.G. (1988) *Antiandrogens*. Moscow: Medicine.

[60] Riera, M.F., Meroni, S.B., Gomez, G.E., Schteingart, H.F., Pellizzari, E.H. and Cigorraga, S.B. (2001) Regulation of lactate production by FSH, IL1 beta, and TNF alpha in rat Sertoli cells. *Gen. Comp. Endocrinol., 122*, 88-97.

[61] Robinson, R., Fritz, I.B. (1981) Metabolism of glucose by Sertoli cells in culture. *Biol. Reprod., Vol. 24*, 1032-1041.

[62] Rodriguez, I., Ody, C., Araki, K., Garcia, I. and Vassalli, P. (1997) An early and massive wave of germinal cell apoptosis is required for the development of functional spermatogenesis. *EMBO J., 16*, 2262-2270.

[63] Russell, L.D. , Ettlin, R.A., Sinha Hikim, A.P., Clegg, E.D. (1990) *Histological and Histopathological Evaluation of the Testis*. Clearwater: Cache River Press.

[64] Ryabov, G.A. (1988) *Hypoxia of critical conditions*. Moscow: Medicine.

[65] Ryzhakov, D.I., Molodyuk, A.V., Artifeksov, S.B. (1980) The functional-morphological study of the sex system of white male rats using electrostimulation technique. *Izvestia Academii Nauk USSR. Seria biologicheskaya. No 1*, 133.

[66] Setchell, B.P. (1970) The secretion of fluid by the testes of rats, ram and goats with some observations on the effect of age, cryptorchidism and hypophysectomy. *Journal of Reproduction and Fertility, Vol. 23*, 79-85.

[67] Setchell, Â.D., Scott, T.W., Voglmayr, J.K., Waites, G.M.H. (1969) Characteristic of testicular spermatozoa and the fluid which transports them into the epididymis. *Biol Reprod. No 1*, 40-66.

[68] Sikka, S.C., Rajasekaran, Ì., Hellstrom, W.J.G. (1995) Role of oxidative stress and antioxidants in male infertility. *J. Androl., Vol.16*, 64-468.

[69] Tepperman, J., Tepperman, H.M. (1989) *Metabolic and endocrine physiology an introductory text (Fifth Edition)*. Chicago, London: Year Book Medical Publishers.

[70] Tramera, F., Roccoa, F., Micalia F. et. al. (1998) Antioxidant systems in rat epididymal spermatozoa. *Biology of Reproduction, Vol. 59*, No 4, 753-758.

[71] Tres, L.L. and Kierszenbaum, A.L. (1999) Cell death patterns of the rat spermatogonial cell progeny induced by sertoli cell geometric changes and Fas (CD95) agonist. *Dev. Dyn., 214*, 361-371.

[72] Yen, S.S.C., Jaffe, R.B. (1998) *Reprodcutive endocrinology (Second edition)*. Philadelphia, London, Toronto, Mexico City, Rio de Janeiro, Sydney, Tokyo, Hong Kong: W.B. Saaunders Company.

In: Cell Differentiation Research Developments
Editor: L. B. Ivanova, pp. 151-170
ISBN: 978-1-60021-939-9

Chapter VI

ROLE OF T CELLS IN HUMAN OSTEOCLASTOGENESIS - BONE DESTRUCTION OF RHEUMATOID ARTHRITIS VIA CYTOKINES FROM T CELLS -

Shigeru Kotake**, Yuki Nanke, Toru Yago, Takefumi Furuya, Manabu Kawamoto, Naomi Ichikawa, Tsuyoshi Kobashigawa and Naoyuki Kamatani

Institute of Rheumatology, Tokyo Women's Medical University, Tokyo 162-0054, Japan.

ABSTRACT

Activated human T cells produce receptor activator NF-κB ligand (RANKL), interleukin 17 (IL-17), and interferon-γ (IFN-γ). All regulate human osteoclastogenesis; RANKL and IL-17 potently induce osteoclastogenesis, while IFN-γ directly inhibits it *in vitro*. However, we have demonstrated that IFN-γ-producing T cells induce human osteoclastogenesis from monocytes through the expression of RANKL. Strongly supporting our findings, it has been reported that peripheral blood T cells from patients with early rheumatoid arthritis (RA) promote osteoclastogenesis in autologus monocytes through the expression of RANKL, although the T cells express IFN-γ. In addition, it has been reported that IFN-γ has indirect pro-osteoclastogenic properties *in vivo*; IFN-γ induces bone resorption in three mouse models of osteoporosis under conditions of estrogen deficiency, infection, and inflammation. On the other hand, the expression or function of some cytokines shows differences between humans and mice in bone cell biology. In the current article, we review recent findings on the role of T cells in human

* Correspondence concerning this article should be addressed to: Dr. Shigeru Kotake, Institute of Rheumatology, Tokyo Women's Medical University, 10-22 Kawada-cho, Shinjuku, Tokyo 162-0054 Japan. Tel: 81-3-5269-1725; Fax: 81-3-5269-1726; e-mail: skotake@ior.twmu.ac.jp.

osteoclastogenesis and bone destruction of rheumatoid arthritis (RA), including the effects of dexamethasone and tacrolimus on the function of T cells.

Keywords: dexamethasone, interferon-γ (IFN-γ), interleukin-17 (IL-17), IL-23, rheumatoid arthritis (RA), osteoclast, osteoprogeterin (OPG), receptor activator of NF-κB ligand (RANKL), single nucleotide polymorphism (SNP), tacrolimus (FK506).

INTRODUCTION

Since 1990, we have investigated human osteoclastogenesis in the pathogenesis of rheumatoid arthritis (RA) by three hypotheses: 1)to investigate the pathogenesis of RA, it is important to study osteoclastogenesis using human rather than mouse cells; 2)osteoclastogenesis in RA is regulated by a mechanism independent of inflammation; and 3) activated human T cells regulate human osteoclastogenesis.

To investigate the pathogenesis of RA, it is critical to investigate osteoclastogenesis using human cells. In bone cell biology, some cytokines show different functions between humans and mice. For example, macrophage-colony stimulating factor (M-CSF) induces colony formation from mouse cells. In human cells, however, M-CSF induces the differentiation of progenitor cells of monocytes rather than colony formation in human cells, although M-CSF is reported to induce colony formation of human macrophages [1]. On the other hand, some cytokines show different expressions between humans and mice; it is reported that the expression of membrane-bound receptor activator NF-κB ligand (RANKL) is limited in human T cells compared with mouse T cells [2]. Moreover, the mouse CD4+CD25+ regulatory subset is isolated from all CD25+ T cells regardless of their level of CD25 expression; however, when similar criteria are followed to isolate these cells from human blood, CD25+ cells (high and low together) did not exhibit an anergic phenotype or significant suppressive function. In humans, CD4+CD25^{hi} exhibit all properties of regulatory T cells [3,4]. In addition, recently it has been reported that a major difference between humans and mice in immunology is the presence of CD4+T cells producing IL-17, Th17 cells, in healthy peripheral blood of humans, but not mice [5]. Thus, in the study of human diseases, it is essential to investigate human osteoclastogenesis using human cells.

Osteoclastogenesis in RA is regulated by a mechanism, at least in part, independent of inflammation. Osteoclastogenesis dependent on and independent of inflammation has been reported in the pathogenesis of RA since RANKL was cloned as an 'osteoclast differentiation factor (ODF)' in 1997. We recently reported that the amount of T cells expressing both interferon-γ (IFN-γ and RANKL in peripheral blood of patients with RA does not correlate with the serum level of C-reactive protein [6]. On the other hand, it has been reported that mature osteoclasts are inactivated by selectively inhibiting cathepsin K, carbonic anhydrase II or vacuolar-H(+) ATPase, which are essential in bone resorption by osteocalsts. In addition, more recently, Geusens et al. showed that, in joint destruction of patients with early active RA, the effects of inflammation (time-averaged ESR) and osteoclast differentiation/activity (circulating OPG:RANKL ratio) are statistically partly independent and may be superimposed [7]. Thus, nowadays it is possible to regulate bone resorption in RA by a mechanism separate from inflammation.

Activated human T cells regulate human osteoclastogenesis, producing RANKL, interleukin-17 (IL-17), or IFN-γ, all of which regulate human osteoclastogenesis. In addition, it has been demonstrated that these cytokines play a critical role in human osteoclastogenesis *in vitro* as well as bone resorption in patients with rheumatoid arthritis (RA). In the current article, we review recent findings on the role of T cells in human osteoclastogenesis and bone destruction of RA.

1. Osteoclasts

Osteoclasts are unique multinucleated cells whose function is specialized to resorb calcified tissues (Figure 1) [6]. On the surface of bone, osteoclasts develop a specialized adhesion structure, 'podosome', which subsequently undergoes reorganization into sealing zones [8]. These ring-like adhesion structures seal osteoclasts to the surface of bone. In the sealed resorption lacuna, localized acidification is driven by carbonic anhydrase II and vacuolar H(+)-ATPase in osteoclasts; carbonic anhydrase II produces protons and vacuolar H(+)-ATPase transfers them into the lacuna. In acidified lacuna, cathepsin-K and matrix metalloproteinase-9 (MMP-9) are released from osteoclasts to degrade calcified tissues [9].

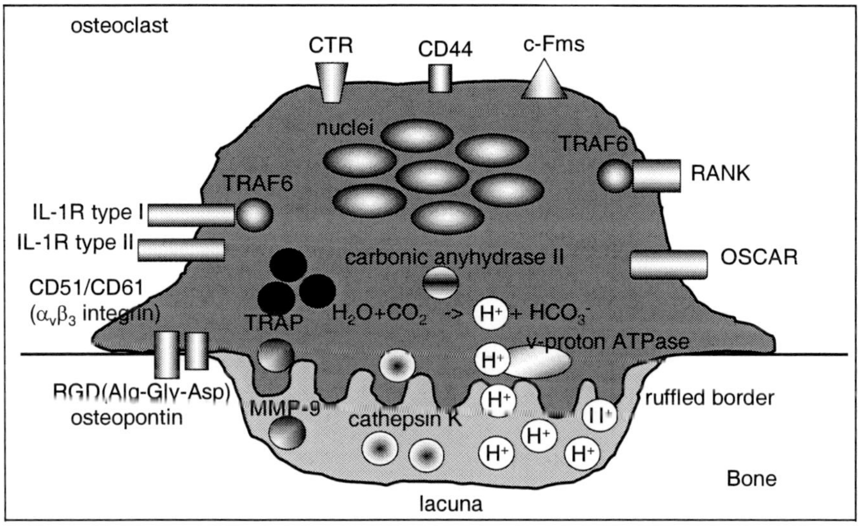

Figure 1. Schematic structure of osteoclasts. CTR, calcitonin receptor; TRAF, TNF receptor-associated factor; TRAP, tartrate-resistant acid phosphatase; MMP, matrix metalloproteinase; OSCAR, osteoclast-associated receptor; c-Fms, M-CSF receptor.

The cooperation of osteoclasts and osteoblasts is critical to maintain skeletal integrity in normal bone. After bone resorption by osteoclasts on normal bone tissues, osteoblasts subsequently rebuild bone in the lacunae resorbed by osteoclasts; this mechanism is called 'bone remodeling'. When the activity or number of osteoclasts is elevated compared with osteoblasts, the bone become fragile, that is, 'osteoporotic'. In addition, bone remodeling is disrupted in all bone diseases associated with changes in bone mass. Thus, bone remodeling is essential to retain both the structure and strength of normal bone.

Osteoclasts also play an important role in the pathogenesis of rheumatoid arthritis (RA). Since 1984, it has been reported that in bone destruction of rheumatoid arthritis (RA), many activated osteoclasts are detected on the surface of eroded bone in the interface with synovial tissues [10]. In addition, we have demonstrated that osteoclasts are detected in synovial tissues as well as eroded bone from patients with RA [11]. In addition, we have reported that the number of precursor cells of osteoclasts increases in bone marrow adjacent to joints with arthritis [12]. Moreover, the amount of cytokines that induce osteoclastogenesis such as IL-1, tumor necrosis factor-α (TNF-α) and IL-6, is elevated in synovial tissues of patients with RA, while the amount of cytokines that inhibit osteoclastogenesis, such as IL-4 and IL-10, is decreased [12-15]. Thus, patients with RA are likely to suffer from joint destruction as well as systemic osteoporosis, in which the number of osteoclasts increases, suggesting that osteoclasts play a critical role in the pathogenesis of RA.

The origin of osteoclasts was unclarified until the 1980s. In 1988, Takahashi et al. established a culture system using mouse spleen cells and osteoblasts to induce osteoclastogenesis *in vitro*, demonstrating that the origin of osteoclasts is hematopoietic cells [16]. The precursor of osteoclasts was then revealed to be colony-forming unit-macrophage (CFU-M) or CFU-granulocyte/macrophage (CFU-GM) in bone marrow or spleen in mice. In 1990, Udagawa et al. demonstrated that osteoclasts are formed from murine macrophages [17]. From these findings, Suda et al. hypothesized that bone marrow hemopoietic cells differentiate into osteoclasts through the stimulation of 'osteoclast-differentiation factor (ODF)' expressed on osteoblasts [18].

Finally, ODF, now termed RANKL, which induces osteoclastogenesis from monocytes or macrophages, was independently cloned by three groups in 1997 (Figure 1) [19]. RANKL is a member of the TNF superfamily of cytokines. The protein constructs a trimeric complex to bind its receptor, receptor activator NF-κB (RANK) (20). A decoy receptor is also cloned, which is designated as 'osteoprotegerin (OPG)' [19] (Figure 2).

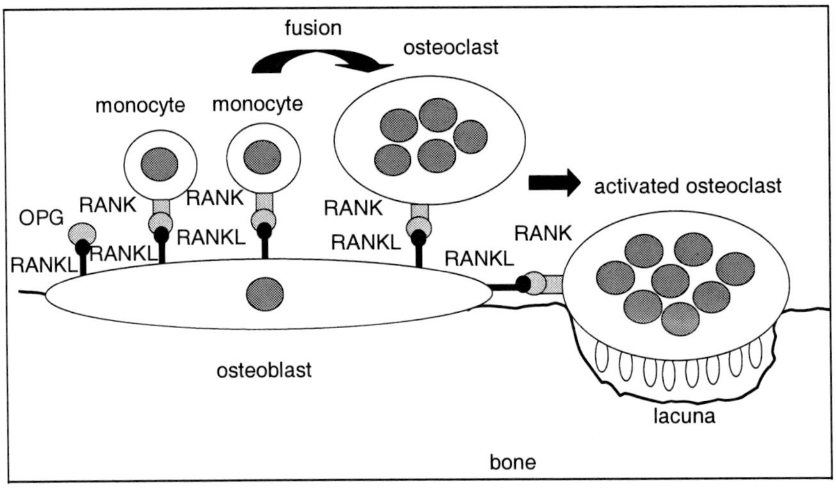

Figure 2. Differentiation and activation of osteoclasts. OPG, osteoprotegerin. A RANK-RANKL system induces both osteoclastogenesis from monocytes and the activation of mature osteoclasts.

2. T cells Producing RANKL

RANKL expressed on T cells as well as on osteoblasts induces osteoclastogenesis. Two groups reported using murine cells that T cells directly induce murine osteoclastogenesis through the expression of RANKL from T cells [21,22]. We have also demonstrated using human T cells and monocytes that human T cells induce osteoclastogenesis from peripheral human monocytes through the expression of RANKL on T cells (Figure 3) [23]. We have also demonstrated that CD4+ T cells in synovial tissues from patients with RA express RANKL, and that the ratio of levels of RANKL to those of OPG in synovial fluids is elevated (Figure 7) [23]. Thus, in synovial fluids of patients with RA, monocytes are likely to differentiate into osteoclasts.

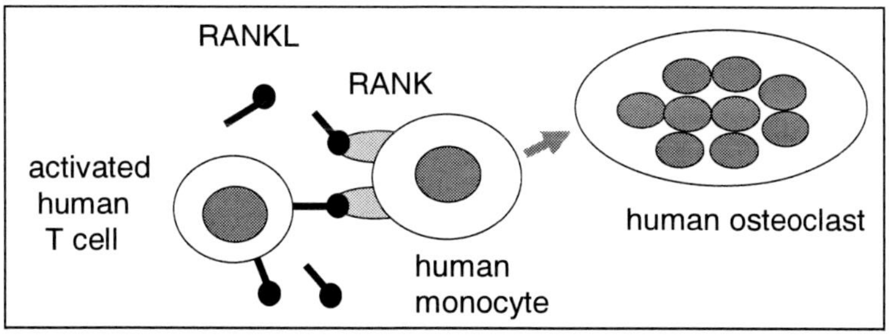

Figure 3. Activated human T cells induce osteoclastogenesis from human monocytes through the expression of RANKL on T cells.

Strongly supporting our findings, two groups have recently demonstrated important results. First, Miranda-Carus et al. have reported that peripheral blood T cells from patients with RA promote osteoclastogenesis from autologous monocytes through the expression of RANKL on T cells, although T cells simultaneously produce IFN-γ [24]. Thus, they concluded that T cells are important contributors to the pathogenesis of bone erosion in RA through interaction with monocytes. Second, Geusens et al. have reported that the ratio of circulating OPG to RANKL in early RA predicts later joint destruction, analyzing 92 patients with early active RA for 5 years [7]. Thus, the reports on osteoclastogenesis and bone destruction in RA patients in these groups support our findings.

In mature osteoclasts, the function of IL-1 and TNF-α, but not RANKL, shows differences between humans and mice. In both humans and mice, RANKL stimulates bone resorption of mature osteoclasts [9]; however, IL-1 and TNF-α show no significant effects on human mature osteoclasts, which contrasts with the potent ability of both cytokines to stimulate murine mature osteoclasts [9]. On the other hand, we have demonstrated that in both humans and mice, TNF-α directly induces osteoclastogenesis [25]. Thus, the differences in species used in the studies are critical to discuss the function of cytokines.

The pathogenesis of periodontitis is similar to that of RA because both diseases show chronic inflammation. The levels of RANKL are elevated in gingival crevicular fluid of patients with periodontitis [26]. In addition, T cells from patients with periodontitis express RANKL [27]. Interestingly, human T cells expressing RANKL transplanted to NOD/SCID mice show alveolar bone destruction [28]. Recently, it has been reported that the induction of an adaptive immune response to orally colonizing non-pathogenic *Pasteurella pneumotropica* by immunization with phylogenetically closely related bacterium, *Actinobacillus actinomycetemcmitans,* results in T cell-derived RANKL-dependent periodontal bone loss in mice [29]. Thus, T cells expressing RANKL play a critical role in bone destruction of perodontitis.

3. Th17 Cells Producing IL-17

1) IL-17

In 1993, IL-17 was initially identified as a transcript selectively expressed by activated T cells encoding a protein with significant sequence identity (~60%) to the Herpesvirus saimiri protein HVS13 [30]. In 1996, IL-17, which induces IL-6, IL-8, and granulocyte colony-stimulating factor (G-CSF) from stromal cells [31], and IL-1 and TNF-α from macrophages [32], was cloned as a unique interleukin showing a function similar to that of IL-1, although IL-17 is produced by T cells [30]. IL-17 receptor (IL-17R), ubiquitously expressed, was reported to be a unique structure different from any receptors of other interleukins. Recently, however, a bioinformatics study predicted a subdomain in IL-17-family receptors with homology to a Toll/IL-1R (TIR) domain, termed the "SEFIR domain" [33]. More recently, Maitra et al. reported that distinct functional motifs within the IL-17 receptor regulate signal transduction and target gene expression [34].

IL-17and IL-17R have been renamed IL-17A and IL-17RA, respectively; five additional IL-17-like ligands (IL-17B-F) and four additional IL-17R-like receptors (IL-17RB-E) have been identified [35]. Interestingly, Toy et al. reported that the biologic activity of IL-17 is dependent on a heteromeric receptor complex composed of IL-17RA and IL-17RC [36].

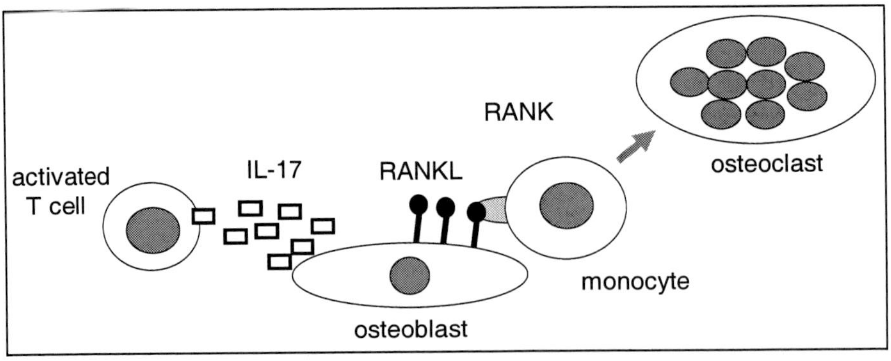

Figure 4. IL-17 from activated T cells induces osteoclastogenesis through the expression of RANKL on osteoblasts.

In 1999, we first demonstrated that IL-17 potently induces mouse osteoclastogenesis from mouse monocytes through the expression of RANKL of osteoblasts (Figure 4) [32,37]. In addition, we recently reported that IL-17 induces human osteoclastogenesis from human monocytes even in the absence of osteoblastic cells or soluble RANKL through both inductively-expressed TNF-α and constitutively-expressed RANKL on human monocytes [37]. In this osteoclastogenesis, the synergistic effect of TNF-α and RANKL plays an important role; the expressed level of each cytokines is too low to induce osteoclastogenesis in each cytokine alone [37]. This synergism has also been reported by other two groups [38,39]. More recently, it has been reported that peripheral blood T cells from patients with early RA promote osteoclastogenesis in autologous monocytes in the absence of exogenous cytokines or osteoblasts, and that osteoclastogenesis is significantly inhibited by neutralizing monoclonal antibodies to IL-17 [24]. Thus, IL-17 induces osteoclastogenesis from monocytes both in the absence and presence of osteoblasts.

Synovial tissues of patients with RA contain an increased number of T cells; however, using the usual ELISA with standard sensitivity, it was very difficult to detect cytokines derived from T cells, such as IL-2 or IFN-γ, in synovial tissues or synovial fluids. Until 1999, this absence of T cell-derived cytokines was an 'enigma' to be solved in the pathogenesis of RA. We first demonstrated that a sufficient amount of a T cell-derived cytokine, IL-17, is detected in synovial fluids from patients with RA using the usual ELISA (Figure 7) [32]. Lubbers et al. reported that treatment with a neutralizing anti-murine IL-17 antibody after the onset of collagen-induced arthritis (CIA) reduces joint inflammation, cartilage destruction, and bone erosion [41]. We also demonstrated that in rats with CIA, peritoneal injection of anti-IL-17 Ab ameliorates synovitis even after the onset of arthritis [42]. In addition, Nakae et al. also reported that IL-17 play an important role in the pathogenesis of a murine arthritis model [43]. Recently, Sarkar et al. reported that dendritic cells genetically modified to express IL-4 exert a therapeutic effect on CIA by targeting IL-17 [44]. Interestingly, Raza et al. reported that early RA of 3-months duration or less is characterized by a distinct and transient synovial fluid cytokine profile of T cells including IL-17, but not IFN-γ [45]. Thus, IL-17 plays an important role in both human RA and murine arthritis models.

It has been reported that interplay between IL-23 and IL-17 production may be a critical immune pathway using gene-targeted mice lacing IL-23 (p19-/-), which is resistant to CIA induction [46]. We recently demonstrated that IL-23 induces osteoclastogenesis in the culture of human PBMC via the expression of IL-17 from T cells [47]. In addition, peritoneal injection of anti-IL-23 Ab ameliorates synovitis of rats with CIA [47]. Thus, IL-23 may induces osteoclastogenesis upstream of IL-17 in the pathogenesis of RA.

2) Th17

In 2005, a novel subset of helper T cells was demonstrated; Th cells expressing IL-17 distinct from Th1 cells or Th2 cells were designated as Th17, Th_{IL-17}, or inflammatory TH (THi). In 1986, Mosmann et al. demonstrated the dichotomy of helper T cells; Th1 cells producing IFN-γ in cellular immunity and Th2 cells producing IL-4 in humoral immunity [48]. The pathogenesis of many diseases has been divided into two categories; for example, RA is Th1 disease, which we confirmed using peripheral helper T cells from patients with RA treated without any disease-modifying anti-rheumatic drugs (DMARDs) or steroids [6]. On the other hand, bronchial asthma and atopic dermatitis are Th2 diseases. The notion that IL-17-producing T cells may be a distinct T cell lineage emerged from gene-targeted mice models with experimental autoimmune encephalomyelitis (EAE) or CIA [49]. In 2005, two laboratories simultaneously reported IL-17-producing CD4+ T cells (Th17, Th_{IL-17}) as a distinct effector lineage [50,51]. In 2006, it was reported that naïve T cells are primed to Th17 by IL-6 and transforming growing factor (TGF)-β, and that IL-23 stimulates the production of IL-17 from primed T cells [52-54]. Human Th17 cells have distinct migratory capacity with the expressions of CCR6 and CCR4 together and antigenic specificities for *Candida albicans* [55]. Reports using infectious mouse models have suggested that Th17 cells develop to mediate protection against extracellular bacteria and *Pneumocystis carinii* [56-58]. In addition, it has been reported that Th17 functions as an osteoclastogenic helper T cell subset that links T cell activation and bone destruction [59]. Hirota et al. reported that autoimmune arthritis of SKG mice is highly dependent on the development of Th17 cells [60]. In addition, IL-17 plays an important role in the pathogenesis of periodontitis [61,62]. Thus, Th17 cells may be involved in bone destruction in RA and periodontitis as well as in the pathogenesis of autoimmune disorders of model mouse such as EAE and autoimmune arthritis, whereas Th17 cells develop to mediate protection against fungi and extracellular bacteria.

4. Th1 Cells Producing IFN-γ

IFN-γ produced by Th1 cells plays an important role in the defense of hosts against viruses or intracellular microorganisms. It has been reported that IFN-γ inhibits osteoclastogenesis [63,64]. In 2000, IFN-γ was shown to inhibit mouse osteoclastogenesis through the degradation of TRAF6 *in vitro* and to repress bone resorption in calvariae from prepubertal mice *in vivo* [65].

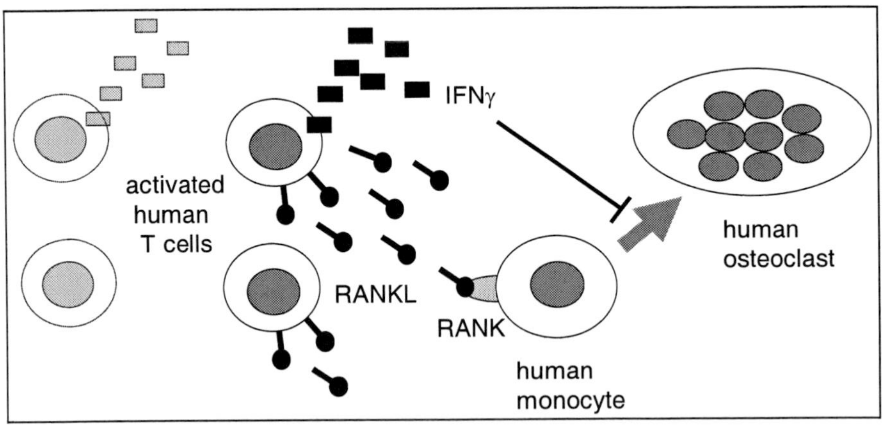

Figure 5. Human activated T cells producing IFN-γ, which simultaneously produce RANKL, induce human osteoclastogenesis from monocytes. IFN-γ from activated T cells inhibits human osteoclastogenesis from monocytes; however, human activated T cells producing IFN-γ, which simultaneously produce RANKL, induce human osteoclastogenesis from monocytes.

We have demonstrated that activated human T cells induce human osteoclastogenesis from monocytes though the expression of RANKL [23] as described above (Figure 3).

Moreover, we recently reported that human IFN-γ-producing T cells (Th1 cells) induce human osteoclastogenesis from monocytes via the production of RANKL from Th1 cells (Figure 5) and that the number of T cells expressing both IFN-γ and RANKL is elevated in the peripheral blood of patients with RA (Figure 6) [6]. Strongly supporting our findings, it has been demonstrated that IFN-γ has indirect pro-osteoclastogenic properties *in vivo*; IFN-γ induces bone resorption in three mouse models with osteoporosis under conditions of estrogen deficiency, infection, and inflammation [66]. In addition, also supporting our findings, Miranda-Carus et al. have reported that peripheral blood T cells from patients with RA promote osteoclastogenesis from autologous monocytes through the expression of RANKL on T cells, although T cells simultaneously produce IFN-γ [24]. Moreover, Stashenko et al. demonstrated using a murine model with periodontitis that Th1 immune response promotes severe bone resorption caused by *Porphyromonas gingivalis* [67]. Thus, IFN-γ or Th1 cells may play a critical role in indirect bone resorption and destruction in a murine model *in vivo* and patients with RA, although IFN-γ directly inhibits murine and human osteoclastogenesis from monocytes *in vitro*.

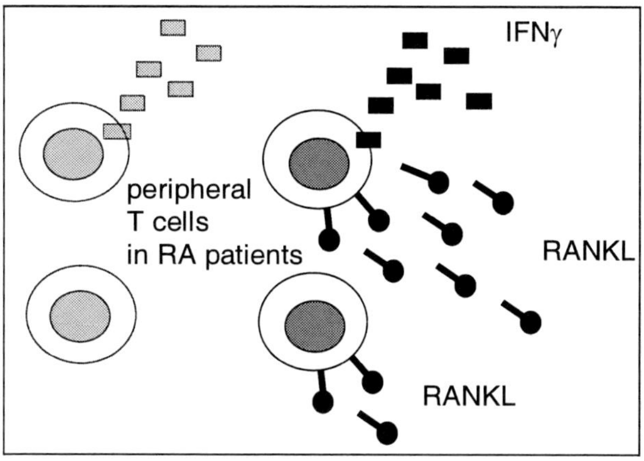

Figure 6. Human helper T cells producing IFN-γ simultaneously produce RANKL in peripheral blood of patients with RA. Theoretically, activated human T cells are divided into 4 types by the production of IFN-γ and RANKL. However, only 2 types, IFN-γ (+), RANKL(+)T cells and IFN-γ (-), RANKL(+)T cells, are detected in peripheral blood of patients with RA.

5. Regulatory T Cells

Regulatory T cells (Tregs) are a subset of CD4+ T cells expressing CD25 with immunosuppressive activity. Recently, it has been reported that Tregs inhibit human osteoclastogenesis from PBMC [68]. This suppression of osteoclastogenesis was blocked by anti-TGF-β antibody or anti-IL-4 antibody, suggesting that TGF-β and IL-4 are key cytokines for the suppressive function of Tregs. On the other hand, it has been reported that TNF downmodulates the function of human CD4+CD25^{hi} Tregs [4]. In fact, CD4+CD25^{hi} Tregs isolated from patients with active RA poorly suppress the proliferation and cytokine secretion of CD4+ effector T cells [4], Thus, the reduced function of Tregs in patients with RA might, at least in part, limit the inhibition of osteoclasts, inducing elevated bone resorption.

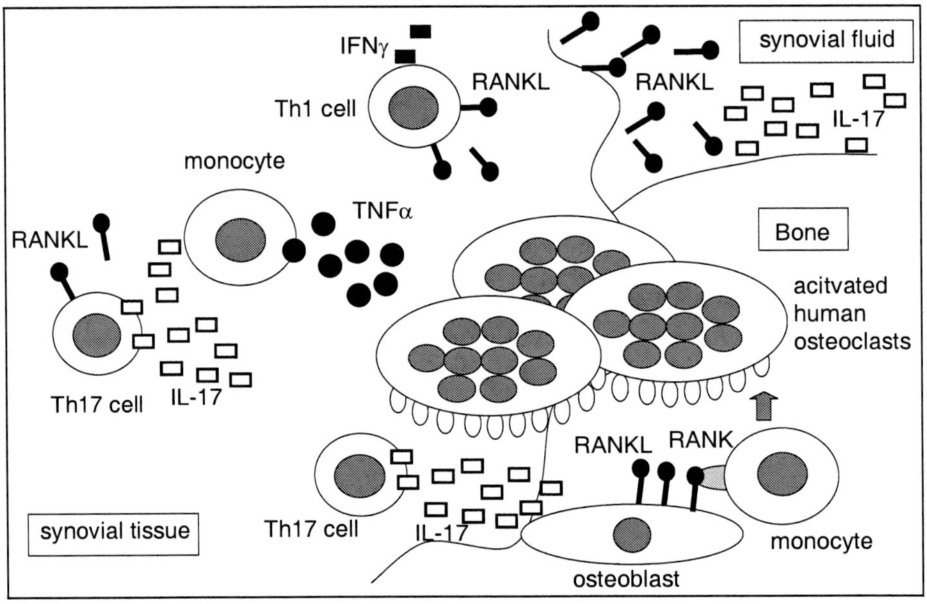

Figure 7. A scheme of the role of T cells in bone resorption in RA. It is speculated that Th17 cells and Th1 cells play an important role in bone resorption in RA through the formation and activation of osteoclasts.

6. Cell-cell Fusion in Osteoclastogenesis

Dendritic cell-specific transmembrane protein (DC-STAMP) is a seven-transmembrane protein originally identified in dendritic cells [69]. Kukita et al. reported that DC-STAMP is essential for the differentiation of osteoclasts, showing that the expression of DC-STAMP is rapidly induced in osteoclast precursor cells by RANKL [70]. Yagi et al. demonstrated that multinucleation of osteoclasts and macrophages is completely abrogated in DC-STAMP -/- mice [71]. Thus, it is speculated that T cells play an important role in inducing DC-STAMP via the expression of RANKL.

In 2006, we reported that antibodies against CD47 expressed on monocytes inhibits human osteoclastogenesis by blocking monocyte fusion [72]. CD47, widely expressed on hematopoietic cells, is a ligand for monocyte/macrophage adhesion and fusion leading to multinucleation. We demonstrated as follows: 1) CD47 is expressed on monocytes but not on osteoclasts; 2) adding anti-CD47 antibody during day 0-3 of the culture period markedly inhibited osteoclastogenesis compared with adding anti-CD47 antibody during days 4-6; and 3)there was a tendency that in total CD14+ monocytes from PBMC, the mean of fluorescence intensity (MFI) on CD47+ cell was higher in patients with RA than in healthy controls [72]. Recently, another group also reported that osteoclast formation is strongly reduced both in vivo and in vitro in the absence of CD47/signal regulatory protein alpha (SIRPα) interaction using blocking antibodies *in vitro* and CD47-/- mice *in vivo* [73]. Thus, these and our findings suggest that CD47 could be a therapeutic target for bone destruction in RA.

7. Dexamethasone

Therapies using systemic glucocorticoid induce osteoporosis as a serious complication; however, the mechanism of glucocorticoid-induced osteoporosis remains unclear. It has been reported that glucocorticoids promote osteoclastogenesis by inhibiting OPG and concurrently stimulating RANKL production by osteoblastic lineage cells [74,75]. In addition, dexamethasone increases the pit area by human osteoclasts induced from monocytes stimulated by RANKL [75]. Thus, glucocorticoids induce osteoporosis both dependently on and independently of osteoblasts.

We hypothesized that activated human T cells are involved in glucocorticoid-induced osteoporosis. We have demonstrated that using human PBMC culture, dexamethasone stimulates osteoclastogenesis by elevating the RANKL-to-IFN-γ ratio expressed on human CD4+ T cells, which is inhibited by adding OPG [76]. Thus, it is suggested that human T cells stimulated by glucocorticoids, at least in part, play a role in human glucocorticoid-induced osteoporosis.

8. Tacrolimus

Tacrolimus (FK506) has been shown to have an optimally anti-inflammatory effect against patients with RA. Tacrolimus diminishes the ability of calcineurin to translocate the nuclear factor of activated T cells (NFAT) that initiates gene transcription for the synthesis of inflammatory cytokines [77]; however, the effect of tacrolimus on human osteoclastogenesis remains unclear.

We have demonstrated that tacrolimus dose-dependently inhibits dexamethasone-induced osteoclastogenesis from human PBMC [78]. As mentioned above, dexamethasone induces human osteoclastogenesis from PBMC, stimulating osteoclastogenesis by elevating the RANKL-to-IFN-γ ratio expressed on human CD4+ T cells [76]. The expression of RANKL stimulated by dexamethasone was markedly reduced by adding tacrolimus [78]. Thus, tacrolimus indirectly inhibited human osteoclastogenesis from PBMC induced by dexamethasone.

On the other hand, tacrolimus directly inhibits osteoclastogenesis. Igarashi et al. reported that FK506 directly induces osteoclast apoptosis in mouse bone marrow cultures [79]. In addition, Miyazaki et al. demonstrated that tacrolimus directly inhibits human osteoclastogenesis via targeting the calcineurin-dependent NFAT pathway and an activation pathway for c-Jun or microphthalmia transcription factor (MITF) in rheumatoid arthritis [80]. Moreover, it has been reported that tacrolimus administration into mice blocks osteclast differentiation, although the administration induces the reduction of bone mass, causing severe impairment of bone formation [81]. Thus, tacrolimus inhibits osteoclastogenesis both directly and indirectly.

9. Single Nucleotide Polymorphism (SNP)

The association of SNPs in RANKL with bone mineral density has been reported [82-84]. In addition, Wu et al. have reported that novel RANKL polymorphisms are associated with an earlier age at RA onset in shared epitope (SE)+, but not SE-, patients and that an interaction between SE-containing HLA-DRB1 and RANKL polymorphisms increases disease penetrance, resulting in a mean age at RA onset that is 18-20 years younger [85]. On the other hand, the association of SNPs in IL-17F with asthma and no significant association in IL-17A and F in osteoarthritis has been reported [86,87], whereas the association of SNPs in IL-17A or F with bone mineral density or RA has not been reported. In addition, a highly significant association between Crohn's disease and the IL-23 receptor gene was found; an uncommon coding variant confers strong protection against Crohn's disease [88]. Replication studies confirmed IL-23 receptor associations in independent cohorts of patients with Crohn's disease or ulcerative colitis [88]. However, the association of SNPs in IL-23 or IL-23 receptor with RA has not been reported.

In 2004, we have demonstrated that SNPs in RANKL and IL-17A genes are significantly associated with radiographic progression after 2 years in patients with early RA [89]. Thus, bone resorption induced by activated T cells may be genetically regulated through SNPs in RANKL and IL-17 in T cells; however, further investigation is necessary to clarify the association using a large number of patients for longer duration.

Conclusion

The role of T cells on human osteoclastogenesis has been clarified in vitro. However, the function of Th1 cells and Th17 cells on bone resorption has not fully been understood in humans including patients with RA, while Th1 cells and Th17 cells have been demonstrated to stimulate bone resorption in a murine model *in vivo*. Thus, further investigation is necessary to clarify the role of Th1 cells and Th17 cells in bone resorption of human diseases including RA.

Acknowledgements

We thank Dr. Nobuo Nara (Tokyo Medical and Dental University) for providing helpful suggestions regarding human hematopoiesis, and Ms. Hanae Kikuchi (Tokyo Women's Medical University) for her excellent experimental technique.

REFERENCES

[1] Motoyoshi, K., Suda, T., Kusumoto, K., Takaku, F., & Miura, Y. (1982). Granulocyte-macrophage colony-stimulating and binding activities of purified human urinary colony-stimulating factor to murine and human bone marrow cells. *Blood.* 1982, 60, 1378-1386.

[2] Kanamaru, F., Iwai, H., Ikeda, T., Nakajima, A., Ishikawa, I., & Azuma, M. (2004). Expression of membrane-bound and soluble receptor activator of NF-kappaB ligand (RANKL) in human T cells. *Immunol Lett.* 94, 239-246.

[3] Baecher-Allan, C., Brown, J.A., Freeman,, G.J., & Hafler, D.A. (2001). CD4+CD25high regulatory cells in human peripheral blood. *J Immunol.* 167, 1245-1253.

[4] Valencia, X., Stephens, G., Goldbach-Mansky, R., Wilson, M., Shevach, E.M., & Lipsky, P.E. (2006). TNF downmodulates the function of human CD4+CD25hi T-regulatory cells. *Blood.* 108, 253-261.

[5] Amadi-Obi, A., Yu, C.R., Liu, X., Mahdi, R.M., Clarke, G.L., Nussenblatt, R.B., Gery, I., Lee, Y.S., & Egwuagu, C.E. (2007). T(H)17 cells contribute to uveitis and scleritis and are expanded by IL-2 and inhibited by IL-27/STAT1. *Nat Med.* 2007 May 13; [Epub ahead of print]

[6] Kotake, S., Nanke, Y., Mogi, M., Kawamoto, M., Furuya, T., Yago, T., Kobashigawa, T., Togari, A., & Kamatani. (2005). IFN-gamma-producing human T cells directly induce osteoclastogenesis from human monocytes via the expression of RANKL. *Eur J Immunol. 35*, 3353-3363.

[7] Geusens, P.P., Landewe, R.B., Garnero, P., Chen, D., Dunstan, C.R., Lems, W.F., Stinissen, P., van der Heijde, D.M., van der Linden, S., & Boers, M. (2006). The ratio of circulating osteoprotegerin to RANKL in early rheumatoid arthritis predicts later joint destruction. *Arthritis Rheum. 54*, 1772-1777.

[8] Luxenburg, C., Geblinger, D., Klein, E., Anderson, K., Hanein, D., Geiger, B., & Addadi L. (2007). The architecture of the adhesive apparatus of cultured osteoclasts: from podosome formation to sealing zone assembly. *PLoS ONE, 2*, e179.

[9] Fuller, K., Kirstein, B., & Chambers, T.J. (2007). Regulation and enzymatic basis of bone resorption by human osteoclasts. *Clin Sci (Lond). 112*, 567-575.

[10] Shimizu, S., Shiozawa, S., Shiozawa, K., Imura, S., & Fujita, T. (1985). Quantitative histologic studies on the pathogenesis of periarticular osteoporosis in rheumatoid arthritis. *Arthritis Rheum. 28*, 25-31.

[11] Kotake, S., Sato, K., Kim, K.J., Takahashi, N., Udagawa, N., Nakamura, I., Yamaguchi, A., Kishimoto, T., Suda, T., & Kashiwazaki, S. (1996). Interleukin-6 and soluble interleukin-6 receptors in the synovial fluids from rheumatoid arthritis patients are responsible for osteoclast-like cell formation. *J Bone Miner Res. 11*, 88-95.

[12] Kotake, S., Higaki, M., Sato, K., Himeno, S., Morita, H., Kim, K.J., Nara, N., Miyasaka, N., Nishioka, K., & Kashiwazaki, S. (1992). Detection of myeloid precursors (granulocyte/macrophage colony forming units) in the bone marrow adjacent to rheumatoid arthritis joints. *J Rheumatol. 19*, 1511-1516.

[13] Kotake, S., Schumacher, H.R. Jr, Yarboro, C.H., Arayssi, T.K., Pando, J.A., Kanik, K.S., Gourley, M.F., Klippel, J.H., & Wilder, R.L. (1997). In vivo gene expression of type 1 and type 2 cytokines in synovial tissues from patients in early stages of rheumatoid, reactive, and undifferentiated arthritis. Proc Assoc Am Physicians. 109, 286-301. *Proc Assoc Am Physicians*. 1997, 109:286-301.

[14] Kotake, S., Schumacher, H.R. Jr, Arayssi, T.K., Gerard, H.C., Branigan, P.J., Hudson A.P., Yarboro, C.H., Klippel, J.H., & Wilder, R.L. (1999). Gamma interferon and interleukin-10 gene expression in synovial tissues from patients with early stages of Chlamydia-associated arthritis and undifferentiated oligoarthritis and from healthy volunteers. *Infect Immun. 67*, 2682-2686.

[15] Kotake, S., Schumacher, H.R. Jr, & Wilder, R.L. (1996). A simple nested RT-PCR method for quantitation of the relative amounts of multiple cytokine mRNAs in small tissue samples. *J Immunol Methods. 199*, 193-203.

[16] Takahashi, N., Akatsu, T., Udagawa, N., Sasaki, T., Yamaguchi, A., Moseley, J.M., Martin, T.J., & Suda, T. (1988). Osteoblastic cells are involved in osteoclast formation. *Endocrinology. 123*, 2600-2602.

[17] Udagawa, N., Takahashi, N., Akatsu, T., Tanaka, H., Sasaki, T., Nishihara, T., Koga, T., Martin, T.J., & Suda, T. (1990). Origin of osteoclasts: mature monocytes and macrophages are capable of differentiating into osteoclasts under a suitable microenvironment prepared by bone marrow-derived stromal cells. *Proc Natl Acad Sci U S A.* 1990. 87, 7260-7264.

[18] Suda, T., Udagawa, N., Nakamura, I., Miyaura, C., & Takahashi, N. (1995). Modulation of osteoclast differentiation by local factors. *Bone. 17*(2 Suppl), 87S-91S.

[19] Udagawa, N., Kotake, S., Kamatani, N., Takahashi, N., & Suda, T. (2002). The molecular mechanism of osteoclastogenesis in rheumatoid arthritis. *Arthritis Res. 4*, 281-289.

[20] Lam, J., Nelson, C.A., Ross, F.P., Teitelbaum, S.L., & Fremont, D.H. (2001). Crystal structure of the TRANCE/RANKL cytokine reveals determinants of receptor-ligand specificity. *J Clin Invest. 108*, 971-979.

[21] Kong, Y.Y., Feige, U., Sarosi, I., Bolon, B., Tafuri, A., Morony, S., Capparelli, C., Li, J., Elliott, R., McCabe, S., Wong, T., Campagnuolo, G., Moran, E., Bogoch, E.R., Van, G., Nguyen, L.T., Ohashi, P.S., Lacey, D.L., Fish, E., Boyle, W.J., & Penninger, J.M. (1999) Activated T cells regulate bone loss and joint destruction in adjuvant arthritis through osteoprotegerin ligand. *Nature. 402*:304-309.

[22] Horwood, N.J, Kartsogiannis, V., Quinn, J.M., Romas, E., Martin, T.J., & Gillespie, M.T. (1999). Activated T lymphocytes support osteoclast formation in vitro. *Biochem Biophys Res Commun. 265*, 144-150.

[23] Kotake, S., Udagawa, N., Hakoda, M., Mogi, M., Yano, K., Tsuda, E., Takahashi, K., Furuya, T., Ishiyama, S., Kim, K.J., Saito, S., Nishikawa, T., Takahashi, N., Togari, A., Tomatsu, T., Suda, T., & Kamatani, N. (2001). Activated human T cells directly induce osteoclastogenesis from human monocytes: possible role of T cells in bone destruction in rheumatoid arthritis patients. *Arthritis Rheum. 44*, 1003-1012.

[24] Miranda-Carus, M.E., Benito-Miguel, M., Balsa, A., Cobo-Ibanez, T., Perez de Ayala, C., Pascual-Salcedo, D., & Martin-Mola, E. (2006). Peripheral blood T lymphocytes

from patients with early rheumatoid arthritis express RANKL and interleukin-15 on the cell surface and promote osteoclastogenesis in autologous monocytes. *Arthritis Rheum. 54*, 1151-1164.

[25] Kobayashi, K., Takahashi, N., Jimi, E, Udagawa, N., Takami, M., Kotake, S., Nakagawa, N., Kinosaki, M., Yamaguchi, K., Shima, N., Yasuda, H., Morinaga, T., Higashio, K., Martin, T.J., & Suda, T. (2000). Tumor necrosis factor alpha stimulates osteoclast differentiation by a mechanism independent of the ODF/RANKL-RANK interaction. *J Exp Med. 191*, 275-286.

[26] Mogi, M., Otogoto, J., Ota, N., & Togari, A. (2004). Differential expression of RANKL and osteoprotegerin in gingival crevicular fluid of patients with periodontitis. *J Dent Res. 83*, 166-169.

[27] Vernal, R., Dutzan, N., Hernandez, M., Chandia, S., Puente, J., Leon, R., Garcia, L., Del Valle, I., Silva, A., & Gamonal, J. (2006). High expression levels of receptor activator of nuclear factor-kappa B ligand associated with human chronic periodontitis are mainly secreted by CD4+ T lymphocytes. *J Periodontol. 77*, 1772-1780.

[28] Teng, Y.T., Nguyen, H., Gao, X., Kong, Y.Y., Gorczynski, R.M., Singh, B., Ellen, R.P., & Penninger, J.M. (2000). Functional human T-cell immunity and osteoprotegerin ligand control alveolar bone destruction in periodontal infection. *J Clin Invest. 106*, R59-67.

[29] Kawai, T., Paster, B.J., Komatsuzawa, H., Ernst, C.W., Goncalves, R.B., Sasaki, H., Ouhara, K., Stashenko, P.P., Sugai, M., & Taubman, M.A. (2007). Cross-reactive adaptive immune response to oral commensal bacteria results in an induction of receptor activator of nuclear factor-kappaB ligand (RANKL)-dependent periodontal bone resorption in a mouse model. *Oral Microbiol Immunol. 22*, 208-215.

[30] Rouvier, E., Luciani, M.F., Mattei, M.G., Denizot, F., & Golstein, P. (1993). CTLA-8, cloned from an activated T cell, bearing AU-rich messenger RNA instability sequences, and homologous to a herpesvirus saimiri gene. *J Immunol. 150*, 5445-5456.

[31] Fossiez, F., Djossou, O., Chomarat, P., Flores-Romo, L., Ait-Yahia, S., Maat, C., Pin, J.J., Garrone, P., Garcia, E., Saeland, S., Blanchard, D., Gaillard, C., Das Mahapatra, B., Rouvier, E., Golstein, P., Banchereau, J., & Lebecque, S. (1996). T cell interleukin-17 induces stromal cells to produce proinflammatory and hematopoietic cytokines. *J Exp Med. 183*, 2593-2603.

[32] Kotake, S., & Kamatani, N. (2002). The Role of IL-17 in Joint Destruction. *Drug News Perspect. 15*, 17-23.

[33] Novatchkova, M., Leibbrandt, A., Werzowa, J., Neubuser, A., & Eisenhaber, F. (2003). The STIR-domain superfamily in signal transduction, development and immunity. *Trends Biochem Sci. 28*, 226-229.

[34] Maitra, A., Shen, F., Hanel, W., Mossman, K., Tocker, J., Swart, D., & Gaffen, S.L. (2007). Distinct functional motifs within the IL-17 receptor regulate signal transduction and target gene expression. *Proc Natl Acad Sci U S A. 104*, 7506-7511.

[35] Kolls, J.K., & Linden, A. (2004). Interleukin-17 family members and inflammation. *Immunity. 21*, 467-476.

[36] Toy, D., Kugler, D., Wolfson, M., van den Bos, T., Gurgel, J., Derry, J., Tocker, J., & Peschon, J. (2006). Interleukin 17 signals through a heteromeric receptor complex. *J Immunol. 177*, 36-39.

[37] Kotake, S., Udagawa, N., Takahashi, N., Matsuzaki, K., Itoh, K., Ishiyama, S., Saito, S., Inoue, K., Kamatani, N., Gillespie, M.T., Martin, T.J., & Suda, T. (1999). IL-17 in synovial fluids from patients with rheumatoid arthritis is a potent stimulator of osteoclastogenesis. *J Clin Invest. 103*, 1345-1352.

[38] Yago, T., Nanke, Y., Furuya, T., Kobashigawa, T., Kamatani, N., & Kotake, S. (2004). Infliximab potently inhibits osteoclastogenesis from human monocytes alone induced by IL-17 in the absence of osteoblasts. *Arthritis Rheum 50 (supple)*, 662, S289 (abst).

[39] Lam, J., Takeshita, S., Barker, J.E., Kanagawa, O., Ross, F.P., & Teitelbaum, S.L. (2000). TNF-alpha induces osteoclastogenesis by direct stimulation of macrophages exposed to permissive levels of RANK ligand. *J Clin Invest. 106*, 1481-1488.

[40] Zou, W., Hakim, I., Tschoep, K., Endres, S., & Bar-Shavit, Z. (2001). Tumor necrosis factor-alpha mediates RANK ligand stimulation of osteoclast differentiation by an autocrine mechanism. *J Cell Biochem. 83*, 70-83.

[41] Lubberts, E., Koenders, M.I., Oppers-Walgreen, B., van den Bersselaar, L., Coenen-de Roo, C.J., Joosten, L.A., & van den Berg, W.B. (2004). Treatment with a neutralizing anti-murine interleukin-17 antibody after the onset of collagen-induced arthritis reduces joint inflammation, cartilage destruction, and bone erosion. *Arthritis Rheum. 50*, 650-659.

[42] Kotake, S. (2002). The role of IL-17 in joint destruction. The 11th International Rheumatology Symposium in Kobe. IS2-1 (abst).

[43] Nakae, S., Saijo, S., Horai, R., Sudo, K., Mori, S., & Iwakura, Y. (2003). IL-17 production from activated T cells is required for the spontaneous development of destructive arthritis in mice deficient in IL-1 receptor antagonist. *Proc Natl Acad Sci USA. 100*, 5986-5990.

[44] Sarkar, S., Tesmer, L.A., Hindnavis, V., Endres, J.L., & Fox, D.A. (2007). Interleukin-17 as a molecular target in immune-mediated arthritis: immunoregulatory properties of genetically modified murine dendritic cells that secrete interleukin-4. *Arthritis Rheum. 56*, 89-100.

[45] Raza, K., Falciani, F., Curnow, S.J., Ross, E.J., Lee, C.Y., Akbar, A.N., Lord, J.M., Gordon, C., Buckley, C.D., & Salmon, M. (2005). Early rheumatoid arthritis is characterized by a distinct and transient synovial fluid cytokine profile of T cell and stromal cell origin. *Arthritis Res Ther. 7*, R784-795.

[46] Murphy, C.A., Langrish, C.L., Chen, Y., Blumenschein, W., McClanahan, T., Kastelein, R.A., Sedgwick, J.D., & Cua, D.J. (2003). Divergent pro- and antiinflammatory roles for IL-23 and IL-12 in joint autoimmune inflammation. *J Exp Med. 198*, 1951-1957.

[47] Yago, T., Nanke, Y., Kawamoto, M., Furuya, T., Kobashigawa, T., Kamatani, N., & Kotake, S. (2005). Interleukin-23 induces human osteoclastogenesis in vitro, and anti-interleukin-23 antibody reduces joint inflammation in rats with collagen-induced arthritis. *Arthritis Rheum. 52* (supple), 1309, S496 (abst).

[48] Mosmann, T.R., Cherwinski, H., Bond, M.W., Giedlin, M.A., & Coffman, R.L. (1986). Two types of murine helper T cell clone. I. Definition according to profiles of lymphokine activities and secreted proteins. *J Immunol. 136*, 2348-2357.

[49] Furuzawa-Carballeda, J., Vargas-Rojas, M.I., & Cabral, A.R. (2007). Autoimmune inflammation from the Th17 perspective. *Autoimmun Rev. 6*, 169-175.

[50] Harrington, L.E., Hatton, R.D., Mangan, P.R., Turner, H., Murphy, T.L., Murphy, K.M., & Weaver, C.T. (2005). Interleukin 17-producing CD4+ effector T cells develop via a lineage distinct from the T helper type 1 and 2 lineages. *Nat Immunol. 6*, 1123-1132.

[51] Park, H., Li, Z., Yang, X.O., Chang, S.H., Nurieva, R., Wang, Y.H., Wang, Y., Hood, L., Zhu, Z., Tian, Q., & Dong, C. (2005). A distinct lineage of CD4 T cells regulates tissue inflammation by producing interleukin 17. *Nat Immunol. 6*, 1133-1141.

[52] Veldhoen, M., Hocking, R.J., Atkins, C.J., Locksley, R.M., & Stockinger, B. (2006). TGFbeta in the context of an inflammatory cytokine milieu supports de novo differentiation of IL-17-producing T cells. *Immunity. 24*, 179-89.

[53] Bettelli, E., Carrier, Y., Gao, W., Korn, T., Strom, T.B., Oukka, M., Weiner, H.L., Kuchroo, V.K. (2006). Reciprocal developmental pathways for the generation of pathogenic effector TH17 and regulatory T cells. *Nature. 441*, 235-238.

[54] Mangan, P.R., Harrington, L.E., O'Quinn, D.B., Helms, W.S., Bullard, D.C., Elson, C.O., Hatton, R.D., Wahl, S.M., Schoeb, T.R., & Weaver, C.T. (2006). Transforming growth factor-beta induces development of the T(H)17 lineage. *Nature. 441*, 231-234.

[55] Acosta-Rodriguez, E.V., Rivino, L., Geginat, J., Jarrossay, D., Gattorno, M., Lanzavecchia, A., Sallusto, F., & Napolitani, G. (2007). Surface phenotype and antigenic specificity of human interleukin 17-producing T helper memory cells. *Nat Immunol.* 8, 639-646.

[56] Ye, P., Rodriguez, F.H., Kanaly, S., Stocking, K.L., Schurr, J., Schwarzenberger, P., Oliver, P., Huang, W., Zhang, P., Zhang, J., Shellito, J.E., Bagby, G.J., Nelson, S., Charrier, K., Peschon, J.J., & Kolls, J.K. (2001). Requirement of interleukin 17 receptor signaling for lung CXC chemokine and granulocyte colony-stimulating factor expression, neutrophil recruitment, and host defense. *J Exp Med. 194*, 519-527.

[57] Happel, K.I., Dubin, P.J., Zheng, M., Ghilardi, N., Lockhart, C., Quinton, L.J., Odden, A.R., Shellito, J.E., Bagby, G.J., Nelson, S., & Kolls, J.K. (2005). Divergent roles of IL-23 and IL-12 in host defense against Klebsiella pneumoniae. *J Exp Med. 202*, 761-769.

[58] Rudner, X.L., Happel, K.I., Young, E.A., & Shellito, J.E. (2007). Interleukin-23 (IL-23)-IL-17 Cytokine Axis in Murine Pneumocystis carinii Infection. *Infect Immun. 75*, 3055-3061.

[59] Sato, K., Suematsu, A., Okamoto, K., Yamaguchi, A., Morishita, Y., Kadono, Y., Tanaka, S., Kodama, T., Akira, S., Iwakura, Y., Cua, D.J., & Takayanagi, H. (2006). Th17 functions as an osteoclastogenic helper T cell subset that links T cell activation and bone destruction. *J Exp Med. 203*, 2673-2682.

[60] Hirota, K., Hashimoto, M., Yoshitomi, H., Tanaka, S., Nomura, T., Yamaguchi, T., Iwakura, Y., Sakaguchi, N., & Sakaguchi, S. (2007). T cell self-reactivity forms a

cytokine milieu for spontaneous development of IL-17+ Th cells that cause autoimmune arthritis. *J Exp Med. 204*, 41-47.

[61] Takahashi, K., Azuma, T., Motohira, H., Kinane, D.F., & Kitetsu, S. (2005). The potential role of interleukin-17 in the immunopathology of periodontal disease. *J Clin Periodontol. 32*, 369-374.

[62] Oda, T., Yoshie, H., & Yamazaki, K. (2003). Porphyromonas gingivalis antigen preferentially stimulates T cells to express IL-17 but not receptor activator of NF-kappaB ligand in vitro. *Oral Microbiol Immunol.*18, 30-36.

[63] Fox, S.W., Fuller, K., Bayley, K.E, Lean, J.M., & Chambers, T.J. (2000). TGF-beta 1 and IFN-gamma direct macrophage activation by TNF-alpha to osteoclastic or cytocidal phenotype. *J Immunol. 165*, 4957-4963.

[64] Udagawa, N., Horwood, N.J., Elliott, J., Mackay, A., Owens, J., Okamura, H., Kurimoto, M., Chambers, T.J., Martin, T.J., & Gillespie, M.T. (1997). Interleukin-18 (interferon-gamma-inducing factor) is produced by osteoblasts and acts via granulocyte/macrophage colony-stimulating factor and not via interferon-gamma to inhibit osteoclast formation. *J Exp Med.185*, 1005-1012.

[65] Takayanagi, H., Ogasawara, K., Hida, S., Chiba, T., Murata, S., Sato, K., Takaoka, A., Yokochi, T., Oda, H., Tanaka, K., Nakamura, K., & Taniguchi, T. (2000). T-cell-mediated regulation of osteoclastogenesis by signalling cross-talk between RANKL and IFN-gamma. *Nature. 408*, 600-605.

[66] Gao, Y., Grassi, F., Ryan, M.R., Terauchi, M., Page, K., Yang, X., Weitzmann, M.N. & Pacifici, R. (2007). IFN-gamma stimulates osteoclast formation and bone loss in vivo via antigen-driven T cell activation. *J Clin Invest.117*, 122-132.

[67] Stashenko, P., Goncalves, R.B., Lipkin, B., Ficarelli, A., Sasaki, H., & Campos-Neto, A. (2007). Th1 immune response promotes severe bone resorption caused by Porphyromonas gingivalis. *Am J Pathol. 170*, 203-213.

[68] Kim, Y.G., Lee, C.K., Nah, S.S., Mun, S.H., Yoo, B., & Moon, H.B. (2007). Human CD4+CD25+ regulatory T cells inhibit the differentiation of osteoclasts from peripheral blood mononuclear cells. *Biochem Biophys Res Commun. 35*7, 1046-1052.

[69] Hartgers, F.C., Vissers, J.L., Looman, M.W., van Zoelen, C., Huffine, C., Figdor, C.G., & Adema, G.J. (2000). DC-STAMP, a novel multimembrane-spanning molecule preferentially expressed by dendritic cells. *Eur J Immunol. 30*, 3585-3590.

[70] Kukita, T., Wada, N., Kukita, A., Kakimoto, T., Sandra, F., Toh, K., Nagata, K., Iijima, T., Horiuchi, M., Matsusaki, H., Hieshima, K., Yoshie, O., & Nomiyama, H. (2004). RANKL-induced DC-STAMP is essential for osteoclastogenesis. *J Exp Med. 200*, 941-946.

[71] Yagi, M., Miyamoto, T., Sawatani, Y., Iwamoto, K., Hosogane, N., Fujita, N., Morita, K., Ninomiya, K., Suzuki, T., Miyamoto, K., Oike, Y., Takeya, M., Toyama, Y., & Suda, T. (2005). DC-STAMP is essential for cell-cell fusion in osteoclasts and foreign body giant cells. *J Exp Med. 202*, 345-351.

[72] Yago, T., Nanke, Y., Kawamoto, M., Furuya, T., Kobashigawa, T., Ichikawa, N., Kamatani, N., & Kotake, S. (2006). Antibodies against CD47 expressed on monocytes in rheumatoid arthritis patients inhibits human osteoclastogenesis by blocking fusion of monocytes. *Arthritis Rheum 50* (supple), 1394, S354 (abst).

[73] Lundberg, P., Koskinen, C., Baldock, P.A., Lothgren, H., Stenberg, A., Lerner, U.H., & Oldenborg, P.A. (2007). Osteoclast formation is strongly reduced both in vivo and in vitro in the absence of CD47/SIRPalpha-interaction. *Biochem Biophys Res Commun. 352*, 444-448.

[74] Hofbauer, L.C., Gori, F., Riggs, B.L., Lacey, D.L., Dunstan, C.R., Spelsberg, T.C., & Khosla, S. (1999). Stimulation of osteoprotegerin ligand and inhibition of osteoprotegerin production by glucocorticoids in human osteoblastic lineage cells: potential paracrine mechanisms of glucocorticoid-induced osteoporosis. *Endocrinology. 140*, 4382-4389.

[75] Sivagurunathan, S., Muir, M.M., Brennan, T.C., Seale, J.P., & Mason, R.S. (2005). Influence of glucocorticoids on human osteoclast generation and activity. *J Bone Miner Res. 20*, 390-398.

[76] Yago, T., Nanke, Y., Kawamoto, M., Furuya, T., Kobashigawa, T., Kamatani, N., & Kotake, S. (2006). Dexamethasone directly induce osteoclastogenesis from human PBMC by elevating the RANKL-to-IFN-γ ratio on T cells even in the absence of osteoblasts. –possible mechanism of corticosteroids in inducing osteoporosis-. *Arthritis Rheum. 50* (supple), 1277, S531 (abst).

[77] Kitahara, K., & Kawai, S. (2007). Cyclosporine and tacrolimus for the treatment of rheumatoid arthritis. *Curr Opin Rheumatol. 19*, 238-245.

[78] Yago, T., Nanke, Y., Kawamoto, M., Furuya, T., Kobashigawa, T., Kamatani, N., & Kotake, S. (2006). Tacrolimus potently inhibits dexamethasone-induced osteoclastogenesis from human PBMC. – possible role of tacrolimus in preventing bone destruction. *Arthritis Rheum. 50* (supple), 1393, S354 (abst).

[79] Igarashi, K., Hirotani, H., Woo, J.T., & Stern, P.H. (2004). Cyclosporine A and FK506 induce osteoclast apoptosis in mouse bone marrow cell cultures. *Bone. 35*, 47-56.

[80] Miyazaki, M., Fujikawa, Y., Takita, C., & Tsumura, H. (2007). Tacrolimus and cyclosporine A inhibit human osteoclast formation via targeting the calcineurin-dependent NFAT pathway and an activation pathway for c-Jun or MITF in rheumatoid arthritis. *Clin Rheumatol. 26*, 231-239.

[81] Koga, T., Matsui, Y., Asagiri, M., Kodama, T., de Crombrugghe, B., Nakashima, K., & Takayanagi, H. (2005). NFAT and Osterix cooperatively regulate bone formation. *Nat Med. 11*, 880-885.

[82] Xiong, D.H., Shen, H., Zhao, L.J., Xiao, P., Yang, T.L., Guo, Y., Wang, W., Guo, Y.F., Liu, Y.J., Recker, R.R., & Deng, H.W. (2006). Robust and comprehensive analysis of 20 osteoporosis candidate genes by very high-density single-nucleotide polymorphism screen among 405 white nuclear families identified significant association and gene-gene interaction. *J Bone Miner Res. 21*, 1678-1695.

[83] Mencej, S., Prezelj, J., Kocijancic, A., Ostanek, B., & Marc, J. (2006). Association of TNFSF11 gene promoter polymorphisms with bone mineral density in postmenopausal women. *Maturitas. 55*, 219-226.

[84] Hsu, Y.H., Niu, T., Terwedow, H.A., Xu, X., Feng, Y., Li, Z., Brain, J.D., Rosen, C.J., Laird, N., & Xu, X. (2006). Variation in genes involved in the RANKL/RANK/OPG bone remodeling pathway are associated with bone mineral density at different skeletal sites in men. *Hum Genet. 118*, 568-577.

[85] Wu, H., Khanna, D., Park, G., Gersuk, V., Nepom, G.T., Wong, W.K., Paulus, H.E., & Tsao, B.P. (2004). Interaction between RANKL and HLA-DRB1 genotypes may contribute to younger age at onset of seropositive rheumatoid arthritis in an inception cohort. *Arthritis Rheum. 50*, 3093-3103.

[86] Kawaguchi, M., Takahashi, D., Hizawa, N., Suzuki, S., Matsukura, S., Kokubu, F., Maeda, Y., Fukui, Y., Konno, S., Huang, S.K., Nishimura, M., & Adachi, M. R. (2006). IL-17F sequence variant (His161Arg) is associated with protection against asthma and antagonizes wild-type IL-17F activity. *J Allergy Clin Immunol. 117*, 795-801.

[87] Southam, L., Heath, O., Chapman, K., & Loughlin, J. (2006). Association analysis of the interleukin 17 genes IL17A and IL17F as potential osteoarthritis susceptibility loci. *Ann Rheum Dis. 65*, 556-557.

[88] Duerr, R.H., Taylor, K.D., Brant, S.R., Rioux, J.D., Silverberg, M.S., Daly, M.J., Steinhart, A.H., Abraham, C., Regueiro, M., Griffiths, A., Dassopoulos, T., Bitton, A., Yang, H., Targan, S., Datta, L.W., Kistner, E.O., Schumm, L.P., Lee, A.T., Gregersen, P.K., Barmada, M.M.,.Rotter, J.I., Nicolae, D.L., & Cho, J.H. (2006). A genome-wide association study identifies IL23R as an inflammatory bowel disease gene. *Science. 314*:1461-1463.

[89] Furuya, T., Ichikawa, N., Kogame, A., Nanke, Y., Higami, K., Hakoda, M., Kamatani, N., & Kotake, S. (2004). Association of polymorphisms in the receptor activator of nuclear factor-kappa B (RANK), RANK ligand, osteoprotegerin, and interleukin 17 genes with disease severity in Japanese patients with rheumatoid arthritis. *Arthritis Rheum. 50* (supple), 853, S354 (abst).

In: Cell Differentiation Research Developments
Editor: L. B. Ivanova, pp. 171-182
ISBN: 978-1-60021-939-9

Chapter VII

Cell Differentiation Research Trade: The Mechanism of Asexual Cell Differentiation in Cultures of Protoplasts, Interspecific Hybrid Embryos and Asexual Tissues

Lanzhuang Chen [1, 2, 3,*] ***and Liming Guan*** [1, 2]
[1]Faculty of Horticulture, Minami Kyushu University, Miyazaki, 884-0003, Japan;
[2]Molecular Biology Division, Frontier Science Research Center, University of Miyazaki, Miyazaki, 889-2192, Japan;
[3]College of Life and Environmental Sciences, Shanghai Normal University, Shanghai, 200234, China.

Abstract

As a sexual reproductive mode, a zygote after cell fusion between sperm and egg, as a destiny, certainly divides into embryo. If it is the dicots, the embryo passes the stages of globular, heart, torpedo and complete one with two cotyledons and hypocotyl that finally becomes a seed. Here, we describe the process of somatic embryogenesis from asexual cells derived from protoplast culture, protoplast fusion, interspecific hybrid embryo culture, and asexual tissue cultures. When the immature globular-stage embryos of tomato and its wild species were taken out and cultured on suitable medium, on which somatic embryos were formed with globular, heart and torpedo stages. When the cultures of cotyledon-derived protoplasts and protoplast fusion of tomato were carried out, somatic embryogenesis was observed, according to the stages from the globular and heart to torpedo. When leaflet tissues from guineagrass and bahiagrass were cultured on suitable medium, embryogenic calli were formed and they were observed by scanning

* Correspondence concerning this article should be addressed to: Lanzhuang Chen, lzchen@nankyudai.ac.jp.

electron microscopy, indicating somatic embryo structure with a scutellum and a coleoptile. The characterizations of the somatic embryo were similar to those previously reported. Multiple shoot and plant regeneration were obtained from somatic embryos. Those somatic embryos above were analyzed with RAPD, RFLP, chromosome, gene cloning and transgenic plants, and the results were also discussed in the text.

1. Introduction

Tomatoes and cereals are becoming increasingly popular ornamentals in Japan, the former as fruiting pot plants, the latter as dry bouquet supplements. Since the 1970's, plant tissue culture techniques have been developed as large-scale clonal propagation means from isolated plant parts. Later, protoplast isolation had also been successfully achieved from different plant tissues (Davey et al. 2005). A unique characteristic of plant tissues and cells is the presence of a cell wall, unlike animal cells, making cell fusion difficult. Successful protoplast isolation allowed for the breakthrough development of cell fusion (Kao and Michayluk 1989), allowing the manipulation of plant cells as easily as those of animals, resulting in a cell fusion boom, e.g. the "pomato" (potato + tomato) by protoplast fusion (Melchers et al. 1978). Cell fusion thus showed a great potential in that it became possible to cross any plant material, overcoming cross incompatibilities. Here, we structure the wider discussion around the knowledge of somatic embryogenesis we accumulated from our studies on *Lycopersicon esculentum* and *L. peruvianum*-complex, and *Panicum maximum* and *Paspalum notatum*, model systems we established, and consider additional factors that should be taken into account to induce somatic embryogenesis in different tissues of different crops (agronomic or horticultural).

2. Somaticembryogenesis from Tomato Tissue Culture

In the genus *Lycopersicon*, *L. peruvianum* is the most remote from the cultivated tomato *L. esculentum* and is regarded as the greatest single reservoir of genetic diversity available in tomato (Rick 1979). The postfertilization abortion (PFA) of hybrid embryos occurred (Barbano and Topoleski 1984, Chen and Imanishi 1991, Chen and Adachi 1992 1995) and hampered the introgression of genes from *L. peruvianum* to the cultivated tomato. And the mechanism of PFA of interspecific hybrid embryos (IHEs), was that the endosperm deterioration in the ovule was the first signal to appear at 7 days after pollination (DAP). However, IHEs continued to develop until to 15 DAP, while the sac was completely filled with proliferating endothelium. Based on observations of IHE PFAs with Nomarski differential interference-contrast microscopy (DIC), an embryo rescue technique (ERT) was developed to culture the IHEs formed before 15 DAP. They were taken out from the ovule and cultured on an improved HLH medium (Neal and Topoleski 1983) supplemented with 2 mg/l 6- benzylaminopurine (BAP) and 1g/l yeast extract (YE). Globular-stage embryos enlarged on it, developed root hairs, showed signs of developing cotyledons, and formed an early torpedo-stage-like structure after 30 days of culture. Green cotyledons were formed on the late torpedo-stage embryos after 3 weeks in culture. After transfer from HLH to MS medium (Murashige and Skoog 1962) supplemented with 1 mg/l zeatin and 0.02 mg/l indolacetic acid (IAA), numerous somatic embryoids and accessory single cotyledons were regenerated after a further 3 weeks followed by the formation of a number of shoots (Figure 1). Shoots were cut individually and transplanted onto hormone-free MS agar for rooting and plantlet regeneration (Chen and Adachi 1992; 1998). Using this method, not only somatic embryogenesis and plant regeneration but also clonal propagation of IHE became possible. The possible cultural stage of IHE established in this study was about 2-3 weeks earlier than previous reports (Smith 1944, Neal and Topoleski 1983).

Figure 1. Somatic embryogenesis with embryo-like structures and single shoot on immature embryo (arrows) of interspecific hybrid between *Lycopersicon esculentum* and *L. peruvianum*. The embryo is aged 13 days after pollination and taken out from immature fruit and cultured on HLH medium.

3. Somatic Embryogenesis from Process of Protoplast Fusion

Cotyledons are an easily reproducible and physiologically stable tissue, giving protoplasts of distinguishable morphology (Koblitz and Koblitz 1982, Shahin 1985). We previously studied conditions affecting the plating efficiency of cotyledon protoplasts (Chen *et al.* 1987 1988). Based on that, we established a culture system of tomato cotyledons (Chen and Adachi 1994). Protoplasts were isolated from cotyledons of 8-10 day old *in vitro*grown seedlings Fusion between leaf and suspension-culture-derived protoplasts has been the major method employed (O'Connell and Hanson 1985 1986). In addition, two types of leaf protoplasts (Wijbrandi *et al.* 1990), or leaf and irradiated leaf protoplasts (Ralushnyak *et al.* 1991) have also been utilized as fusion partners. Two ways of protoplast fusion have mainly been conducted with chemicals, e.g. polyethylene glycol (PEG) and electro-fusion. Two methods have been reported for the selection of parasexual hybrids: (1) prefusion treatments of protoplasts, like iodoacetamide (IOA) treatment of one partner (Ozias-Akins *et al.* 1986, Terada *et al.* 1987) or gamma-irradiation (Ralushnyak *et al.* 1991); (2) postfusion selection of fused protoplasts by antibiotics, such as adding kanamycin to culture media for hybrid selection (Wijbrandi *et al.* 1990), or DNA analysis of calli or leaves of parasexual hybrids (Derks *et al.* 1991, Wachocki *et al.* 1991). Chen and Adachi (1998) and Chen et al. (2001, 2002) have established a culture system for plant regeneration via somatic embryogenesis from a new combination of protoplast fusion partners from cotyledon protoplasts of *L. esculentum* and suspension-culture-derived protoplasts of the *L. peruvianum*-complex and obtained somatic hybrids with a multiple step selection protocol. Cotyledon protoplasts treated with IOA (10 mM, 15 min at 4°C) gave an optimal effect to prevent cell division when used as a fusion partner. Since two kinds of protoplasts, either white in color (wild species) or green (tomato) were used, we were able to visually identify heterokaryon formation. Pre-selection could be based on the collapse of cotyledon protoplasts after IOA treatment before fusion or after PEG treatment alone. The TM medium for fusion culture is suitable for cotyledon protoplast culture, but for the suspension culture- derived protoplasts, only fused protoplasts can divide. During culture, two types of somatic embryogenesis were observed under microscope. The first type on the surface of calli are globular structured and cup-shaped somatic embryos with tube-like protuberances. Embryo developmental stages including globular, heart, early torpedo and late torpedo were observed (Figure 2). The other type was characterized by the production of multiple or single somatic embryos at various stages, light purple and/or deep-green in color, which grow towards the medium (Chen and Adachi 1998, Chen *et al.* 1998 2002a 2002b). Embryo ontogeny was asynchronous; globular-through cotyledon-stage embryos were often observed on a single callus. A dome-shaped shoot meristem and double cotyledons were often seen together. When the somatic embryos were placed onto fresh TM-4 medium, they grew into shoots with double cotyledons, a hypocotyl and a shoot meristem. In the process of plant regeneration from protoplast fusion, a typical development of sexual embryos under natural conditions *in vivo* was observed with a series of embryogenesis steps: globular, heart, early torpedo, and late torpedo stages. Therefore, this established system can be used not only in plant regeneration by somatic embryogenesis but also in genetic transformation and clonal propagation of rare genetic resources through plant tissue culture.

4. Somatic Embryogenesis from Guineagrass Tissue Culture

A method to transform rice by *Agrobacterium* was reported by Hiei *et al.* (1994). This method had the following advantages compared with direct gene transformation: no need for special instruments, simple operation, and high repeatability. This method thus showed efficiency for monocot plants. Several explants had been applied for plant regeneration, including immature embryos in guinea grass (Akashi and Adachi 1991), ovary culture (Chen *et al.* 2002c 2002d), immature inflorescence in finger millet (George and Eapen1990), protoplasts in barley (Lazzeri and Lörz 1990), or mature seeds in rice (Hiei *et al.* 1994). These culture systems usually gave regeneration via somatic embryogenesis. However, there are other easy-to-use reproductive tissues, e.g. ovary, as well as vegetative tissues, which can be collected in all seasons, but such protocols have not yet been well established. Here, we describe some regeneration systems through somatic embryogenesis established from ovaries and leaflets of guinea grass (*Panicum maximum*).

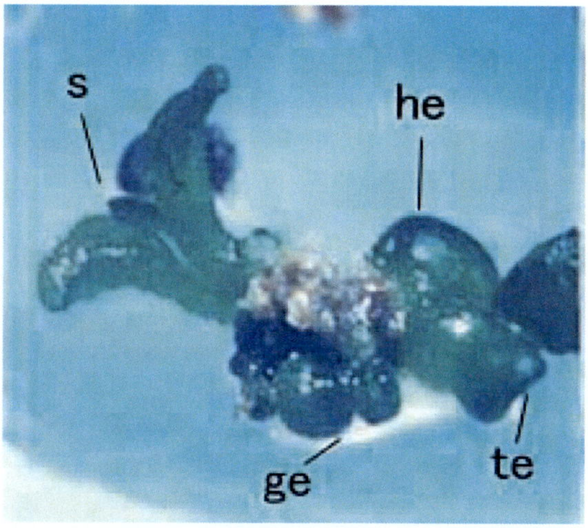

Figure 2. Somatic embryogenesis with different stages from fused protoplasts between *Lycopersicon esculentum* and *L. peruvianum*, displaying young shoot (s), globular-stage embryo (ge), Heart-stage embryo (he) and torpedo-stage embryo (te).

We were interested in establishing a system of somatic embryogenesis and plant regeneration from leaflets of warm-season forage grasses, which could be used as materials in any season and at any time for genetic transformation. A simple and efficient culture of leaflets for plant regeneration in guinea grass (Chen *et al.* 2001a 2001b 2002c 2002d) was established. For the material preparation of guinea grass, leaflet tissue was taken from terminal tillers arising at the end of stolons. Leaflets (2-3 cm) were isolated by trimming the leaves and excising the parts of the leaf-stem tissue 1 cm above the root base which contains the basal meristem and basal leaf tissue. The outer 4-5 leaves were removed. The remaining tissues were cut into 5- mm lengths and washed in streaming tap water for 20-30 min. The explants were transferred to 70% ethanol for 1 min, sterilized in 0.15% sodium hypochlorite solution (Antiformin) for 15 min and rinsed in sterile distilled water three times, 5 min each time. The explants were then placed on MS medium supplemented with 5 mg/l 2,4-D, 30 g/l sucrose, and 2.5 g/l Gellan gum (Wako Pure Chemical Industries, Ltd., Japan) for callus formation with one of the cut surfaces in contact with the medium. After 3 weeks culture, calli were formed in a yellow liquid paste, appearing contaminated. However, culturing for a further 3 resulted in the loss of the yellow color, which may be a characteristic of bahia grass. During the 6th-9th week of culture, formation of nodular (NO) and white color embryogenic calli on the surface of the yellowish-white color calli was observed. In this stage, the structures of embryogenic calli were confirmed with SEM, indicating NO tissue, and SC, CO and germination of the plumule (P) (Figures 3, 4). Those structures were similar to previous reports of *P. maximum* (Lu and Vasil 1985, Chen *et al.* 2002c 2002d) and *P. notatum* (Marousky and West 1990). In preliminary experiments, when the NO and white color calli were moved onto regeneration medium for culture, somatic embryos were initiated. So the calli were termed embryogenic calli. Small size and white color calli were selected and propagated inthe same medium by subculture every two weeks. The embryogenic calli formed anthocyanin pigment and green spots emerged from them. The P grew to multiple shoots after another 3 weeks. The complete plants were regenerated from the shoots on hormone-free MS medium in a total of 4 months (Chen *et al.* 2001a 2001b). A simple and efficient culture system of leaflets from guinea grass was also established in similar manner to bahia grass (Chen *et al.* 2002c 2002d). Recently, attempts at inducing an apomixis-specific gene (Chen *et al.* 1995 1999 2005, Seo *et al.* 2004) by *Agrobacterium* transformation into the selected yellowish calli have been carried out (Chen *et al.* 2003).

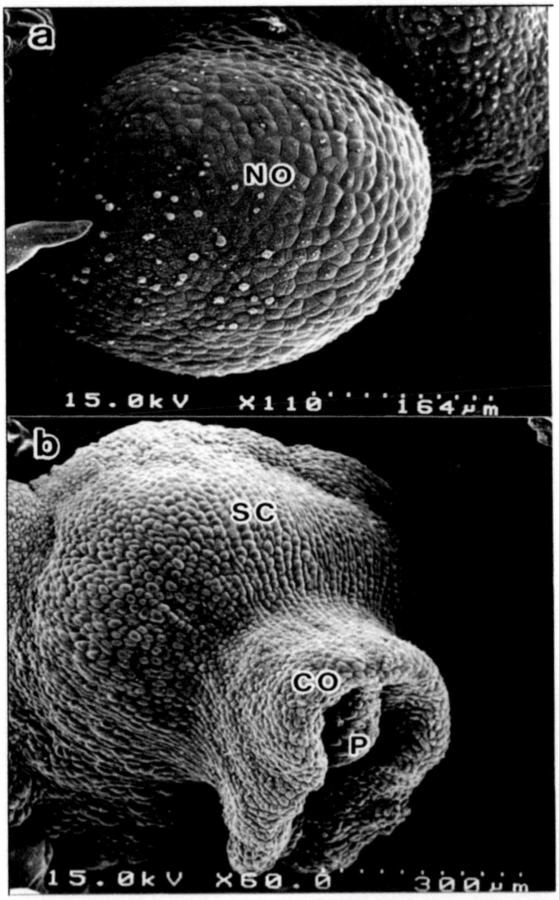

Figure 3. Somatic embryogenesis from leaflet culture of guinea grass (*Panicum maximum*), observed with scanning electron microscopy (SEM). a. Nodular embryogenic tissue (NO) occurred at earlier culture stages; b. Scutellum (SC) and coleoptile (CO) and germination of plumule (P) were generated on the culture.

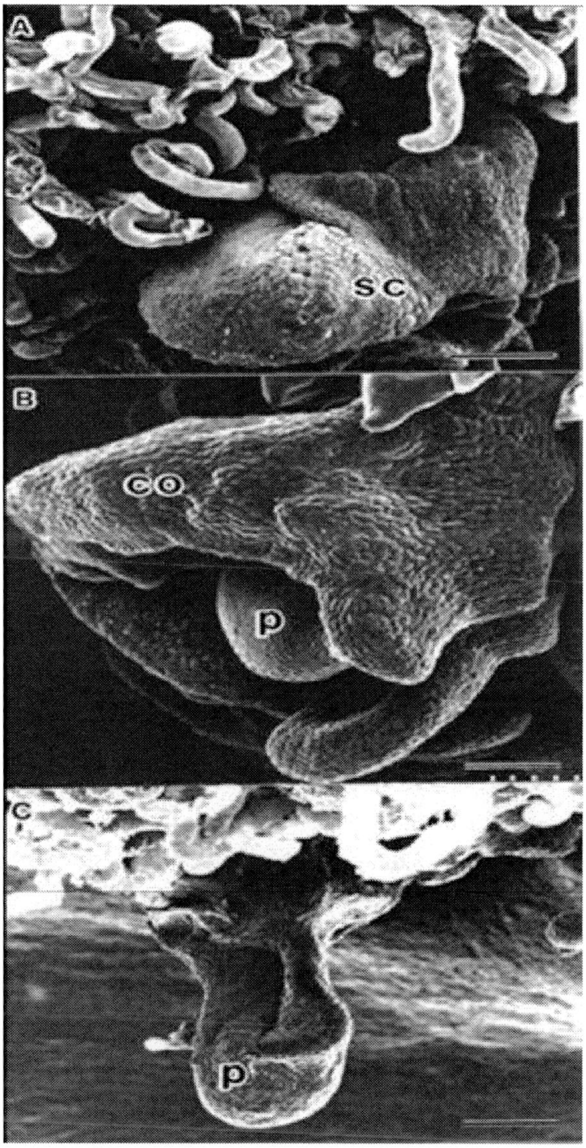

Figure 4. Somatic embryogenesis from leaflet culture of bahia grass (*Paspalum notatum*), observed with scanning electron microscopy (SEM). Scutellum (SC) (A), coleoptile (CO), and plumule (P) (B) and germination of plumule (P) (C) were observed as the callus growing and differentiation.

5. Detection of Interspecific Hybrids and Somatic Hybrids

For the detection of interspecific hybrids, morphological comparison of hybrid and parental leaves and flowers in *L. esculentum* X *L. peruvianum* have already been made by Thomas and Pratt (1981) and Imanishi (1988). Our results indicated that the regenerated plants showed morphological characteristics more similar to the male parent, *L. peruvianum*, than the female one, *L. esculentum*, but in any case revealed the hybrid nature of the interspecific crosses. It is found that no changes of chromosome number between hybrids and parents were observed. From that, it is suggested that the hybrids were derived directly from the embryo without passing through callus formation. The absence of the callus formation during the embryo-to-plant *in vitro* development was also checked under a dissecting microscope. *L. peruvianum* was superior to *L. esculentum* in shoot regeneration ability (Koornneef et al. 1087). Our observation was used to determine the hybrid nature by comparing their shoot regeneration ability, indicating that the number of shoots regenerated from hybrids ranged between those of their parents in the interspecific cross. Based on this method, the regenerants were true hybrids. This method, therefore, might be used as a fast and simple, but reliable method for the characterization of interspecific hybrids, not only in tomato but also in other crops.

For identification of somatic hybrids, the first choice is to compare the morphology as it is easy and visual able method. This method has been used in L. esculentum (+) L. peruvianum (Kinsara et al. 1986; San et al. 1990; Sakata et al. 1991; Giddings and Rees 1992a, b; Sakata and Monma 1993). In our study, the leaves of the somatic hybrids showed few wrinkles, no leaflets and long hairs. The chromosome number of hybrids showed expected $2n = 4x = 48$ in most of regenerants, but in few plants, their number were between 69-72 chromosomes. RAPD analysis was popular in detection of hybrids of interspecific and somatic hybridization (Xu et al. 1993; Hossain et al. 1994; Chen et al. 1998). The fact that different band patterns were observed when the same one primer, indicates a wide range of genetic changes among the hybrids. That may be due to somaclonal variation (Xu et al. 1993), and IOA treatment (Ozias-Akins et al. 1986), and re-arrangement and amplification of organelle DNA (Rode et al. 1988). RFLP analysis for identification of hybrids is considered as most believable methods. When we sued *Bam* HI to digest DNA and nuclear rRNA gene, *pRR*217 as a hybridization probe to determine the hybrids, all the putative somatic hybrids tested showed the fragments from both parents, as they carried 48 chromosomes ($2n = 4x$). From RFLP analysis, it is considered that tetraploid hybrids contain two partners' fragments, only different partner protoplasts are fused, and pre-selection using IOA treatment and culture medium regulation are efficient to allow somatic hybrids develop.

6. Conclusion

A practical culture system was established with somatic culture of interspecific hybrid immature embryos, cotyledon protoplast and protoplast fusion between *Lycopersicon esculentum* and '*Lycopersicon peruvianum*-complex' in order to obtain interspecific hybrids and somatic hybrids. On the other hand, to establish an efficient culture system for genetic transformation, somatic embryogenesis and plant regeneration have been successfully established from the culture of immature ovaries, leaflets and leaflet culture-derived suspension cultures in monocots (guineagrass and bahiagrass).

ACKNOWLEDGEMENTS

We are indebted to Tetsuro Kinoshita, Makoto Inoue, Kaoru Murai, Eiko Anami, Sachiko Amano, Reiji Okabe, Takuro Hamaguchi, and Taiji Adachi, for their advice, technique assistance and encouragement during the researches on somatic embryogenesis of tomato, guinea grass and bahia grass. Research in LZ C'slaboratory on apomixis has been partly supported by Science Research Grant № 10660010, 14560005 and 17380005, from the Ministry of Education, Science, Sports and Culture of Japan.

REFERENCES

Akashi R, Adachi T (1991) High frequency somatic embryo formation in cultures of immature embryos of guineagrass, *Panicum maximum Jacq. Japanese Journal of Breeding 41*, 85-93

Barbano PP, Topoleski LD (1984) Postfertilization hybrid seed failure in *Lycopersicon esculentum* x *Lycopersicon peruvianum* ovules. *Journal of the American Society for Horticultural Science 109*, 95-100

Chen LZ, Adachi T (1992) Embryo abortion and efficient rescue in interspecific hybrids, *Lycopersicon esculentum* and the '*peruvianum*-complex'. *Japanese Journal of Breeding 42*, 65-77

Chen LZ, Adachi T (1994) Plant regeneration via somatic embryogenesis from cotyledon protoplasts of tomato (*Lycopersicon esculentum* Mill.). *Breeding Science 44*, 257-262

Chen LZ, Adachi T (1995) Mechanism of abortion of postfertilization hybrid embryo in interspecific backcross, *Lycopersicon esculentum* x (*L. esculentum* x *L. peruvianum*). *Cytologia 60*, 123-131

Chen LZ, Adachi T (1996) Efficient hybridization between *Lycopersicon esculentum* and *L. peruvianum* via 'embryo rescue' and *in vitro* propagation. *Plant Breeding 115*, 251-256

Chen LZ, Adachi T (1998) Protoplast fusion between *Lycopersicon esculentum* (Mill.) and *Lycopersicon peruvianum*-complex: Somatic embryogenesis, plant regeneration and morphology. Plant Cell Reports *17*, 508-514

Chen *et al.* Somatic embryogenesis, plant regeneration and clonal propagation in plants morphology. *Plant Cell Reports 17*, 508-514

Chen LZ, Amano S, Guan LM, Adachi T (2005) Preliminary isolation of viable embryo sacs and protoplasts from viable egg cells in facultative apomictic guineagrass (*Panicum maximum*). *Cytologia 70*, 171-179

Chen LZ, Anami E, Guan LM, Adachi T (2001a) Somatic embryogenesis and plant regeneration from leaflets of "Nanou" bahiagrass. *Plant Biotechnology 18*, 58-65

Chen LZ, Anami E, Guan LM, Adachi T (2001b) Establishment of embryogenic suspension culture derived from leaflets of sexual bahiagrass (*Paspalum notatum*) with regeneration ability in long term. *Plant Biotechnology 18*, 209-214

Chen LZ, Guan LM, Seo MK, Hoffmann F, Adachi T (2005) Developmental expression of ASG-1 during gametogenesis in apomictic guinea grass (*Panicum maximum*). *Journal of Plant Physiology 162*, 1141-1148

Chen LZ, Imanishi S, Hiura I (1987) Culture of cotyledon protoplast of *Lycopersicon esculentum* – Effect of pre-treatment before the isolation of protoplasts on the yield and cell division of protoplasts. *Journal of Yamagata Agricuture and Forest Society 44*, 65-69

Chen LZ, Imanishi S, Hiura I (1988) Effects of temperature, isolation method and medium on tomato cotyledon protoplast division and colony formation. *Journal of Yamagata Agriculture and Forest Society 45*, 19-23

Chen LZ, Imanishi S (1991) Cross-compatibility between the cultivated tomato *Lycopersicon esculentum* and the wild species *L. peruvianum, L. chilense* assessed by ovule culture in vitro. *Japanese Journal of Breeding 41*, 223-230

Chen LZ, Kinoshita T, Kishima Y, Akashi R, Adachi T (1998) Plant regeneration via somatic embryogenesis after somatic hybridization between *Lycopersicon esculentum* and *L. peruvianum*. *SABRAO Journal of Breeding and Genetics 30*, 25-34

Chen LZ, Kozono T (1994) Cytology and quantitative analysis of aposporous embryo sac development in guineagrass (*Panicum maximum* Jacq.). *Cytologia* 59, 253-260

Chen LZ, Kozono T (1994) Cytological evidence of seed-forming embryo development in polyembryonic ovules of facultatively apomictic guineagrass (*Panicum maximum* Jacq.). *Cytologia 59*, 351-359

Chen LZ, Miyazaki C, Kojima A, Saito A, Adachi T (1999) Isolation and characterization of a gene expressed during early embryo sac development in apomictic guineagrass (*Panicum maximum*). *Journal of Plant Physiology 154*, 55-62

Chen LZ, Murai K, Inoue M, Kaneko Y, Sato Y, Adachi T (2002a) Somatic hybrids between *Lycopersicon esculentum* and *Lycopersicon chmielewskii*. *Plant Biotechnology 19*, 389-396

Chen LZ, Murai K, Inoue M, Kaneko Y, Tsutsumi T, Adachi T (2002b) Somatic hybrids with low temperature tolerance between *Lycopersicon esculentum* and *Lycopersicon hirsutum*. *SABRAO Journal of Breeding and Genetics 34*, 83-94

Chen LZ, Okabe R, Guan LM, Adachi T (2002c) A simple and efficient culture of leaflets for plant regeneration in guineagrass (*Panicum maximum*). *Plant Biotechnology 19*, 63-68

Chen LZ, Okabe R, Hamaguchi T, Guan LM, Adachi T (2002d) Effect of harvest seasons on the efficiency of ovary culture in *Panicum maximum*. *Plant Biotechnology 19*, 173-179

Chu CC, Wang CC, Sun CS, Hsu C, Yin KC, Chu CY, Bi FY (1975) Establishment of an efficient medium for anther culture of rice through comparative experiments on the nitrogen source. *Science Sinica 18*, 659-668

Davey MR, Anthony P, Power JB, Lowe KC (2005) Plant protoplast: status and biotechnological perspectives. *Biotechnology Advance 23*, 131-171

Derks FMM, Wijbrandi J, Koornneef M, Colinhooymans CM (1991) Organelle analysis of symmetric and asymmetric hybrids between *Lycopersicon peruvianum* and *Lycopersicon esculentum*. *Theoretical and Applied Genetics 81*, 199-204

Formm ME, Morrish F, Armstrong C, Williams R, Thomas J, Klein TM (1990) Inheritance and expression of chimeric genes in the progeny of transgenic maize plants. *Biotechnology 8*, 833-839

George J, Eapen S (1990) High frequency plant-regeneration through direct shoot development and somatic embryogenesis from immature inflorescence culture of finger millet (*Eleusine coracana* Gaertn). *Euphytica 48*, 269-274

Gradziel TM, Robinson RW (1989) Breakdown of self-incompatibility during pistil development in *Lycopersicon peruvianum* by modified bud pollination. *Sexual Plant Reproduction 2*, 38-42

Gradziel TM, Robinson RW (1991) Overcoming unilateral breeding barriers between *Lycopersicon peruvianum* and cultivated tomato *Lycopersicon esculentum. Euphytica 54*, 1-9

Haccius B (1978) Question of unicellular origin of nonzygotic embryos in callus culture. *Phytomorphology 28,* 74-81

Hiei Y, Ohta S, Komari T, Kumashiro T (1994) Efficient transformation of rice (*Oryza sativa* L.) mediated by Agrobacterium and sequence analysis of the boundaries of the T-DNA. *The Plant Journal 6*, 271-282

Hogenboom NG (1972a) Breaking breeding barriers in *Lycopersicon*. I. The genus *Lycopersicon*, its breeding barriers and the importance of breaking these barriers. *Euphytica 21*, 221-227

Hogenboom NG (1972b) Breaking breeding barriers in *Lycopersicon*. III. Inheritance of self-incompatibility in *Lycopersicon peruvianum* (L.) Mill. *Euphytica 21*, 244-256

Imanishi S, Suto Y (1987) Plant regeneration from cotyledon protoplasts of *Lycopersicon pimpinellifolium*. *Japanese Journal of Breeding 37*, 199-202

Kao KN, Michayluk MR (1989) Fusion of plant protoplasts - Techniques. In: Bajaj YPS (ed) *Biotechnology in Agriculture and Forestry (vol 8) Plant Protoplasts and Gnetic Engineering I*, Springer-Verlag, Berlin, pp 276-288

Koblitz H, Koblitz D (1982) Experiments on tissue culture in the genus Lycopersicon Miller-mesophyll protoplast regeneration to plants in *Lycopersicon esculentum* cv. 'Nadja'. *Plant Cell Reports 1*, 143-146

Komatsu T, Yamakata M, Hakuzan R, Doi O, Ueno K, Nakatani T, Mochidome N, Tsuyushige M, Tsurumi Y, Houman M, Shinya A (1991) A new variety 'nan-ou' bahia grass (*Papspalum notatum* Flugge). *Kyushu Agricultural Research* p 159

Lazzeri PA, Lörz H (1990) Regenerable suspension and protoplast culture of barley and stable transformation via DNA uptake into protoplasts. In: Lycett GW, Grieson D (eds) *Genetic Engineering in Crop Plants*, Butterworths, London, pp 231-238

Lu C, Vasil IK (1985) Histology of somatic embryogenesis in *Panicum maximum* (guineagrass). *American Journal of Botany 72*, 1908-1913

Marousky FJ, West SH (1990) Somatic embryogenesis and plant regeneration from cultured mature caryopses of bahiagrass (*Paspalum notatum* Flugge.) *Plant Cell, Tissue and Organ Culture 20*, 125-129

Melchers G, Sacristan MD, Holder AA (1978) Somatic hybrid plants of potato and tomato regenerated from fused protoplasts. *Carlsberg Research Communication 43*, 203-208

Murashige T, Skoog F (1962) A revised medium for rapid growth and bioassays with tobacco tissue culture. *Physiologia Plantarum 15*, 473-479

Neal CA, Topoleski LD (1983) Effects of the basal medium on growth of immature tomato embryos *in vitro*. *Journal of American Society of Horticultural Science 108*, 434-438

O'Connell MA, Hanson MR (1985) Somatic hybridization between *Lycopersicon esculentum* and *Lycopersicon pennellii. Theoretical and Applied Genetics 70*, 1-12

O'Connell MA, Hanson MR (1986) Regeneration of somatic hybrid plants formed between *Lycopersicon esculentum* and *Solanum rickii. Theoretical and Applied Genetics 72*, 59-65

Ozias-Akins P, Ferl RJ, Vasil IK (1986) Somatic hybridization in the Gramineae: *Pennisetum aericanum* (L.) K. Schum. (peal millet) + *Panicum maximum* Jacq. (guinea grass). *Molecular General Genetics 203*, 365-370

Ralushnyak YI, Latypov SA, Samoylov AM, Piven NM, Gleba YY (1991) Introgressive hybridization of tomatoes by 'gamma-fusion' of *Lycopersicon esculentum* Mill. And *Lycopersicon peruvianum* var. dentatum Dun. protoplasts. *Plant Science 73*, 65-78

Rick CM (1979) Biosystemic studies in *Lycopersicon* and closely related species of Solanum. In Hawkes JG, Lester RN, Skelding AD (eds), *The Biology and Taxonomy of the Solanaceae,* Academic Press London, pp 667-677

Seo MK, Chen LZ, Lee HY, Adachi T (2004) Efficient *Agrobacterium*-mediated transformation of ASG-1 gene from guinea grass (*Panicum maximum*) to rice. *Applied Biological Sci*ence *8*, 34-46

Shahin EA (1985) Totipotency of tomato protoplasts. *Theoretical and Applied Genetics 69*, 235-240

Shatters RG, Wheeler RA, West SH (1994) Somatic embryogenesis and plant regeneration from callus cultures of "Tifton 9" bahiagrass. *Crop Science 34*, 1378-1384

Smith PG (1944) Embryo culture of a tomato species hybrid. *Proceedings of American Society of Horticultural Science 44*, 413-416

Terada R, Yamashita Y, Nishibayashi S, Shimamoto K (1987) Somatic hybrids between *Brassica oleracea* and *B. campestris*: selection by the use of iodoacetamide inactivation and regeneration ability. *Theoretical and Applied Genetics 73*, 379-384

Tan MLMC, Colijn-Hooymans CM, Lindhout WH, Kool AJ (1987) A comparison of shoot regeneration from protoplasts and leaf discs of different genotypes of the cultivated tomato. *Theoretical and Applied Genetics 75*, 105-108

Vasil V, Castillo AM, Fromm ME, Vasil IK (1992) Herbicide resistant fertile transgenic wheat plants obtained by microprojectile bombardment of regenerable embryogenic callus. *Biotechnology 1*0, 667-674

Wachocki SE, Bonnema AB, O'Connell MA (1991) Comparison of the organization of the mitochondrial genome in tomato somatic hybrids and cybrids. *Theoretical and Tpplied Genetics 81*, 420-427

Wijbrandi J, Posthuma A, Kok JM, Rijken R, Vos JGM, Koornneef M (1990) Asymmetric somatic hybrids between *Lycopersicon esculentum* and irradiated *Lycopersicon peruvianum*. *Theoretical and Applied Genetics 80*, 305-312

Zapata FJ, Evans PK, Power JB, Cocking EC (1977) The effect of temperature on the division of leaf protoplasts of *Lycopersicon esculentum* and *Lycopersicon peruvianum. Plant Science 8*, 119-124

Zhong H, Bolyard MG, Srinivasan C, Sticklen MB (1993) Transgenic plants of turfgrass (*Agrostis palustric* Huds.) from microprojectile bombardment of embryogenic callus. *Plant Cell Reports 13*, 1-6.

In: Cell Differentiation Research Developments
Editor: L. B. Ivanova, pp. 183-199
ISBN: 978-1-60021-939-9

Chapter VIII

Migration, Differentiation and Functional Role of Monocyte Subsets

Frank Tacke*

Medical Clinic III, University Hospital Aachen, Germany.

Abstract

Monocytes represent about 5-10% of blood leukocytes in mice or men and are established circulating precursors for tissue macrophages and dendritic cells (DCs). Monocyte-derived macrophages and DCs fulfill critical roles in innate and adaptive immunity during inflammation and inflammatory disorders, and it is believed that monocytes also maintain these populations in some peripheral tissues during homeostasis. Recent identification of mouse monocyte subsets that closely resemble human monocyte subsets has inspired a variety of techniques wherein monocytes can be readily traced *in vivo* to address these critical questions. There are two major monocyte subsets that vary in chemokine receptor (CCR) and adhesion molecule expression, migratory and differentiation properties. In humans, 'classical' $CD14^+$ $CD16^-$ monocytes express CCR2, CD64, CD62L, whereas 'non-classical' $CD14^{low}$ $CD16^+$ monocytes lack CCR2. Their counterparts in mice are $CCR2^+$ Gr-1^{hi} and $CCR2^-$ Gr-1^{low} monocytes, respectively. Gr-1^{hi} ($Ly6C^{hi}$) monocytes are recruited to inflammatory sites, e.g. inflamed skin or acutely inflamed peritoneum. Gr-1^{hi} ($Ly6C^{hi}$) monocytes are precursors for the epidermal DCs, Langerhans cells, after skin inflammation. Gr-1^{low} monocytes have been proposed as precursors for steady state DCs, but experimental evidence is as of yet limited. In atherosclerosis, a chronic inflammatory disease of the blood vessels, both subsets differ in their migratory capacity, chemokine receptor pathways and differentiation towards macrophages or DCs. The rapid progress in recent years led to a better understanding of monocyte biology in the steady state and inflammation.

* Correspondence concerning this article should be addressed to: Dr. Frank Tacke, M.D., PhD, RWTH-University Hospital Aachen, Medical Clinic III, Pauwelsstr. 30, 52074 Aachen, Germany. Phone: +49 241-8035848; Fax: +49 241-8082455; Email: frank.tacke@gmx.net.

INTRODUCTION

In humans and mice, monocytes represent about 5-10% of peripheral blood leukocytes. They originate from a myeloid precursor in the bone marrow [1], are released in the circulation and then enter tissues. The half-life of monocytes in blood is believed to be relatively short, about one day in mice [2] and 3 days in humans [3]. This short half-life in blood has fostered the concept that blood monocytes may continuously repopulate macrophage or dendritic cell (DC) populations to maintain homeostasis and, during inflammation, fulfill critical roles in innate and adaptive immunity [4]. However, a body of evidence suggests that macrophages in several different organs self-renew without input from blood precursors [5]. For example, pulmonary and airway macrophages are not replaced by blood monocytes following procedures involving irradiation if care is taken to protect the viability of local, radiosensitive precursors [6], and correspondingly, studies in parabiotic mice indicate failure of blood monocytes to give rise to lung macrophages in the steady state [7]. Moreover, depletion of blood monocytes does not affect alveolar macrophage counts [8]. Indeed, replenishment of pulmonary macrophages from blood monocytes is mainly only observed after whole body irradition that would eliminate a radiosensitive, local precursor [5]. Nonetheless, the fact that monocytes give rise to macrophages in any virtually any organ in the context of inflammation is undisputed.

Recent research has focused on the mechanisms and molecules involved in monocyte recruitment as well as on their differentiation in various tissues [5,9]. Two major fates of monocyte differentiation are easily identified and separable: many monocytes develop into macrophages and others become dendritic cells (DCs) [10]. Functionally, macrophages robustly degrade material that they engulf [11] and apparently fail to present antigens or initiate a T cell response [12-14], release a variety of cytokines upon activation and are thought to be important for the local clearance of dead cells in inflamed and non-inflamed tissues [15]. In contrast, DCs have a poor proteolytic capacity [11], but are able to phagocytize material from other cells, process and intracellularly retain peptides / antigens, that will be subsequently presented by MHC-molecules together with appropriate costimulation for the priming of T cells [16]. After DC maturation, the plasma membrane of DCs turns over slowly allowing long duration of MHC/peptide complexes that facilitate long signaling interactions with T cells [17]. In contrast, activated macrophages very rapidly renew plasma membrane proteins, including MHC II, such that particular MHC/peptide complexes would be expected to a have a rather short half-life at the cell surface [18]. Moreover, macrophages and DCs have distinct migratory properties: whereas macrophages largely remain in tissues, DCs are capable of migrating from peripheral tissues to lymphoid organs in order to interact with T cells and induce an immune response [19].

HUMAN MONOCYTE SUBSETS

Over fifteen years ago, heterogeneity among human monocytes was recognized [20]. The differential expression of CD14 (part of the receptor for lipopolysaccharide) and CD16 (also known as FcγRIII) were traced initially to define two major subsets in peripheral blood: the so called 'classical' $CD14^+CD16^-$ monocytes, typically representing up to 95% of the monocytes in a healthy individual, and the 'non-classical' $CD14^{low}CD16^+$ monocytes comprising the remaining fraction of monocytes [20,21]. These subsets differ in many respects, including adhesion molecule and chemokine receptor (CCR) expression. $CD14^+CD16^-$ monocytes express CCR2, CD62L (L-Selectin) and FCγRI (CD64), whereas $CD14^{lo}CD16^+$ monocytes lack CCR2, and have higher levels of MHC-II and FCγRII (CD32) [5,21-24]. Both subsets express the receptor for fractalkine, CX_3CR1, but $CD14^{lo}CD16^+$ monocytes characteristically express higher levels [23,24]. It is important to note that considerable heterogeneity is known to exist in the minor $CD16^+$ monocyte fraction. An "intermediate" population of monocytes that is $CD14^+$ (as opposed to $CD14^{lo}$) and $CD16^+$ differentiates *in vitro* differently than $CD14^{lo}CD16^+$ monocytes [25] and probably is the subset of the $CD16^+$ monocytes that expresses CCR5 [24]. Within the $CD16^+$ monocyte fraction, there is also differential expression of a relative of PSGL-1 called M-DC8 [26-28].

Interest in $CD16^+$ monocytes has been fueled in part by observations that this subset was elevated in blood during inflammatory conditions and carcinoma, including among many others HIV dementia [29], atherosclerosis [30], rheumatoid arthritis [31], and cancer [32]. Frequently, the occurrence of elevations in circulating $CD16^+$ monocytes was interpreted to signify a role for $CD16^+$ monocytes in the disease process where the elevation was observed. Hence, in a number of studies, $CD16^+$ monocytes have been called inflammatory monocytes in the context of chronic inflammation. However, this designation appears to be inappropriate until we learn more about the *in vivo* function of this subset, especially given the observations that the mouse counterparts to $CD16^+$ human monocytes, the $Ly\text{-}6C^-/Gr\text{-}1^{lo}$ monocytes [23,33,34], do not migrate substantially to sites of inflammation *in vivo* compared with the $CCR2^+$ subset [23,33-37].

Several studies have analyzed the differentiation of human monocyte subsets *in vitro* and these studies showed in some cases a predisposition of $CD16^+$ monocytes to differentiate into antigen-presenting dendritic cells [25,38], e.g. when freshly isolated blood monocytes are tested in an *in vitro* model of transendothelial migration. In this model, human umbilical vein endothelial cells (HUVEC) are grown as a monolayer on an endotoxin-free collagen matrix (Figure 1). Freshly isolated human blood monocytes, incubated for 1.5 h in this model, migrate into the collagen and undergo differentiation. Whereas the majority of monocytes remain in the subendothelial matrix and display a macrophage-like phenotype, some reversely transmigrate the endothelial layer in an ablumenal-to-lumenal direction within 2 days and differentiate into antigen-presenting cells (Figure 1) [39]. In this model, the $CD16^+$ subset of human monocytes preferentially becomes migratory DCs, indicating that the $CD14^{lo}CD16^+$ cells might be precursors of DCs [38].

Without knowing more about the tissues where these subsets will migrate *in vivo*, these *in vitro* experiments are difficult to put into context. The extent to which human monocyte subsets differentiate into DCs *in vivo* is indeed too complex to predict, as both the phenotype of the cultured monocytes and factors present in the *in vitro* culture are expected to have a strong influence on the differentiation. For instance, $CD14^+CD16^-$ monocytes differentiate into Langerhans cells, the typical epidermal DC population, with potent antigen-presentation functions in a modified tissue-engineered model of human epidermal equivalents [40]. However, in the absence of the epidermal environment, but in the presence of vascular endothelial cells, CD16[+] monocytes are more prone to become DCs [38]. To relate the latter findings to what occurs *in vivo*, it will be essential to identify the trafficking behavior of CD16[+] monocytes or their murine equivalents *in vivo*.

The most popular model to study monocyte-derived DCs is to culture blood monocytes of either subset [41] in the presence of GM-CSF and IL-4 [10,42-44] or other related cytokine cocktails [45]. These DCs have been considered to the be the "gold standard" DC [46] to assess maturation and many aspects of the biology of human DCs. Addition of TGFβ1 to the culture has revealed how they can be directed to generate Langerhans cells *in vitro* [47]. Recently, three papers have begun to question the possibility that GM-CSF/IL-4-treated monocyte-derived DCs have true counterparts *in vivo*. Two of these studies investigated the phenotype of DCs in human lymph nodes and failed to find critical markers on lymph node DCs that are abundantly expressed on the culture-derived DCs [48,49]. Finally, in the epidermal constructs where $CD14^{+}CD16^{-}$ monocytes differentiated into Langerhans cells [40], resembling a similar outcome *in vivo* [37], IL-4/GM-CSF-derived DCs failed to do so [40]. Thus, whereas the study of IL-4/GM-CSF-cultured monocyte-derived DCs is undisputedly useful for studying DC biology, it is important to keep in mind that the pathway of differentiation that leads to the generation of these cells may not be recapitulated *in vivo*.

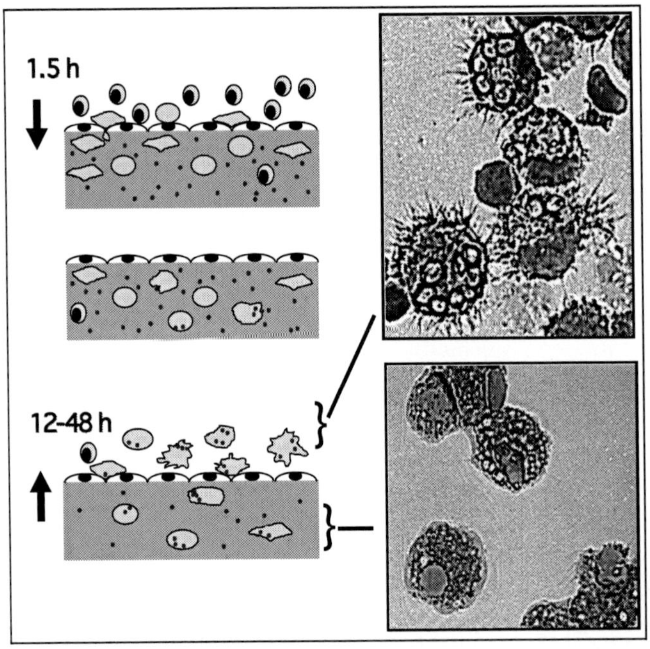

Figure 1. Differentiation of human monocyte subsets in an *in vitro* model of transendothelial migration. Human monocytes (yellow) or other leukocytes (pink) are applied for 1.5 h to a monolayer of human umbilical vein cells, grown on an endotoxin-free collagenous matrix, and non adherent / transmigrating cells are washed off thereafter. The collagenous matrix contains small particles (red dots) that are phagocytized by entering monocytes (12 h). Between 12-48 h, some monocytes are reversely transmigrating the endothelium in an ablumenal-to-lumenal direction. The photographs show monocyte-derived cells in the subendothelial layer with a typical macrophage morphology (lower picture) and reversely transmigrating with a typical dendritic cell (DC) morphology (upper picture). Human CD16[+] monocytes are more prone to differentiate into DCs than CD14[+] monocytes in this model.

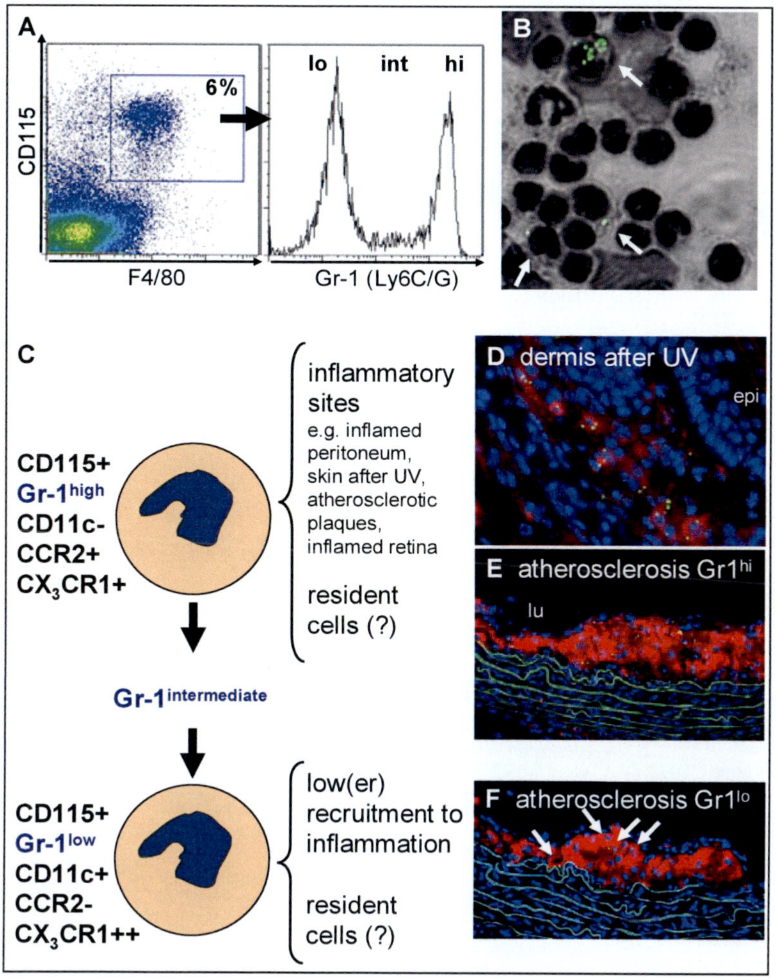

Figure 2. Mouse monocyte subsets and their migratory properties. A) Mouse blood monocytes can be identified by flow-cytometric analysis as low SSC cells that stain positive for CD115 (M-CSF receptor), F4/80 antigen and CD11b (Mac1, not shown). Monocytes can be divided according to their Gr-1 (Ly6C/G) expression into Gr-1 high (hi), intermediate (int) and low (lo) cells. B) Mouse blood monocyte subsets can be selectively labeled by the intravenous (i.v.) injection of fluorescent latex particles (see main text for details). Cytospin analysis of blood 2 days after injection of latex i.v. shows latex$^+$ monocytes (arrows) with typical mononuclear morphology. C) Schematic summary of the current concept of trafficking of monocyte subsets. Gr-1^{hi} monocytes are preferentially recruited to sites of inflammation, but their contribution to the turnover of steady state monocyte-derived cells in tissues is as yet not clear. Gr-1^{lo} monocytes originate from Gr-1^{hi} monocytes during maturation through a subset with intermediate Gr-1 expression (Gr-1^{int}). Gr-1^{lo} monocytes are by far less efficiently recruited to inflammatory sites, but have been proposed to contribute to the turnover of resident cells in the steady state. D-F) Monocyte subsets were specifically labeled *in vivo* with fluorescent latex particles to trace their trafficking pattern. Gr-1^{hi} monocytes enter the skin after local UV irradiation-induced inflammation (D) or atherosclerotic plaques in apoE$^{-/-}$ mice (E), but Gr-1^{lo} monocytes can also be detected in atherosclerotic plaques in apoE$^{-/-}$ mice after their specific labeling (F). Representative immunohistochemical analysis shown; green, latex particles in monocyte-derived cells; red, CD68 (macrophage marker); blue, nuclei counterstained with DAPI. Epi, epidermis; lu, lumen.

Mouse Monocyte Subsets

The mouse counterparts of the human monocyte subsets have been identified more recently, prompting intensive investigation in order to determine their functional roles in homeostasis and in inflammatory conditions [5,23,33,34]. Mouse blood monocytes can be identified by flow-cytometric analysis as cells with a low side scatter profile that coexpress CD115 (M-CSF receptor), CD11b (Mac 1) and F4/80 antigen (Figure 2A) [34]. In mice that have green fluorescent protein (gfp) knocked into the fractalkine-receptor (CX_3CR1)-gene [50], two major monocyte populations are apparent with low or high gfp intensity [23,51]. These monocyte subsets can also be distinguished by high or low expression level of surface Gr-1 (most antibodies to Gr-1 recognize two molecules, Ly6C and Ly6G), previously considered to mark only neutrophils [23,33,34]. In fact, monocytes express Ly6C [33], and neutrophils Ly6G and Ly6C. Similar to the human $CD14^+$ 'classical' monocytes, Gr-1^{hi} monocytes are $CCR2^+CX_3CR1^+CD62L^+$, whereas Gr-1^{lo} monocytes, similar to human $CD16^+$ 'nonclassical' monocytes, are $CCR2^-CX_3CR1^{++}CD62L^-$ (Table 1) [23,33,34,51]. In fact, even a small third subset with intermediate expression of Gr-1 and 'transitional' patterns of chemokine receptors exists, likely analogous to $CD14^+CD16^+$ "intermediate" monocytes (Table 1) [33,34]. The terms "transitional" or "intermediate" derive from the fact that Gr-1^{hi} monocytes are precursors for Gr-1^{lo} monocytes in the blood [33,34,52,53], despite the original claim that the two subsets are independently derived [23]. The transition from Gr-1^{hi} monocytes to Gr-1^{lo} monocytes goes through a stage wherein there is formation of a transient population of cells (Gr-1^{int}) that expresses intermediate (and unique) phenotypic features.

Table 1. Mouse monocyte subsets. Characteristics of the three identified mouse blood monocyte subsets, separated by high, intermediate (int) or low expression of surface Gr-1, are summarized

	'classical' monocytes	'intermediate' subset	'nonclassical' monocytes
Gr-1 (Ly6C/G) expression	Gr-1 high	Gr-1 int	Gr-1 low
surface molecule phenotype	CD115+, F4/80+, CD11b+		
	CD62L+		CD62L-
chemokine receptor expression	CCR2+	CCR2+	CCR2-
	CX_3CR1+	CX_3CR1+	CX_3CR1++
		CCR7+	CCR5+
		CCR8+	
human monocyte equivalent	CD14+ CD16- CD64+	CD14+ CD16+ CD64+	CD14lo CD16+ CD64-

The identification of the mouse monocyte subsets opened the possibility to study their trafficking and differentiation pattern *in vivo*. Different techniques have been developed to trace monocyte subsets in mice. For instance, the different expression level of CX_3CR1 in Gr-1^{hi} (=CX_3CR1^+) and Gr-1^{lo} (=CX_3CR1^{++}) monocytes [23] corresponds to a different gfp intensity in the CX_3CR1^{gfp} knock-in mice, allowing one to trace monocytes for a period of time without adoptive transfer [51]. However, once recruited to tissues, depending on the course of differentiation, monocytes from $CX_3CR1^{gfp/+}$ mice can lose or increase their gfp expression as they differentiate into tissue macrophages or DCs [23,34,50,54] so that tracing their fate in long-term studies is not possible.

For the purposes of tracing monocyte fate, one can also purify monocyte populations from blood for adoptive transfer experiments [23,53,55]. However, the adoptive transfer of monocytes can potentially introduce several artifacts as the isolation and *ex vivo* manipulations significantly alter the activation status, differentiation, viability, and migratory properties of monocytes [23,33], especially if antibodies are bound to the monocyte surface. Nonetheless, adoptive transfer methodologies yield results consistent with other methods to track monocytes. Some additional limitations of adoptive transfer of monocytes include the technical difficulty associated with the procedure (often requiring many donor mice per recipient) and the difficulty in quantifying monocyte behaviors, since only a small fraction of transferred monocytes is typically recovered for analysis.

Our group has developed an alternative technique to stably and specifically label either of the two major monocyte subsets with a phagocytic tracer directly *in vivo* without adoptive transfer (Figure 2B) [52], a method that in principle resembles a technique previously applied in rats [56,57]. Gr-1^{lo} monocytes can be labeled by the intravenous (i.v.) injection of 0.5-μm fluorescent latex particles. After i.v. latex administration, about 10-15% of the monocytes become latex$^+$, and about 90% of the latex$^+$ cells are Gr-1^{lo} monocytes by 24 h. Latex$^+$ Gr-1^{lo} monocytes can be traced for over seven days in wild-type mice, and, as expected, the numbers of latex$^+$ monocytes decrease over time due to the rapid turnover of blood monocytes [52]. However, the disappearance of these labeled monocytes from the circulation is not as rapid as would be predicted based on the reported half-life of blood monocytes [52,58], except in the case where the mice have an inflammatory disease such as atherosclerosis [59]. On the other hand, Gr-1^{hi} monocytes can be stably labeled using a modification of this technique, in which monocytes are first transiently depleted by i.v. application of liposomes loaded with the bisphosphonate clodronate [60] prior to the introduction of latex beads [52]. The i.v. administration of latex particles during monocyte depletion labels bone marrow monocytes, through cell transfer [52], and about 15% of monocytes later entering the circulation are latex$^+$ and uniformly Gr-1^{hi}. Monocytes remain Gr-1^{hi} for about 5 days before they start to convert to the Gr-1^{lo} subset [52]. Again, the persistence of the labeled Gr-1^{hi} monocytes in the blood is longer than expected [58], but is notably decreased in mice with inflammatory disease [59] as was noted above for the Gr-1^{lo} monocytes. These methods do not affect the migration patterns of labeled monocytes, do not significantly affect monocyte gene expression patterns, and do not induce inflammatory cytokines that can be detected in serum [59].

The transient depletion of blood monocytes by the systemic administration of clodronate-loaded liposomes has revealed interesting insights into the relationship between the subsets (Figure 2C) [23,33,34,52]. In the bone marrow, the majority of $CD115^+$ (pro)monocytes are Gr-1^{hi}, and CCR2 has been identified as a crucial factor for the exit of (Gr-1^{hi}) monocytes from bone marrow into blood [61]. After depletion of all blood monocytes by clodronate-loaded liposomes, returning circulating monocytes are exclusively Gr-1^{hi} and remain so for at least 5 days. By 5 day, monocytes begin to convert to the other subsets as some downregulate Gr-1 expression. By 7-9 days, the expected ratio of the mouse monocyte subsets has equilibrated [33]. If latex particles are injected i.v. during monocyte depletion, all of the monocyte initially labeled as Gr-1^{hi} monocytes have fully converted to Gr-1^{lo} monocytes by day 7-9 [52]. This rather slow conversion is observed only in the setting where monocytes have been first depleted. The conversion from Gr-1^{hi} to Gr-1^{int} and Gr-1^{lo} monocytes is more rapid in unperturbed mice and the conversion occurs within a few hours when Gr-1^{hi} monocytes encounter a phagocytic tracer, as the injection of latex beads (into unperturbed mice) promotes conversion to the Gr-1^{lo} subset within 24 h [52]. The relatively slower Gr-1^{hi}>Gr-1^{lo} conversion of monocytes that enter blood during restoration of the monocyte pool after depletion is consistent with the idea that conversion requires a certain period of monocyte maturation in the circulation.

The latex bead-labeling technique is readily applicable to projects that aim to trace the fate of monocyte subsets and has already been successfully applied to the study of Langerhans cell precursors [37]. In this study, the latex-labeling technique was used as one of three methods, including bone marrow chimeras and adoptive transfer, to trace monocyte precursors in the epidermis. All methods yielded findings that were in agreement, illustrating that the various methods, each with some caveats, are useful, reliable, and internally in agreement. Latex-labeled monocytes typically carry only 1-3 particles per cell (Figure 2B) [52]; this extent of labeling does not apparently alter their homing capacity, since the frequency of labeled monocytes recruited to the peritoneum after thioglycollate injection is similar to their frequency in the blood of the labeled mice [59]. The technique is more quantitative and is not as demanding technically as adoptive transfer; it also requires fewer mice than adoptive transfer, since endogenous monocytes are traced. Furthermore, in an adoptive transfer experiment, the transferred monocytes have to compete with endogenous monocytes which may likely alter their trafficking or differentiation, unlike after direct labeling of endogenous monocytes. However, the possible adverse sequelae associated with administration of the particles or the clodronate-loaded liposomes should not be dismissed, although these effects may not be as significant as those encountered during the preparation of cells for adoptive transfer. Death by apoptosis in response to the uptake of clodronate-loaded liposomes [60] would lead to similar sequelae as occurs in transgenically engineered mice where cell populations, such as dendritic cells [13] or macrophages [62], are induced to undergo selective suicide by apoptosis. In the case of clodronate-loaded liposomes, apoptosis of the targeted cell population is restricted to the locale of liposome administration [60], but it is systemic in the transgenic suicide models [63]. Also, the transgenic suicide models may in certain cases not be as specific as anticipated, as for instance the CD11c-DTR mice, ablating CD11c+ DCs after diphtheria toxin administration, also show depletion of marginal zone spleen and sinusoidal lymph nodes macrophages after induction [64]. Possible similar effects on other tissue macrophages may exist in this model, and have not been examined [64].

TRAFFICKING PATTERN AND CHEMOKINE RECEPTOR USAGE OF MONOCYTE SUBSETS *IN VIVO*

As Gr-1^{hi} monocytes express CCR2, a molecule well known to be involved in inflammatory monocyte recruitment [9], it was anticipated that Gr-1^{hi} monocytes are rapidly recruited to sites of inflammation (Figure 2C). In fact, the recruitment of Gr-1^{hi} monocytes into inflamed tissue has been convincingly shown in a variety of experimental models, such as the acutely inflamed peritoneum [23,33,59], the inflamed retina [55] and infection with Toxoplasma gondii [65] or Listeria monocytogenes [35] at the site of inoculation. Furthermore, severe inflammation of the skin, as seen after local ultraviolet (UV) irradiation, also results in the massive recruitment of Gr-1^{hi} monocytes (Figure 2D) [37]. In typical models of acute inflammation, Gr-1^{hi} monocyte recruitment is critically dependent on CCR2 [66,67], but not on CX_3CR1 [50]. In addition to CCR2, CCR6/CCL20 is involved in the recruitment of Gr-1^{hi} monocytes (and other DC precursors?) into epithelial tissues after adjuvants application to the skin or buccal mucosa [36,68]. This CCR6/CCL20 dependent mechanism of Gr-1^{hi} monocyte recruitment into the dermis and epithelium is essential for CD8+ T cell priming against innocuous protein antigen administered with adjuvant [36].

We and others have also investigated the role of monocyte recruitment in atherosclerosis, a more complex chronic inflammatory disease model. It is well established that monocytes critically participate in the development and progression of atherosclerotic lesions, as monocyte-derived cells are major components of these plaques [69]. The various techniques to study mouse monocyte trafficking, e.g. bead-labeling, adoptive transfer or genetic labeling have been applied in atherosclerosis-prone apo$E^{-/-}$ mice in order to unravel the rate of recruitment of Gr-1^{hi} (and Gr-1^{lo}) monocytes into atherosclerotic lesions and the chemokine receptors that these subsets utilize for entering plaques [59,70,71]. These studies revealed a robust recruitment of Gr-1^{hi} monocytes into atherosclerotic plaques, where they differentiate primarily into macrophages (Figure 2E) [59,71]. Gr-1^{hi} monocytes required CCR2, CCR5, and, unexpectedly, CX_3CR1 for emigration into atherosclerotic lesions [59]. In contrast, the migratory pattern of Gr-1^{hi} monocytes in the absence of inflammation is not well understood. This may differ dependent on the specific environment in different organs. For instance, Gr-1^{hi} or Gr-1^{lo} monocytes do not seem to contribute to DC populations in the spleen in the steady state, but following infection or sterile inflammation, they become a specialized type of splenic DCs and/or develop into DCs in other inflamed organs [53,72]. In contrast, Gr-1^{hi} monocytes can give rise to DC populations in the lung after adoptive transfer even in the absence of overt inflammation [73].

Even less is known about the trafficking and fate of Gr-1^{lo} monocytes. In contrast to Gr-1^{hi} monocytes, they migrate scarcely or not at all to inflamed tissue in mice, including the acutely inflamed peritoneum [23,33,59], or to skin after intracutaneous injection of latex beads [34], administration of vaccine formulations [36], or epicutaneous UV exposure [37]. The only exception recognized to date is atherosclerosis as a model of chronic vascular inflammation. Gr-1^{lo} monocytes are capable of entering atherosclerotic plaques, although less efficiently than their Gr-1^{hi} counterparts [59]. Due to their low capacity to traffic to inflamed tissues, it has therefore been hypothesized that Gr-1^{lo} monocytes may fulfill critical roles in replacing resident macrophages or DCs in the steady state [23].

It has further been suggested that the Gr-1^{lo} monocytes utilize CX_3CR1 to migrate into noninflamed tissue for replacing resident macrophages or DCs, given their higher surface expression of CX_3CR1 and the CX_3CL1-dependent transendothelial migration of human CD16$^+$ monocytes *in vitro* [23,24]. However, the experimental evidence for the trafficking pattern and the potential role of Gr-1^{lo} monocytes as precursors for resident tissue cells is rather limited. After adoptive transfer, Gr-1^{lo} monocytes can be found in blood, spleen, liver, lung, and brain of recipients for up to 3 days after transfer, and then at very diminished numbers by day 4 [23]. Given the fact that the human counterparts to Gr-1^{lo} monocytes, the CD16$^+$ monocytes, express low levels of CCR5 [22,24], this chemokine receptor should be considered a potential mediator of the trafficking of this subset. One study in mice appears consistent with the possibility that CCR2$^-$ monocytes (Gr-1^{lo} subset) express elevated CCR5 levels [74]. In atherosclerosis, Gr-1^{lo} monocytes emigrate into lesions independent of CCR2 and CX_3CR1, but require CCR5 [59]. Gr-1^{lo} monocytes have been also found to migrate to the lung in the steady state and appear to give rise to CD11b^{hi} pulmonary DCs [53,73]. CX_3CR1 appears to be involved in this trafficking pattern (unpublished observations).

Relevance of Monocyte-Derived Dendritic Cells *In Vivo*

While the differentiation potential of human and murine monocytes into DCs *in vitro* is undoubtedly clear [10], the extent to which monocyte differentiate *in vivo* into DCs is more challenging to address. It will be important to trace differentiation and migration in several different tissues and conditions. Until now, it has often been assumed that the precursor for DCs in spleen will be relevant in all lymphoid and peripheral tissues [1,44,75]. This is unlikely to be true.

Some of the Gr-1^{lo} monocytes that emigrated to the spleen after adoptive transfer expressed DC markers like CD11c or MHC II [23], but interpreting these data is difficult, because other experiments by the same group revealed recently that monocytes, irrespective of the subset, are not DC precursors in spleen [1]. Overall, the extent to which Gr-1^{lo} monocytes replenish tissue-resident macrophages and DC populations in the absence of inflammation, or the extent to which monocytes contribute to the steady state DC pool *in vivo* in general, is still unknown. The puzzling nature of the behavior of this subset *in vivo*, particularly its apparent resistance to extravascular recruitment, raises more questions than answers at present about the importance of this subset *in vivo*. Several situations have been identified in which monocyte subsets give rise to DCs *in vivo*, as discussed below.

After intradermal (i.d.) injection of fluorescent latex particles in skin, mild inflammation is observed that results in the recruitment of phagocytic CD11b$^+$ monocytes. About 25% of these latex-bearing monocytes migrate thereafter into the draining lymph node where they differentiate into MHCII$^+$DEC205$^+$ DCs that variably express CD11c [76,77]. Interestingly, neither CCR2 nor CX_3CR1 is required for monocyte recruitment, migration or differentiation towards DCs in this model [34]. Gr-1^{lo} monocytes are not recruited in this model, but Gr-1^{hi} and the minor Gr-1^{int} population of monocytes are. Indeed, the small Gr-1^{int} fraction appears to be the population most “predisposed” to give rise to DCs in this model, and these cells express mRNAs for chemokine receptors that regulate migration to the lymph nodes, CCR7 and CCR8 (Table 1) [34].

Although it is undisputed that monocytes can give rise to DCs under inflammatory conditions, the extent to which they do so *in vivo* in clinically relevant settings and especially for the turnover of DCs in the steady state remains unknown. Recently, a clonogenic bone marrow progenitor that is $CD117^+$ (c-kit), $CD115^+$ and CX_3CR1^+ has been identified that gives rise to monocytes, macrophages and resident $CD8\alpha^+/CD8\alpha^-$ spleen DCs [1]. These cells are upstream precursors of bona fide monocytes, but they may circulate somewhat independently of monocytes and have a greater plasticity for differentiation than the conventional blood monocyte.

Direct evidence that a circulating monocyte subset gives rise to a resident peripheral DC population, again especially clear under inflammatory conditions, is provided for the differentiation of $Gr\text{-}1^{hi}$ monocytes into epidermal Langerhans cells [37]. In the steady state, Langerhans cells arise from a local, radioresistant precursor [67]. This precursor is dependent upon CD115 expression for differentiation or survival [37]. However, severe skin inflammation induced by local UV irradiation results in the full depletion of Langerhans cells from skin, apparently including loss of the local precursor also [67]. In this setting, $Gr\text{-}1^{hi}$, in contrast to $Gr\text{-}1^{lo}$, monocytes are recruited to the inflamed skin in a CCR2- and CCR6-dependent manner, proliferate locally and differentiate into functional Langerhans cells or dermal macrophages *in vivo* [37,67,68]. This work reveals the differentiation potential of monocytes and shows how they fulfill their predicted role as circulating precursors for DCs and macrophages, especially during inflammatory conditions in which there is increased demand for DCs and macrophages. Similarly, in the context of vaccination through skin and other mucosal routes, $Gr\text{-}1^{hi}$ monocytes appear to be the most relevant DC precursors for cross-priming that results in $CD8^+$ T cell response. The recruitment of these monocytes is dependent upon CCR6 [36].

Monocytes also have important roles as DC precursors in models of infectious diseases. In spleen, $Gr\text{-}1^{hi}$ monocytes are linked to the formation of so-called "Tip-DCs", a population of antigen-presenting cells expressing tumor necrosis factor and inducible nitric oxide synthase. These cells are essential for the killing of Listeria monocytogenes-infected cells within the spleen [78]. Another interesting role for $Gr\text{-}1^{hi}$ monocytes in the context of immunization is the "priming" of naïve B cells that will ultimately facilitate their differentiation into Ig-producing cells. This process was recently found to be dependent upon $Gr\text{-}1^+F4/80^+$ cells that very well may be $Gr\text{-}1^{hi}$ monocytes [79]. Interestingly, these $Gr\text{-}1^+$ cells appeared to serve as the source for IL-4 that in turn mediates the B cell priming [79].

Recent studies raised the possibility that DCs in peripheral organs, including but not limited to lung, may arise at least to some extent from monocytes in the steady state. After DCs are artificially ablated in the airway and lung, adoptively transferred monocytes can contribute to their repopulation [53]. Further work revealed that monocytes became pulmonary DCs in resting mice even in the absence of ablation or overt inflammation [73]. Future studies will show if this is a general mechanism or specific to certain organs such as the lung.

In closing, after only a few years following the identification of monocyte subsets in mice [23], we have come closer to understanding their roles *in vivo*, as the cited examples above illustrate. The rate of progress in this area is rapidly picking up pace, giving rise to the expectation that we will soon know much more about the biology of monocytes in the steady state and inflammation. Further work *in vivo* is essential for placing *in vitro* work with human monocytes into context and providing input for the generation of improved *in vitro* models. The parallels between Langerhans cell precursors in the study by Schaerli *et al.*, who employed an *in vitro* model of human epidermis [40], with the recent work of Ginhoux *et al.* *in vivo* [37] indicate that studies in mice will indeed likely be relevant to gaining a better understanding of the biology of human monocytes.

REFERENCES

[1] Fogg DK, Sibon C, Miled C, Jung S, Aucouturier P, Littman DR, Cumano A, Geissmann F. A clonogenic bone marrow progenitor specific for macrophages and dendritic cells. *Science* 2006;311:83-87.

[2] Van Furth R, Cohn ZA. Origin and Kinetics of Mononuclear Phagocytes. *J Exp Med* 1968;128:415-435.

[3] Whitelaw DM. Observations on Human Monocyte Kinetics After Pulse Labeling. *Cell and Tissue Kinetics* 1972;5:311-&.

[4] Ziegler-Heitbrock HW. Definition of human blood monocytes. *J Leukoc Biol* 2000;67:603-606.

[5] Gordon S, Taylor PR. Monocyte and macrophage heterogeneity. *Nat Rev Immunol* 2005;5:953-964.

[6] Tarling JD, Lin H, Hsu S. Self-Renewal of Pulmonary Alveolar Macrophages - Evidence from Radiation Chimera Studies. *J Leukoc Biol* 1987;42:443-446.

[7] Sawyer RT. The Ontogeny of Pulmonary Alveolar Macrophages in Parabiotic Mice. *J Leukoc Biol* 1986;40:347-354.

[8] Sawyer RT, Strausbauch PH, Volkman A. Resident Macrophage Proliferation in Mice Depleted of Blood Monocytes by Sr-89. *Lab Invest* 1982;46:165-170.

[9] Imhof BA, Aurrand-Lions M. Adhesion mechanisms regulating the migration of monocytes. Nat Rev Immunol 2004;4:432-444.

[10] Leon B, Lopez-Bravo M, Ardavin C. Monocyte-derived dendritic cells. *Semin Immunol* 2005;17:313-318.

[11] Delamarre L, Pack M, Chang H, Mellman I, Trombetta ES. Differential lysosomal proteolysis in antigen-presenting cells determines antigen fate. *Science* 2005;307:1630-1634.

[12] Albert ML, Sauter B, Bhardwaj N. Dendritic cells acquire antigen from apoptotic cells and induce class I restricted CTLs. *Nature* 1998;392:86-89.

[13] Jung S, Unutmaz D, Wong P, Sano GI, los Santos K, Sparwasser T, Wu SJ, Vuthoori S, Ko K, Zavala F, Pamer EG, Littman DR, Lang RA. In vivo depletion of CD11c(+) dendritic cells abrogates priming of CD8(+) T cells by exogenous cell-associated antigens. *Immunity* 2002;17:211-220.

[14] Zammit DJ, Cauley LS, Pham QM, Lefrancois L. Dendritic cells maximize the memory CD8 T cell response to infection. I*mmunity* 2005;22:561-570.

[15] Savill J, Dransfield I, Gregory C, Haslett C. A blast from the past: Clearance of apoptotic cells regulates immune responses. *Nat Rev Immunol* 2002;2:965-975.
[16] Trombetta ES, Ebersold M, Garrett W, Pypaert M, Mellman I. Activation of lysosomal function during dendritic cell maturation. *Science* 2003;299:1400-1403.
[17] Cella M, Engering A, Pinet V, Pieters J, Lanzavecchia A. Inflammatory stimuli induce accumulation of MHC class II complexes on dendritic cells. *Nature* 1997;388:782-787.
[18] Baron CL, Raposo G, Scholl SM, Bausinger H, Tenza D, Bohbot A, Pouillart P, Goud B, Hanau D, Salamero J. Modulation of MHC class II transport and lysosome distribution by macrophage-colony stimulating factor in human dendritic cells derived from monocytes. *J Cell Sci* 2001;114:999-1010.
[19] Randolph GJ, Angeli V, Swartz MA. Dendritic-cell trafficking to lymph nodes through lymphatic vessels. *Nat Rev Immunol* 2005;5:617-628.
[20] Passlick B, Flieger D, Ziegler-Heitbrock HW. Identification and characterization of a novel monocyte subpopulation in human peripheral blood. *Blood* 1989;74:2527-2534.
[21] Ziegler-Heitbrock HWL, Fingerle G, Strobel M, Schraut W, Stelter F, Schutt C, Passlick B, Pforte A. The Novel Subset of CD14+ CD16+ Blood Monocytes Exhibits Features of Tissue Macrophages. *Eur J Immunol* 1993;23:2053-2058.
[22] Weber C, Belge KU, von Hundelshausen P, Draude G, Steppich B, Mack M, Frankenberger M, Weber KSC, Ziegler-Heitbrock HWL. Differential chemokine receptor expression and function in human monocyte subpopulations. *J Leukoc Biol* 2000;67:699-704.
[23] Geissmann F, Jung S, Littman DR. Blood monocytes consist of two principal subsets with distinct migratory properties. *Immunity* 2003;19:71-82.
[24] Ancuta P, Rao R, Moses A, Mehle A, Shaw SK, Luscinskas FW, Gabuzda D. Fractalkine preferentially mediates arrest and migration of CD16+ monocytes. *J Exp Med* 2003;197:1701-1707.
[25] Grage-Griebenow E, Zawatzky R, Kahlert H, Brade L, Flad HD, Ernst M. Identification of a novel dendritic cell-like subset of CD64(+)/CD16(+) blood monocytes. *Eur J Immunol* 2001;31:48-56.
[26] Schakel K, Mayer E, Federle C, Schmitz M, Riethmuller G, Rieber EP. A novel dendritic cell population in human blood: one-step immunomagnetic isolation by a specific mAb (M-DC8) and in vitro priming of cytotoxic T lymphocytes. *Eur J Immunol* 1998;28:4084-4093.
[27] Siedlar M, Frankenberger M, Ziegler-Heitbrock LHW, Belge KU. The M-DC8-positive leukocytes are a subpopulation of the CD14(+) CD16(+) monocytes. *Immunobiology* 2000;202:11-17.
[28] Schakel K, Kannagi R, Kniep B, Goto Y, Mitsuoka C, Zwirner J, Soruri A, von Kietzell M, Rieber EP. 6-sulfo LacNAc, a novel carbohydrate modification of PSGL-1, defines an inflammatory type of human dendritic cells. I*mmunity* 2002;17:289-301.
[29] Pulliam L, Gascon R, Stubblebine M, McGuire D, McGrath MS. Unique monocyte subset in patients with AIDS dementia. *Lancet* 1997;349:692-695.
[30] Schlitt A, Heine GH, Blankenberg S, Espinola-Klein C, Dopheide JF, Bickel C, Lackner KJ, Iz M, Meyer J, Darius H, Rupprecht HJ. CD14+CD16+ monocytes in

coronary artery disease and their relationship to serum TNF-alpha levels. *Thromb Haemost* 2004;92:419-424.

[31] Baeten D, Boots AMH, Steenbakkers PGA, Elewaut D, Bos E, Verheijden GFM, Verbruggen G, Miltenburg AMM, Rijnders AWM, Veys EM, de Keyser F. Human cartilage gp-39+, CD16+monocytes in peripheral blood and synovium - Correlation with joint destruction in rheumatoid arthritis. *Arthritis Rheum* 2000;43:1233-1243.

[32] Saleh MN, Goldman SJ, LoBuglio AF, Beall AC, Sabio H, McCord MC, Minasian L, Alpaugh RK, Weiner LM, Munn DH. CD16+ monocytes in patients with cancer: spontaneous elevation and pharmacologic induction by recombinant human macrophage colony-stimulating factor. *Blood* 1995;85:2910-2917.

[33] Sunderkotter C, Nikolic T, Dillon MJ, van Rooijen N, Stehling M, Drevets DA, Leenen PJM. Subpopulations of mouse blood monocytes differ in maturation stage and inflammatory response. *J Immunol* 2004;172:4410-4417.

[34] Qu CF, Edwards EW, Tacke F, Angeli V, Llodra J, Sanchez-Schmitz G, Garin A, Haque NS, Peters W, van Rooijen N, Sanchez-Torres C, Bromberg J, Charo IF, Jung S, Lira SA, Randolph GJ. Role of CCR8 and other chemokine pathways in the migration of monocyte-derived dendritic cells to lymph nodes. *J Exp Med* 2004;200:1231-1241.

[35] Drevets DA, Dillon MJ, Schawang JS, van Rooijen N, Ehrchen J, Sunderkotter C, Leenen PJM. The Ly-6C(high) monocyte subpopulation transports Listeria monocytogenes into the brain during systemic infection of mice. *J Immunol* 2004;172:4418-4424.

[36] Le Borgne M, Etchart N, Goubier A, Lira SA, Sirard JC, van Rooijen N, Caux C, Ait-Yahia S, Vicari A, Kaiserlian D, Dubois B. Dendritic cells rapidly recruited into epithelial tissues via CCR6/CCL20 are responsible for CD8+ T cell crosspriming in vivo. *Immunity* 2006;24:191-201.

[37] Ginhoux F, Tacke F, Angeli V, Bogunovic M, Loubeau M, Dai XM, Stanley RE, Randolph GJ, Merad M. Langerhans cells arise from monocytes in vivo. *Nat Immunol* 2006;7:265-273.

[38] Randolph GJ, Sanchez-Schmitz G, Liebman RM, Schakel K. The CD16(+) (FcgammaRIII(+)) subset of human monocytes preferentially becomes migratory dendritic cells in a model tissue setting. *J Exp Med* 2002;196:517-527.

[39] Randolph GJ, Beaulieu S, Lebecque S, Steinman RM, Muller WA. Differentiation of monocytes into dendritic cells in a model of transendothelial trafficking. *Science* 1998;282:480-483.

[40] Schaerli P, Willimann K, Ebert LM, Walz A, Moser B. Cutaneous CXCL14 targets blood precursors to epidermal niches for Langerhans cell differentiation. *Immunity* 2005;23:331-342.

[41] Sanchez-Torres C, Garcia-Romo GS, Cornejo-Cortes MA, Rivas-Carvalho A, Sanchez-Schmitz G. CD16(+) and CD16(-) human blood monocyte subsets differentiate in vitro to dendritic cells with different abilities to stimulate CD4(+) T cells. *Int Immunol* 2001;13:1571-1581.

[42] Sallusto F, Lanzavecchia A. Efficient Presentation of Soluble-Antigen by Cultured Human Dendritic Cells Is Maintained by Granulocyte-Macrophage Colony-Stimulating

Factor Plus Interleukin-4 and Down-Regulated by Tumor-Necrosis-Factor-Alpha. *J Exp Med* 1994;179:1109-1118.

[43] Zhou LJ, Tedder TF. CD14(+) blood monocytes can differentiate into functionally mature CD83(+) dendritic cells. *Proc Natl Acad Sci U S A* 1996;93:2588-2592.

[44] Shortman K, Liu YJ. Mouse and human dendritic cell subtypes. *Nat Rev Immunol* 2002;2:151-161.

[45] Piemonti L, Bernasconi S, Luini W, Trobonjaca Z, Minty A, Allavena P, Mantovani A. IL-13 supports differentiation of dendritic cells from circulating precursors in concert with GM-CSF. *European Cytokine Network* 1995;6:245-252.

[46] Pulendran B, Palucka K, Banchereau J. Sensing pathogens and tuning immune responses. *Science* 2001;293:253-256.

[47] Geissmann F, Prost C, Monnet JP, Dy M, Brousse N, Hermine O. Transforming growth factor beta 1 in the presence of granulocyte/macrophage colony-stimulating factor and interleukin 4, induces differentiation of human peripheral blood monocytes into dendritic Langerhans cells. *J Exp Med* 1998;187:961-966.

[48] Granelli-Piperno A, Pritsker A, Pack M, Shimeliovich I, Arrighi JF, Park CG, Trumpfheller C, Piguet V, Moran TM, Steinman RM. Dendritic cell-specific intercellular adhesion molecule 3-grabbing nonintegrin/CD209 is abundant on macrophages in the normal human lymph node and is not required for dendritic cell stimulation of the mixed leukocyte reaction. *J Immunol* 2005;175:4265-4273.

[49] Krutzik SR, Tan B, Li HY, Ochoa MT, Liu PT, Sharfstein SE, Graeber TG, Sieling PA, Liu YJ, Rea TH, Bloom BR, Modlin RL. TLR activation triggers the rapid differentiation of monocytes into macrophages and dendritic cells. *Nat Med* 2005;11:653-660.

[50] Jung S, Aliberti J, Graemmel P, Sunshine MJ, Kreutzberg GW, Sher A, Littman DR. Analysis of fractalkine receptor CX(3)CR1 function by targeted deletion and green fluorescent protein reporter gene insertion. *Mol Cell Biol* 2000;20:4106-4114.

[51] Palframan RT, Jung S, Cheng G, Weninger W, Luo Y, Dorf M, Littman DR, Rollins BJ, Zweerink H, Rot A, von Andrian UH. Inflammatory chemokine transport and presentation in HEV: a remote control mechanism for monocyte recruitment to lymph nodes in inflamed tissues. *J Exp Med* 2001;194:1361-1373.

[52] Tacke F, Ginhoux F, Jakubzick C, van Rooijen N, Merad M, Randolph GJ. Immature monocytes acquire antigens from other cells in the bone marrow and present them to T cells after maturing in the periphery. *J Exp Med* 2006;203:583-597.

[53] Varol C, Landsman L, Fogg DK, Greenshtein L, Gildor B, Margalit R, Kalchenko V, Geissmann F, Jung S. Monocytes give rise to mucosal, but not splenic, conventional dendritic cells. *J Exp Med* 2007;204:171-180.

[54] Niess JH, Brand S, Gu XB, Landsman L, Jung S, McCormick BA, Vyas JM, Boes M, Ploegh HL, Fox JG, Littman DR, Reinecker HC. CX(3)CR1-mediated dendritic cell access to the intestinal lumen and bacterial clearance. *Science* 2005;307:254-258.

[55] Xu HP, Manivannan A, Dawson R, Crane IJ, Mack M, Sharp P, Liversidge J. Differentiation to the CCR2(+) inflammatory phenotype in vivo is a constitutive, time-limited property of blood monocytes and is independent of local inflammatory mediators. *J Immunol* 2005;175:6915-6923.

[56] Ryan GB. Origin and Sequence of Cells Found in Acute Inflammatory Response. *Australian Journal of Experimental Biology and Medical Science* 1967;45:149-162.

[57] Matsuno K, Ezaki T, Kudo S, Uehara Y. A life stage of particle-laden rat dendritic cells in vivo: their terminal division, active phagocytosis, and translocation from the liver to the draining lymph. *J Exp Med* 1996;183:1865-1878.

[58] Van Furth R. Human monocytes and cytokines. *Research in Immunology* 1998;149:719-720.

[59] Tacke F, Alvarez D, Kaplan TJ, Jakubzick C, Spanbroek R, Llodra J, Garin A, Liu J, Mack M, van Rooijen N, Lira SA, Habenicht AJ, Randolph GJ. Monocyte subsets differentially employ CCR2, CCR5, and CX3CR1 to accumulate within atherosclerotic plaques. *J Clin Invest* 2007;117:185-194.

[60] van Rooijen N, Sanders A. Liposome-Mediated Depletion of Macrophages - Mechanism of Action, Preparation of Liposomes and Applications. *J Imunol Methods* 1994;174:83-93.

[61] Serbina NV, Pamer EG. Monocyte emigration from bone marrow during bacterial infection requires signals mediated by chemokine receptor CCR2. *Nat Immunol* 2006;7:311-317.

[62] Duffield JS, Forbes SJ, Constandinou CM, Clay S, Partolina M, Vuthoori S, Wu SJ, Lang R, Iredale JP. Selective depletion of macrophages reveals distinct, opposing roles during liver injury and repair. *J Clin Invest* 2005;115:56-65.

[63] Saito M, Iwawaki T, Taya C, Yonekawa H, Noda M, Inui Y, Mekada E, Kimata Y, Tsuru A, Kohno K. Diphtheria toxin receptor-mediated conditional and targeted cell ablation in transgenic mice. *Nat Biotechnol* 2001;19:746-750.

[64] Probst HC, Tschannen K, Odermatt B, Schwendener R, Zinkernagel RM, Van Den Broek M. Histological analysis of CD11c-DTR/GFP mice after in vivo depletion of dendritic cells. *Clin Exp Immunol* 2005;141:398-404.

[65] Robben PM, LaRegina M, Kuziel WA, Sibley LD. Recruitment of Gr-1(+) monocytes is essential for control of acute toxoplasmosis. *J Exp Med* 2005;201:1761-1769.

[66] Boring L, Gosling J, Chensue SW, Kunkel SL, Farese RV, Broxmeyer HE, Charo IF. Impaired monocyte migration and reduced type 1 (Th1) cytokine responses in C-C chemokine receptor 2 knockout mice. *J Clin Invest* 1997;100:2552-2561.

[67] Merad M, Manz MG, Karsunky H, Wagers A, Peters W, Charo I, Weissman IL, Cyster JG, Engleman EG. Langerhans cells renew in the skin throughout life under steady-state conditions. *Nat Immunol* 2002;3:1135-1141.

[68] Merad M, Hoffmann P, Ranheim E, Slaymaker S, Manz MG, Lira SA, Charo I, Cook DN, Weissman IL, Strober S, Engleman EG. Depletion of host Langerhans cells before transplantation of donor alloreactive T cells prevents skin graft-versus-host disease. *Nat Med* 2004;10:510-517.

[69] Breslow JL. Mouse models of atherosclerosis. *Science* 1996;272:685-688.

[70] Swirski FK, Pittet MJ, Kircher MF, Aikawa E, Jaffer FA, Libby P, Weissleder R. Monocyte accumulation in mouse atherogenesis is progressive and proportional to extent of disease. *Proc Natl Acad Sci U S A* 2006;103:10340-10345.

[71] Swirski FK, Libby P, Aikawa E, Alcaide P, Luscinskas FW, Weissleder R, Pittet MJ. Ly-6C monocytes dominate hypercholesterolemia-associated monocytosis and give rise to macrophages in atheromata. *J Clin Invest* 2007;117:195-205.

[72] Naik SH, Metcalf D, van Nieuwenhuijze A, Wicks I, Wu L, O'Keeffe M, Shortman K. Intrasplenic steady-state dendritic cell precursors that are distinct from monocytes. *Nat Immunol* 2006;7:663-671.

[73] Landsman L, Varol C, Jung S. Distinct differentiation potential of blood monocyte subsets in the lung. *J Immunol* 2007;178:2000-2007.

[74] Mack M, Cihak J, Simonis C, Luckow B, Proudfoot AEI, Plachy J, Bruhl H, Frink M, Anders HJ, Vielhauer V, Pfirstinger J, Stangassinger M, Schlondorff D. Expression and characterization of the chemokine receptors CCR2 and CCR5 in mice. *J Immunol* 2001;166:4697-4704.

[75] del Hoyo GM, Martin P, Vargas HH, Ruiz S, Arias CF, Ardavin C. Characterization of a common precursor population for dendritic cells. *Nature* 2002;415:1043-1047.

[76] Randolph GJ, Inaba K, Robbiani DF, Steinman RM, Muller WA. Differentiation of phagocytic monocytes into lymph node dendritic cells in vivo. *Immunity* 1999;11:753-761.

[77] Rotta G, Edwards EW, Sangaletti S, Bennett C, Ronzoni S, Colombo MP, Steinman RM, Randolph GJ, Rescigno M. Lipopolysaccharide or whole bacteria block the conversion of inflammatory monocytes into dendritic cells in vivo. *J Exp Med* 2003;198:1253-1263.

[78] Serbina NV, Salazar-Mather TP, Biron CA, Kuziel WA, Pamer EG. TN/iNOS-producing dendritic cells mediate innate immune defense against bacterial infection. *Immunity* 2003;19:59-70.

[79] Jordan MB, Mills DM, Kappler J, Marrack P, Cambier JC. Promotion of B cell immune responses via an alum-induced myeloid cell population. *Science* 2004;304:1808-1810.

In: Cell Differentiation Research Developments
Editor: L. B. Ivanova, pp. 201-215
ISBN: 978-1-60021-939-9

Chapter IX

HEMATOPOIETIC DIFFERENTIATION OF MOUSE EMBRYONIC STEM CELLS, A VALUABLE SYSTEM FOR IN VITRO STUDY OF EMBRYONIC HEMATOPOIESIS AND LEUKEMIAGENESIS

Gang-Ming Zou*
Department of Pathology, Johns Hopkins University School of Medicine, Baltimore, MD 21231, USA.

ABSTRACT

Mouse embryonic stem (ES) cells are derived from the inner cell mass of the 3.5-day-old blastocyst. These cells retain the differentiation ability after many passages in the presence of leukemia inhibitory factor. Removal of leukemia inhibitory factor results in mouse ES cells differentiation into embryoid bodies. Mouse ES cells have been regarded as a versatile biological system which has made significant advance in cell and developmental biology. They possess developmental potentials to differentiate into hematopoietic cells. Using the mouse ES/EB system, a precursor that responds to vascular endothelial growth factor and generates colonies consisting of undifferentiated blast cells was identified. These VEGF-responsive blast cell colonies were shown to contain endothelial, primitive erythroid, and various definitive hematopoietic precursors. These cells are referred to as blast colony-forming cells and are thought to be cells represent the hemangioblast precursor of blood and endothelial lineages. Other hematopoietic cells, such as progenitor B cells, mature B cells, dendritic cells, mast cells, neutrophils, and T cells can also be generated from mouse ES cell differentiation. This ES cell differentiation toward to the generation of hematopoietic cells might be a useful system to study gene function and regulation in hematopoietic development. Using this system, we recently identified protein tyrosine phosphatase Shp-2 is essential for mouse

* Correspondence concerning this article should be addressed to: Gang-Ming Zou, MD, PhD., Department of Pathology, Johns Hopkins University School of Medicine, 1550 Orleans St. CRB2, M341, Baltimore, MD 21231. Tel: 410-955 3511; Fax: 410-614 0671; Email: gzou1@jhmi.edu.

ES cell-derived hematopoietic differentiation and Ape1 regulate hematopoietic differentiation of mouse ES cells through its redox function domain. In this chapter, we review the induction of mesoderm cells, particularly hematopoietic cells from mouse ES cell differentiation. We also summarize and discuss the recent advance in the hematopoietic cell development using mouse ES cell differentiation system.

Keywords: ES cell, stem cells, embryonic hematopoisis, leukemia, Shp-2, Ape1.

ABBREVIATIONS

AGM	Aorta-gonad-mesonephros;
BFU-E	Erythroid burst-forming cells;
BL-CFC	Blast colony-forming cells;
CFU-E	Erythroid colony-forming cells;
CML	Chronic myeloid leukemia;
EB	Embryoid body;
ES	Embryonic stem cells;
Ery-D	Definitive erythroid progenitor cells;
Ery-P	Primitive erythroid progenitor cell;
GM-CSF	Granulocyte-Monocyte-Colony Stimulating Factor;
HPC	Hematopoietic Progenitor cells;
HSC	Hematopoietic stem cells;
LIF	Leukemia Inhibitory Factor;
LPS	Lipopolysaccharide;
mIL-3	Murine Interleukin-3;
RAG2	Recombination-activating gene-2;
Rb	Retinoblastoma;
SCF	Stem Cell Factor;
SiRNA	Small interfering RNA.

INTRODUCTION

Cell differentiation in developmental biology describes the process by which cells acquire a “type’. The morphology of a cell may change during differentiation. It is the process by which unspecialized cells become specialized into one of the many cells that make up the body, such as a heart, liver, brain, muscle, or blood cells. During differentiation, certain genes have been turned on or become activated, whereas some other genes have been switched off or inactivated. A cell that is able to differentiate into many cell types is known as pluripotent. These cells are named stem cells in animal and meristematic cells in higher plants. Differentiation is an important phenomenon in embryogenesis. The fertilized egg, the first entity of the life, has the ability to generate an entire organism. The fertilized egg commit to cell differentiation which result in the formation of a blastocyst. Blastocyst composes outer trophoblast cells and undifferentiated inner cell mass. Mouse ES cells are derived from the inner cell mass of the 3.5-day-old blastocyst (Evans *et al.*, 1981; Martin. 1981). ES cells retain the differentiation ability of the inner cell mass after many passages in the presence of leukemia inhibitory factor (LIF) (Sato and Nakano, 2001). The pluripotency of these cells as demonstrated in vivo by the introduction of ES cell into blastocyst. The formation of mouse chimeras demonstrated that ES cells can contribute to all cell lineages including the germ line (Bradley et al, 1984). Removal of LIF results in ES cell differentiation into embryoid bodies (EB) (Zou et al, 2003) (also see Figure 1). Mouse ES cells exhibit the ability in vitro to reproduce the various somatic cell types (See Figure 2) and also the cells of germ line, therefore, become an effective in vitro system to study cell differentiation and regulation in the embryo stage.

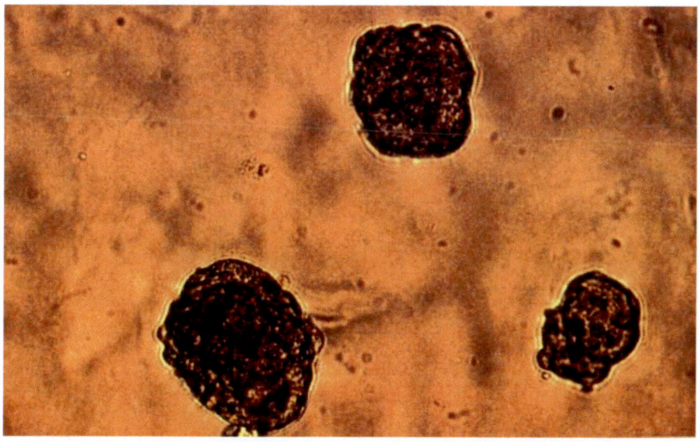

Figure 1. Photograph of three EBs on day 10 differentiated from mouse ES cells.

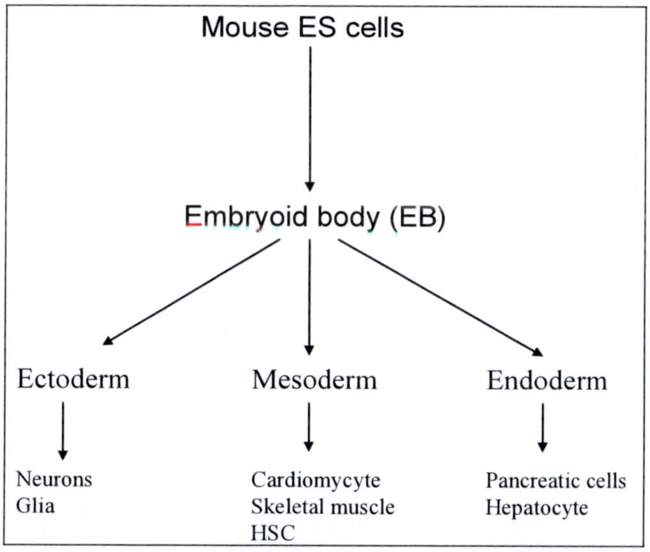

Figure 2. Differentiation of mouse ES cells to ectoderm, mesoderm and endoderm.

THE HEMATOPOIETIC HIERARCHY

Hematopoiesis is defined as the process by which all the different cell lineage that from the blood and immune system are generated from a common pluritopotent hematopoietic stem cells. During the lifespan, two separate hematopoietic systems exist, both arising during embryonic development but only one persisting in the adult stage. The primitive system is derived from the extraembryonic yolk sac and consists mainly of nucleated erythroid cells, which carry oxygen to the developing embryonic tissues (Robertson et al, 1999). When the embryo increases in size, this early circulatory system is superseded by the more complex definitive hematopoietic system, which originates within the embryo itself and continues throughout adult life. This definitive hematopoietic system produces all adult blood cell types including erythrocytes and cells of the myeloid and lymphoid lineages. All these cells are derived from pluripotent hematopoietic stem cells (HSCs) through a succession of precursors with progressively limited potential under the control of specific cytokines such as interleukins and granulocyte/monocyte-stimulating factors. In most cases, the cytokines that determine differentiation to a particular lineage are well defined. However, the factors that regulated HSC generation and maintenance of pluripotency are not fully understood currently. The adult hematopoietic tissues (i.e. bone marrow, thymus and spleen) are seeded by multilineage blood cells derived from the fetal liver. During embryogenesis, however, fetal liver hematopoiesis is preceded by the emergence of pluripotent HSCs in a region of the paro-aortic splanchnopleural mesoderm containing the dorsal aorta, gonadal ridge and mesonephros, named the aorta-gonad-mesonephros (AGM) region. In vitro studies and repopulation analyses in myeloablated recipient mice suggest this region as a major source of long-term repopulating-HSCs between 8.5-11.5 days post coitum (dpc) in the mouse, prior to the onset of liver hematopoiesis. These cell clusters express CD45 as well as several markers in common with adjacent endothelial cells including CD34, which is commonly used to identify HSCs in bone marrow and peripheral blood. CD34-positive cells isolated from murine AGM at 10.5 dpc can give rise to cells of all mematopoietic lineages in vitro. These intro-aortic cell clusters constitute the first site of definitive HSC generation during development and represent the origins of adult hematopoiesis. Within the embryonic AGM region, the eventual fate of hemangioblasts is determined by factors that bind at the cell surface, triggering downstream signaling pathways that culminate in the activation of hematopoietic of other lineage-specific genes. A stromal layer underlying the ventral floor of the dorsal aorta within the AGM has been suggested to represent a niche. This HSC niche supports HSC development. This highly defined ventral region, coupled with the absence of contaminating committed hematopoietic progenitors within the AGM at this stage of development, provides an ideal environment in which to investigate the factors involved in the generation and regulation of HSCs in vivo.

Early Embryonic Hematopoiesis is Recapitulated in EBs Derived from Mouse ES Cells

In vitro differentiation of murine ES cells generates relatively large numbers of early embryonic hematopoietic cells and their immediate precursors which can be isolated and manipulated for further study (Weiss and Orkin, 1996). Hematopoiesis in EBs also occurs in distinct waves. The two step replating assay demonstrates that separate populations of erythroid progenitors appear in a characteristic temporal fashion. Primitive erythroid progenitors which arise at day 4 or 5 of EB development and then rapidly disappear yield small colonies that synthesize embryonic globins. Definitive erythroid progenitors appear later and generate larger colonies that contain adult globins. These two precursor populations have distinct growth factor requirements: the former are erythropoietin responsive while the latter require Epo and SCF for full development, consistent with diminished requirements for SCF in primitive as compared with definitive hematopoiesis in vivo.

Hemangioblast from Mouse ES Cell Differentiation

The hemangioblast is a bipotential cell that gives rise to hematopoietic and endothelial cells (Chung et al, 2002; Perlingeiro et al, 2003). Although the existence of the hemangioblast was first postulated early last century, a cell with this activity has yet to be unequivocally identified in mammals. In the last decade, gene targeting experiments in the mouse have uncovered genes which are required for development of both the hematopoietic and endothelial lineages, Using the ES/EB system, a precursor that responds to VEGF and generates colonies consisting of undifferentiated blast cells was identified (Kennedy et al, 1997). These VEGF-responsive blast cell colonies were shown to contain endothelial, primitive erythroid, and various definitive hematopoietic precursors. These cells are referred to as blast colony-forming cells (BL-CFC) and are thought to be cells represent the hemangioblast precursor of blood and endothelial lineages. This system is now being used to dissect the molecular regulation of hemangioblast development and differentiation (Forrai and Robb, 2003). ES cells deficient for the transcription factor scl/tal-1 are unable to generate hemangioblasts, while those deficient for Runx1 generate reduced numbers of these precursors (Lacaud et al, 2001). These findings suggest that both genes play pivotal roles at the earliest stages of hematopoietic and endothelial development. Ape1 (also named Ref-1) is a dual functional protein which plays a role both in DNA repair and redox regulation (Evans et al, 2000; Kelley and Parsons, 2001). Recently, we used siRNA to knock down Ape1 gene expression in EB cells and discovered that reduction of Ape1 by siRNA results in a significant decrease in the frequency of hemangioblast formation. These studies suggest that Ape1 may play a critical role in embryonic hematopoiesis (Zou et al, 2007).

HSC from Mouse ES Cell Differentiation

HSCs are precursors that give rise to all red, white blood cells, and platelet. Mouse HSCs may exhibit a different phenotype compared with human HSCs, for example, human HSCs do not express Sca-1 (see Table 1). C-kit^{+}Sca-1^{+} murine bone marrow cells are representative for mouse HSCs. Mouse ES cell differentiated in vitro will yield a multitude of hematopoietic derivatives, whereas progenitors displaying true stem cell activity remain difficult to obtain. Possible causes are a biased differentiation to primitive yolk sac type hematopoiesis, and a variety of developmental or functional deficiencies. Recent studies in the Zebrafish have identified the caudal homeobox transcription factors (cdx1/4) and posterior hox genes (hoxa9a, hoxb7a) as key regulators for blood formation during embryonic development (Lengerke et al, 2007). Activation of Cdx and Hox genes during the in vitro differentiation of mouse ES cells followed by co-culture on supportive stromal cells generates ESC-derived HSCs capable of multilineage repopulation of lethally irradiated adult mice. One approach to obtain definitive HSCs from ES cells is to enforce expression of gene that stimulates hematopoiesis. Using a murine ES cell line, Wang et al (2005) demonstrated that Cdx promote commitment to hematopoietic mesoderm; stimulate hematopoietic progenitor formation from murine ES cells. Brief pulses of ectopic Cdx4 or HoxB4 expression are sufficient to enhance hematopoiesis during ES cell differentiation, presumably by acting as developmental switches to activate posterior Hox genes (Lengerke et al, 2007).

Table 1. Cell surface marker of hematopoietic stem cells

Mouse HSC	Human HSC
CD34$^{low/-}$	CD34^{+}
CD38+	CD59^{+}
Thy1$^{+/low}$	Thy1^{+}
SCA-1^{+}	CD38 $^{low/-}$
C-kit^{+}	c-kit $^{low/-}$
lin^{-}	lin^{-}

Erythroid Progenitor from Mouse ES Cell Differentiation

Erythrocytes develop from multipotent hematopoietic progenitor cells through a series of differentiation, including a common myeloid progenitor and the progenitors that have been already to the red cell lineage, such as BFU-E and CFU-E. BFU-E differentiates into CFU-E on stimulation by erythropoietin, and then further differentiates into erythroblasts when stimulated by other factors. In 1990, Lindenbaum and Grosveld, two scientists in England, first reported about erythroid progenitor differentiation of mouse ES cells (Lindenbaum and Grosveld, 1990). They used mouse ES cells to study globin gene expression and switching in vitro. They demonstrated that ES-derived EBs express the full complement of mouse embryonic globin genes in the correct temporal order and that on further differentiation, a switch occurs to the fetal/adult genes. NF-E1, the erythroid-specific transcription factor, was expressed coordinately with that of globin in EBs. Their experiments indicate that the ES cell system provides a valuable model to study erythroid hematopoietic development. We recently found that protein tyrosine phosphatase Shp-2 is essential for both Ery-P and Ery-D colony formation in mouse ES cell differentiation (Zou et al, 2006). Furthermore, we have identified that Apel redox function is required for Ery-P and Ery-D colony formation using mouse ES cell differentiation system (Zou et al, 2007a).

MYELOID PROGENITOR CELLS FROM MOUSE ES CELL DIFFERENTIATION

Myeloid progenitor can be generated from mouse ES cell differentiation. When cultured in semisolid medium, ES cell lines efficiently generated EBs containing blood islands in which hematopoietic cells from all six myeloid lineages could be detected. Erythropoietin significantly increased blood island formation. Colony-forming cells of all myeloid lineages as well as multipotent progenitors can be identified. Hematopoietic differentiation of ES cells under these conditions reflects formation of the complete range of blood cells found in the yolk sac of the early fetus. Therefore this system can be a unique model in which to study the earliest events of hematopoietic development in vitro (Burkert et al, 1991). Raf kinases play an integral role in embryonic myeloid hematopoiesis, as in vitro murine ES cell-derived myeloid progenitor development is quantitatively impaired in the absence of B-Raf. This phenotype is caused by the loss of B-Raf expression and associated ERK1/2 activation during hematopoietic progenitor cell formation (Kamata et al, 2005). Shp-2 is a member of cytoplasmic Src homology 2 (SH2) domain-containing protein tyrosine phosphatases. It is a gene product of *PTPN11* gene. Although Shp-2 has been shown to be necessary for hematopoiesis using a mouse model expressing a mutant residual protein (Shp-$2^{\Delta/\Delta}$), it might not the direct evidence to prove HSCs is the direct target for Shp-2 in Shp-2-mediated hematopoietic regulation since embryonic mesoderm development is defect in this Shp-2 mutant mice. To better address this question, we recently used siRNA to knockdown *PTPN11* gene expression in EB cells and examined the consequences on murine ES-derived primitive, and definitive hematopoietic development. We have found that myeloid progenitor formation was decreased significantly after transfection with *PTPN11* siRNA (Zou et al, 2006). Apel (also named Ref-1) is a molecule with dual functions in DNA repair and redox regulation of a number of transcription factors. This molecule is required for normal embryo development (Xanthoudakis et al, 1996). To explore the role of Apel in regulating embryonic hematopoiesis, we recently used the murine ES cell differentiation system and a siRNA approach to knockdown Apel gene expression. We have found that myeloid progenitor development from murine ES cell differentiation in vitro was reduced significantly when *Apel* gene expression was knocked down by *Apel*-specific siRNA. To explore the mechanism whereby Apel regulates hematopoiesis, we found that inhibition of the redox activity of Apel with E3330, a specific Apel redox inhibitor, but not Apel DNA repair activity, which was blocked using the small molecule methoxyamine, affected cytokine-mediated myeloid progenitor development in vitro. These studies reveal the physiological role of Apel normal embryonic myeloid hematopoiesis and that the redox function, but not the repair endonuclease activity, of Apel is critical in regulation of normal embryonic hematopoietic development in mouse (Zou et al, 2007a).

Lymphoid Progenitor from Mouse ES Cell Differentiation

An antibody-secreting B cell is derived from a lymphoid stem cell through a series of developmental stages: progenitor B cell (pro-B cell), precursor B cell (pre-B) cell, immature B cell, and mature B cell stage. Pro-B cell development, like proliferation or differentiation, is regulated by various cytokines, such as IL-3, IL-7 etc. IL-3 stimulates in vitro growth of primary c-kit$^+$B220$^+$ pro-B cells sorted from murine fetal liver (Winkler et al, 1995). Por-B cells can be generated from mouse ES cell differentiation through coculture with OP9 stromal cells (Cho et al, 1999). Pu.1 plays an essential role in the development of all lymphoid lineages; however, it also regulates commitment to other blood cell lineages. We recently provided evidence for early B cell lineage commitment as determined by coexpression of CD19 and B220 when Pu.1 expression was knocked down in HPCs by Pu.1 siRNA. The expression of early B cell factor and paired box protein 5 transcription factors was induced when cells were treated by Pu.1 siRNA. We also identified that knockdown of Pu.1 expression was more efficient in generating progenitor B cells compared with the more common in vitro method of B lymphoid development by means of coculture of CD34$^+$ EB cells with OP9 stromal cells. So, in early B cell development, constitutive Pu.1 expression inhibits the earliest B cell development through repressing early B cell factor and paired box protein 5 expressions (Zou et al, 2005).

Dendritic Cell Generation from Mouse ES Cell Differentiation

Dendritic cells (DCs) are antigen presenting cells that are potent stimulators of both T and B cell-mediated immune responses. They are normally present in very small numbers in the peripheral blood. They can be differentiated in vitro from various cellular sources, including bone marrow, cord blood and PB mononuclear cells. The majority of research and clinical protocols to date differentiate DCs from precursors using GM-CSF in combination with either TNF-α or IL-4 (Zou and Tam, 2002). Since the limitation in number of dendritic cells in blood, investigators have studied to generate DCs from ES cells and suggest the potential clinical application of DC derived from ES cells (Fairchild et al, 2005; Senju et al, 2003). Senju et al (2003) developed a method to generate DCs from mouse ES cells. They cultured ES cells for 10 days on feeder cell layers of OP9, in the presence of GM-CSF in the latter 5 days. Then the ES cell-derived cells were transferred to bacteriologic Petri dishes without feeder cells and further cultured for 7 days, irregularly shaped floating cells with protrusions appeared and these expressed MHC class II, CD11c, CD80, and CD86, with the capacity to stimulate primary mixed lymphocyte reaction and to process and present protein antigen to T cells. The functions of these ES- derived DCs (esDCs) were comparable with those of DCs generated from bone marrow cells. Upon transfer to new dishes and stimulation with IL-4 plus TNF-α, combined with anti-CD40 monoclonal antibody or LPS, the esDCs completely became mature DCs, characterized by a typical morphology and higher capacity to stimulate MLR. Pro-inflammatory Cytokine LIGHT is a type II transmembrane protein belonging to the TNF family that was originally identified as a weak inducer of apoptosis. It is also named CD258. The receptor for LIGHT is HVEM and LT-βR. LIGHT plays a role in inducing maturation of dendritic cells, such as upregulating CD80, CD86 expression on dendritic cells (Zou et al, 2004). We recently used esDCs to examine the regulatory role of LIGHT in chemokine production in DCs. We found that esDCs express CCR7 and CCR10 (the receptor of CCL27) upon the LIGHT stimulation. LIGHT also upregulates CCL27, but not CCL19 and CCL21 expression in those cells. LIGHT stimulation increases the esDC migration potential. Further investigation show that LIGHT activated DCs autocrine CCL27 which regulate their migration as Blockage of CCL27 on esDCs using neutralizing antibody reduces migration potential. In signaling study, we identified that LIGHT activated NF-κB in esDCs and inhibition of NF-κB activation by specific inhibitor can partly attenuate the effect of LIGHT in regulation of CCL27 expression. We demonstrate that pro-inflammatory LIGHT stimulation enhances CCL27 expression through activation of NF-κB in DCs (Zou et al, 2007b).

T Cell Differentiation from Mouse ES Cells

T lymphocyte plays a central role in cell-mediated immunity. Mouse ES cell differentiation also generates T cells (de Pooter et al, 2003). Recombination-activating gene-2 (RAG2) is required for the rearrangement of antigen receptor genes. Mice homozygous for a loss-of-function mutation of the RAG2 do not produce mature B and T lymphocytes. But chimeric mice that result from injection of normal ES cells into blastocysts from RAG2-deficient mice develop normal mature lymphocyte populations, all of which are derived from the injected ES cells; this process is called RAG2-deficient blastocyst complementation. Homozygous Retinoblastoma (Rb) mutant ES cells can develop into phenotypically normal, mature B and T lymphocytes in the RAG-2-deficient background. Rb-deficient B and T cells do not have major defects in either activation or function (Chen et al, 1993). Mature T cell can also be generated from implanted ES-derived EBs in nude mice (Chen et al, 1995). Recently, it has been suggested that directed differentiation of mouse ES cells into T cells required the engagement of Notch receptors by Delta-like 1 ligand expressed on the OP9-DL1 stromal cell line. ES-derived T cell progenitors effectively reconstituted the T cell compartment of immunodeficient mice, enabling an effective response to a viral infection (Schmitt et al, 2004). A transient Flk1^{+}CD45^{-} subset of EB cells are able to differentiate into T lymphocytes in vitro; and the use of reaggregate thymic organ cultures greatly enhanced T lymphocyte frequency. These findings strongly reveal that ES cells can exhibit in vitro T-cell potential.

B Lymphocyte Differentiation from Mouse ES Cells

B lymphocyte plays a key role in the humoral immune response. They are an essential component of the adaptive immune system. Mature B lymphocyte can be generated from the differentiation of mouse ES cells. B lymphocytes generated in vitro from mouse ES cells are functional. These ES-derived B lymphocytes secrete Ig upon stimulation with LPS at a mature mitogen-responsive stage. The ES cell-derived A-MuLV-transformed pre-B (EAB) cells are phenotypically and functionally indistinguishable from standard A-MuLV-transformed pre-B cells derived from infection of mouse fetal liver or bone marrow. EAB cells possess functional V (D)J recombinase activity (Cho et al, 1999).

Mast Cells

Mast cell (or mastocyte) is a resident cell of areolar connective tissue that contains many granules rich in histamine and heparin. Beside their role in allergy and anaphylaxis, they also play an important protective role as well, being intimately involved in wound healing and defense against pathogens. Mast cells differentiate from multipotent hematopoietic progenitors in the bone marrow. Cytokine SCF is a major chemotactic factor for mast cells and their progenitors. It elicits cell-cell and cell-substratum adhesion, facilitates the proliferation, and sustains the survival, differentiation, and maturation, of mast cells (Okayama and Kawakami 2006). Mouse ES cell differentiation produces mast cells. To determine the importance of the prosurvival factors Bcl-2 and Bcl-XL for mast cell development and survival, bcl-$x^{-/-}$ and bcl-$2^{-/-}$ mouse ES cells were maintained in medium supplemented with either IL-3 or IL-3 in combination with SCF to favor mast cell development. Deficiency in either bcl-x or bcl-2 completely inhibited the development of ES-derived mast cells. ES-derived mast cells lacking bcl-x or bcl-2 exhibited strong expression of A1, another prosurvival Bcl-2 family member. Both bcl-x and bcl-2 are indispensable for survival of mast cells derived from mouse ES cell differentiation (Moller et al, 2007).

Neutrophils

Neutrophils are a type of white blood cell, specifically a form of granulocyte, These cells contain neutrally-staining granules, tiny sacs of enzymes that help the cell to kill and digest microorganisms it has engulfed by phagocytosis. The mature neutrophils have a segmented nucleus while the immature neutrophil has band-shape nucleus. The lifespan of neutrophil is about 3 days. Neutrophilia, an increased proportion of neutrophils in the blood, is a common finding with acute bacterial infections. A nouse ES cell/OP9 coculture system is effective to produce functional neutrophils. The OP9 stromal cell line significantly enhances the production of neutrophils from ES cell differentiation. Morphologic and functional evaluation of these ES-derived neutrophils indicates that most of them are mature neutrophils. Their morphology, ability to produce superoxides, flux calcium, undergo chemotaxis in response to MIP-2, and contain the neutrophil-specific markers Gr-1 suggests that they are comparable with purified mouse bone marrow neutrophils (Lieber et al, 2004). Successful differentiation of ES cells into neutrophils provides a useful model system for studying neutrophil differentiation and function and the factors that regulate them.

Endothelial Cell Differentiation from Mouse ES Cells

Endothelial cells are the thin and flattened cells, and a layer of them lines the inside surfaces of body cavities, blood vessels, and lymph vessels, making up the endothelium. They are involved in many aspects of vascular biology, such as vasoconstriction and vasodilation, and hence the control of blood pressure, blood clotting, atherosclerosis, angiogenesis, inflammation and swelling (oedema). Vascular endothelial cells regulate both normal hematopoiesis (Ohneda et al, 1997) and malignant hematopoiesis (Hatfield et al, 2006). Endothelial cells are also one of cell member of HSC niche (Zou, 2007). Hematopoietic cells and endothelial cells can be derived from a common precursor cell named hemangioblast in embryonic stage. Moreover, some evidence suggested that hematopoietic cells can emerge from endothelial cells (Li et al, 2005). Endothelial cell can be generated from mouse ES cell differentiation (Marchetti et al, 2002; Sauer et al, 2005; Gimond et al, 2006). Marchetti et al (2002) described a novel genetic approach that selects for endothelial cells from differentiating ES cells. The strategy they used is based on the establishment of ES-cell clones that carry an integrated puromycin resistance gene under the control of a vascular endothelium-specific promoter, tie-1. Using EGFP as a reporter gene, they have confirmed the endothelial specificity of the tie-1 promoter in the embryoid body model. They suggested that application of endothelial lineage selection to differentiating ES cells will be a useful approach for future pro-angiogenic and endothelial cell replacement therapies. To explore the molecular mechanisms driving angiogenesis in tissues derived from ES cells are currently unknown. Sauer et al (2005) recently investigated the effects of direct current electrical field treatment on endothelial cell differentiation of mouse ES cells. Treatment of ES cell-derived embryoid bodies with field strengths ranging from 250 V/m to 750 V/m, dose-dependently increased the capillary area staining positive for CD31. Pretreatment with the JNK inhibitor SP600125 resulted in a significant decrease in capillary areas, whereas the p38 inhibitor SB203580 was without effects. Endothelial cells derived, purified and expanded in vitro from ES cells sustain an important endothelial cell function, the ability to undergo vasculogenesis in collagen gels. These studies indicate that endothelial products derived in vitro from stem cells could be useful in regenerative medicine applications (McCloskey et al, 2005). Using ES-derived endothelial cells, it is suggested that PDK1 is essential for the motility of vascular endothelial cells (Primo et al, 2007)

Application of Mouse ES Cell Differentiation to Study Leukemiagenesis

Cancers are suggested to originate in stem cells through the accumulation of multiple mutations. Some of these mutations result in a loss of heterozygosity (LOH). A recent report demonstrates that exposure of mouse ES cells to nontoxic amounts of mutagens increase the frequency of LOH. Therefore, mutagen induction of LOH in ES cells suggests a new pathway to account for the multiple homozygous mutations in human tumors. This induction is able to mimic early mutagenic events that generate cancers in human tissue stem cells (Bielas et al, 2007). Mouse ES differentiation generates both primitive erythroid and definitive myeloid lineages in a process that mimics hematopoiesis in the mammalian yolk sac. To investigate leukemic transformation of these embryonic hematopoietic progenitors, mouse ES cells were transfected with the Chronic Myeloid Leukemia-specific *BCR/ABL* gene. Following the culture for differentiation, two transformed subclones, EB57 and EB67, were isolated. They retained characteristics of embryonic hematopoietic progenitors and induced a fatal leukemia in mice characterized by massive splenomegaly and granulocytosis. EB57 secretes IL-3 and unidentified cytokines that can stimulate autocrine and paracrine cell proliferation, presumably accounting for the reactive granulocytosis in diseased mice (Peters et al, 2001). In BCR-ABL transfected mouse CCE ES cells, the major effect of BCR-ABL expression was the persistence of primitive morphology of ES cells despite LIF deprivation, correlated with a constitutive activation of STAT3, the major self-renewal factor of ES cells, but no evidence of activation of STAT5. To study the relevance of STAT3 activation by BCR-ABL in human CML, Coppo et al (2003) demonstrated an increased levels of STAT3 proteins phosphorylated both on tyrosine and serine residues on the $CD34^+$ cells, purified from CML patients at different stages of their disease. These studies indicate the potential functional link between BCR-ABL oncogene and a self-renewal in the context of ES cells through constitutive activation of STAT3. Thus, the BCR-ABL ES cell model suggests a role for STAT3 in the pathogenesis of human CML.

Conclusions

It is obvious that mouse ES cell differentiation take advantage in study the molecular event in regulation of embryonic hematopoiesis. Most type of hematopoietic cells can be generated from mouse ES cell differentiation in vitro. ES cell differentiation can be used to test candidate growth factors and cytokines for initiation and support of embryonic hematopoiesis. ES-derived EBs also provides a valuable source for new cytokines that stimulate primitive and/or definitive hematopoiesis. Moreover, this stem cell differentiation model also exhibits the implication in the study of leukemiagenesis.

References

Bielas JH, Venkatesan RN, Loeb LA. LOH-proficient embryonic stem cells: a model of cancer progenitor cells? *Trends Genet.* 2007. 23:154-157.

Bradley A, Evans M, Kaufman MH, Robertson E. Formation of germ-line chimaeras from embryo-derived teratocarcinoma cell lines. *Nature*. 1984. 309:255-256.

Burkert U, von Ruden T, Wagner EF. Early fetal hematopoietic development from in vitro differentiated embryonic stem cells. *New Biol.* 1991. 3:698-708.

Chen J, Gorman JR, Stewart V, Williams B, Jacks T, Alt FW. Generation of normal lymphocyte populations by Rb-deficient embryonic stem cells. Curr Biol. 1993. 3:405-413.

Chen U, Mok H. Development of mouse embryonic stem (ES) cells: IV. Differentiation to mature T and B lymphocytes after implantation of embryoid bodies into nude mice. *Dev Immunol.* 1995. 4:79-84.

Cho SK, Webber TD, Carlyle JR, Nakano T, Lewis SM, Zuniga-Pflucker JC. Functional characterization of B lymphocytes generated in vitro from embryonic stem cells. *Proc Natl Acad Sci U S A*. 1999. 96:9797-9802.

Chung YS, Zhang WJ, Arentson E, Kingsley PD, Palis J, Choi K. Lineage analysis of the hemangioblast as defined by FLK1 and SCL expression. *Development.* 2002. 129:5511-5520.

Coppo P, Dusanter-Fourt I, Millot G, Nogueira MM, Dugray A, Bonnet ML, Mitjavila-Garcia MT, Le Pesteur D, Guilhot F, Vainchenker W, Sainteny F, Turhan AG. Constitutive and specific activation of STAT3 by BCR-ABL in embryonic stem cells. *Oncogene*. 2003. 22:4102-4110.

de Pooter RF, Cho SK, Carlyle JR, Zuniga-Pflucker JC. In vitro generation of T lymphocytes from embryonic stem cell-derived prehematopoietic progenitors. *Blood*. 2003. 102:1649-1653.

Evans MJ, Kaufman MH. 1981. Establishment in culture of pluripotential cells from mouse embryos. *Nature*. 292:154-156.

Evans AR, Limp-Foster M, Kelley MR. Going APE over ref-1. *Mutat Res*. 2000. 461:83-108. Review.

Fairchild PJ, Nolan KF, Cartland S, Waldmann H. Embryonic stem cells: a novel source of dendritic cells for clinical applications. *Int Immunopharmacol*. 2005. 5:13-21.

Forrai A, Robb L. The hemangioblast--between blood and vessels. *Cell Cycle*. 2003. 2:86-90. Review.

Gimond C, Marchetti S, Pages G. Differentiation of mouse embryonic stem cells into endothelial cells: genetic selection and potential use in vivo. *Methods Mol Biol*. 2006. 330:303-329.

Hatfield K, Ryningen A, Corbascio M, Bruserud O. Microvascular endothelial cells increase proliferation and inhibit apoptosis of native human acute myelogenous leukemia blasts. *Int J Cancer*. 2006. 119:2313-2321.

Kamata T, Kang J, Lee TH, Wojnowski L, Pritchard CA, Leavitt AD. A critical function for B-Raf at multiple stages of myelopoiesis. *Blood*. 2005. 106:833-840.

Kelley MR, Parsons SH. Redox regulation of the DNA repair function of the human AP endonuclease Apel/ref-1. *Antioxid Redox Signal.* 2001. 3:671-683

Kennedy M, Firpo M, Choi K, Wall C, Robertson S, Kabrun N, Keller G. A common precursor for primitive erythropoiesis and definitive haematopoiesis. *Nature.* 1997. 386:488-493.

Lacaud G, Robertson S, Palis J, Kennedy M, Keller G. Regulation of hemangioblast development. *Ann N Y Acad Sci.* 2001. 938:96-107.

Lengerke C, McKinney-Freeman S, Naveiras O, Yates F, Wang Y, Bansal D, Daley GQ. The Cdx-Hox pathway in hematopoietic stem cell formation from embryonic stem cells. *Ann N Y Acad Sci.* 2007. 1106:197-208

Li W, Ferkowicz MJ, Johnson SA, Shelley WC, Yoder MC. Endothelial cells in the early murine yolk sac give rise to CD41-expressing hematopoietic cells. *Stem Cells Dev.* 2005 14:44-54.

Lieber JG, Webb S, Suratt BT, Young SK, Johnson GL, Keller GM, Worthen GS. The in vitro production and characterization of neutrophils from embryonic stem cells. *Blood.* 2004. 103:852-859.

Lindenbaum MH, Grosveld F. An in vitro globin gene switching model based on differentiated embryonic stem cells. *Genes Dev.* 1990. 4(12A):2075-2085.

Marchetti S, Gimond C, Iljin K, Bourcier C, Alitalo K, Pouyssegur J, Pages G. Endothelial cells genetically selected from differentiating mouse embryonic stem cells incorporate at sites of neovascularization in vivo. *J Cell Sci.* 2002. 115:2075-2085.

Martin GR. 1981. Isolation of a pluripotent cell line from early mouse embryos cultured in medium conditioned by teratocarcinoma stem cells. Proc. *Natl. Acad. Sci. U S A.* 78:7634-7638.

McCloskey KE, Gilroy ME, Nerem RM. Use of embryonic stem cell-derived endothelial cells as a cell source to generate vessel structures in vitro. *Tissue Eng.* 2005. 11:497-505.

Moller C, Karlberg M, Abrink M, Nakayama KI, Motoyama N, Nilsson G. Bcl-2 and Bcl-XL are indispensable for the late phase of mast cell development from mouse embryonic stem cells. *Exp Hematol.* 2007. 35:385-393.

Ohneda O, Bautch VL. Murine endothelial cells support fetal liver erythropoiesis and myelopoiesis via distinct interactions. *Br J Haematol.* 1997. 98:798-808.

Okayama Y, Kawakami T. Development, migration, and survival of mast cells. *Immunol Res.* 2006. 34:97-115.

Perlingeiro RC, Kyba M, Bodie S, Daley GQ. A role for thrombopoietin in hemangioblast development. *Stem Cells.* 2003. 21:272-280.

Peters DG, Klucher KM, Perlingeiro RC, Dessain SK, Koh EY, Daley GQ. Autocrine and paracrine effects of an ES-cell derived, BCR/ABL-transformed hematopoietic cell line that induces leukemia in mice. *Oncogene.* 2001. 20:2636-46.

Primo L, di Blasio L, Roca C, Droetto S, Piva R, Schaffhausen B, Bussolino F. Essential role of PDK1 in regulating endothelial cell migration. *J Cell Biol.* 2007. 176:1035-1047.

Robertson S, Kennedy M, Keller G. Hematopoietic commitment during embryogenesis. *Ann N Y Acad Sci.* 1999. 872:9-15.

Sato M, Nakano T. Embryonic stem cell. *Intern. Med.* 2001. 40:195-200.

Sauer H, Bekhite MM, Hescheler J, Wartenberg M. Redox control of angiogenic factors and CD31-positive vessel-like structures in mouse embryonic stem cells after direct current electrical field stimulation. *Exp Cell Res.* 2005. 304:380-390.

Schmitt TM, de Pooter RF, Gronski MA, Cho SK, Ohashi PS, Zuniga-Pflucker JC. Induction of T cell development and establishment of T cell competence from embryonic stem cells differentiated in vitro. *Nat Immunol.* 2004. 5:410-417.

Senju S, Hirata S, Matsuyoshi H, Masuda M, Uemura Y, Araki K, Yamamura K, Nishimura Y. Generation and genetic modification of dendritic cells derived from mouse embryonic stem cells. *Blood.* 2003. 101:3501-3508.

Wang Y, Yates F, Naveiras O, Ernst P, Daley GQ. Embryonic stem cell-derived hematopoietic stem cells. *Proc Natl Acad Sci U S A.* 2005. 102:19081-19086.

Weiss MJ, Orkin SH. In vitro differentiation of murine embryonic stem cells. New approaches to old problems. *J Clin Invest.* 1996. 97:591-595. Review.

Winkler TH, Melchers F, Rolink AG. Interleukin-3 and interleukin-7 are alternative growth factors for the same B-cell precursors in the mouse. *Blood.* 1995. 85:2045-2051.

Xanthoudakis S, Smeyne RJ, Wallace JD, Curran T. The redox/DNA repair protein, Ref-1, is essential for early embryonic development in mice. *Proc Natl Acad Sci U S A.* 1996. 93:8919-8923.

Zou GM. Protein tyrosine phosphatase Shp-2 in Cytokine signalings and stem cell regulations. In *"Stem Cell Research Developments"*. Calvin A. Fong (Editor). Nova Science Publishers, Inc. Hauppauge. NY. 2007. pp.107-134.

Zou GM, Chen JJ, Yoder MC, Wu W, Rowley JD. Knockdown of PU.1 by siRNA on CD34+ EB cells derived from mouse embryonic stem cell turn cell fate determination to pro-B cells. *Proc Natl Acad Sci USA.* 2005. 102:13236-13241

Zou GM, Luo MH, Reed A, Kelley MR, Yoder MC. Ape1 regulates hematopoietic differentiation of embryonic stem cells through its redox functional domain. *Blood.* 2007a. 109:1917-1922.

Zou GM, Chan RJ, Shelley WC, Yoder MC. Knockdown of Shp-2 by siRNA on mouse embryonic stem cell derived EB cells affects hematopoietic development. *Stem Cells.* 2006. 24:587-594.

Zou GM, Martinson J, Hu WY, Tam Y, Klingemann HG. The effect of LIGHT in inducing maturation of monocyte-derived dendritic cells from MDS patients. *Cancer Immunol Immunother.* 2004. 53:681-689.

Zou GM, Hu WY, Wu W. TNF family molecule LIGHT regulates chemokine CCL27 expression on mouse embryonic stem cell-derived dendritic cells through NF-kB activation. *Cell Signal.* 2007b. 19:87-92.

Zou GM, Tam YK. Cytokines in the generation and maturation of dendritic cells: recent advances. *Eur Cytokine Netw.* 2002. 13:186-199.

Zou GM, Wu W, Chen J, Rowley JD. Duplexes of 21-nucleotide RNAs mediate RNA interference in differentiated mouse ES cells. *Biol Cell.* 2003. 95:365-371.

In: Cell Differentiation Research Developments
Editor: L. B. Ivanova, pp. 217-231
ISBN: 978-1-60021-939-9

Chapter X

ROLE OF FSH AND TESTOSTERONE IN MALE GERM CELL DIFFERENTIATION

Andrés Aragón[1,*], María Elena Ayala[2], Simón Martínez[3] and Nazario Pescador[1]

[1]Laboratorio de Biología de la Reproducción, Facultad de Medicina Veterinaria y Zootecnia, Universidad Autónoma del Estado de México, México;
[2]Unidad de Biología de la Reproducción, Laboratorio de Pubertad, Facultad de Estudios Superiores Zaragoza, Universidad Nacional Autónoma de México, México City, México;
[3]Centro de Investigacion y Estudios Avanzados en Salud Animal, Facultad de Medicina Veterinaria y Zootecnia, Universidad Autónoma del Estado de México, México.

ABSTRACT

Follicle stimulating hormone (FSH) and testosterone have been described as necessary factors for proliferation of Sertoli and germ cells, as well as having a role in apoptosis. However, their role in germ cell differentiation has not been studied in detail. FSH is a proteic hormone produced in the anterior hypophysis and has receptors located in Sertoli cells, which in turn function as nurse cells for germ cells in the seminiferous epithelium. The expression of FSH receptors changes during the cycle of seminiferous epithelium. Germ cells in consecutive steps of differentiation are located in each progressive stage of the seminiferous epithelium, indicating a role for FSH signaling in the differentiation of germ cells. CREB (CRE-protein binding) is a transcription factor involved in differentiation of germ cells and it is modulated by FSH. Some CREM-regulated products have been shown to participate in germ cell differentiation. Recent studies have focused on the intracellular signaling after FSH joins with its receptor, and how it influences germ cell differentiation. Testosterone-stimulated germ cell development occurs via paracrine communication with Sertoli cells. The androgen

* *Correspondence concerning this article should be addressed to: Andrés Aragón, email: andresammx@yahoo.com.

receptor is expressed in a stage-specific manner during the seminiferous epithelium cycle. Recent studies on gene expression show a number of genes modulated by testosterone, many of which do not follow the traditional pathway of androgen/receptor/gene expression, but are activated by another signalization pathway. This chapter begins with the background necessary to understand the general role of FSH and testosterone in seminiferous epithelium and then addresses the molecular signaling initiated by these chemical messengers.

INTRODUCTION

Cell differentiation occurs as a response to the signals of neighbor cells. This interaction permits the differentiating cell to express its specialized character in the overall context of location within a specific tissue. Spermatogenesis is a process constant of proliferation and meiosis, in which every daughter cell is compromised toward germ cell differentiation (spermatogonia, spermatocytes and spermatids) or to function in the renewal of its own population (spermatogonia).

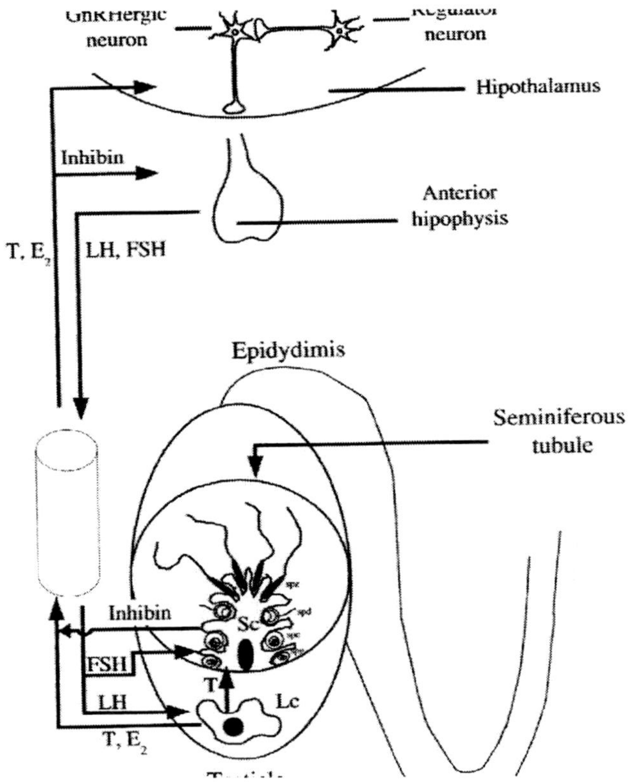

Figure 1. Schematic representation of the principal chemical messengers regulating spermatogenic function. T, testosterone; E_2, estradiol; LH, luteinizing hormone; FSH, follicle- stimulating hormone; Sc, Sertoli cell; Lc, Leydig cell; spg, spermatogonium; spc, spermatocyte; spd, spermatid; spz, spermatozoon.

It is widely accepted that the spermatogenic process is regulated by the gonadotropins follicle-stimulating hormone and luteinizing hormone which, in turn, are released by the gonadotropes-gonadotropin release hormone (GnRH). GnRHergic neurons located in the hypothalamus are regulated by other neurons, of which the nervous endings contact GnRHergic neurons (Figure 1). There is feedback among the sexual steroids produced by the testicles (progesterone, testosterone, and estradiol), the gonadotropes in the anterior hypophysis, and the hypothalamus.

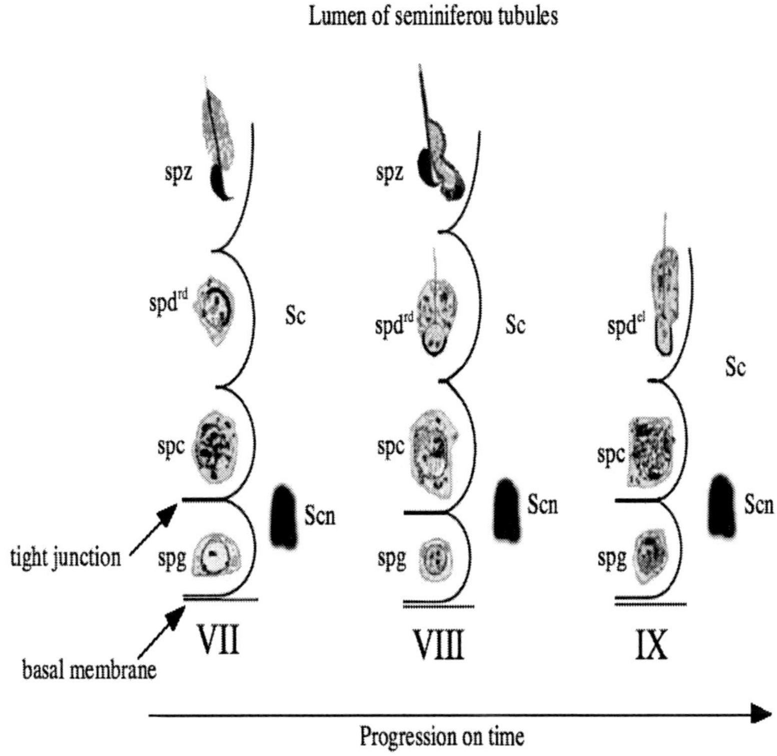

Figure 2. Germ cell associations in stages of mouse seminiferous epithelium. Germ cells with distinct morphology in stages VII-IX are depicted. All transformations (dynamic differentiation) occur in a segment of the seminiferous tubule. spg, spermatogonium; spc, spermatocyte; spdrd, round spermatid, spd^{el}, elongated spermatid; spz, spermatozoid; Sc, Sertoli cell; Nsc, Sertoli cell nucleus.

Spermatogenesis occurs into the seminiferous tubules, out of which Leydig cells and blood vessels are located. In the seminiferous epithelium, the spermatogonium, located in the basal compartment, is the most primitive germ cell. Proliferation of spermatogonies occurs by mitosis. By the process of meiosis, the spermatocytes, located in adluminal compartment, give origin to the spermatids. Spermatids undergo a dramatic process of differentiation that converts them into the highly specialized sperm cells. Histologically, spermatogenesis progression is observed as a series of consecutives stages, each one characterized by a particular association of germ cells. A complete series of changes in cell associations (stages) arranged in the logical sequence of developmental progression is called the cycle of seminiferous epithelium. These associations develop and remain in one region of a seminiferous tubule. A segment of a tubule contains (in rodents) a single association; whereas, a complete series of segments constitutes a wave of seminiferous epithelium [1]. The number of stages and the type of associations is specie-specific. For example, there are twelve stages in mouse, fourteen in rat, and six in man. This classification divides a continuous process into discrete segments (Figure 2), then makes it possible to investigate the expression and role of individual factors along the spermatogenic process. Such an arrangement of the seminiferous epithelium means that inducing-differentiation factors are present in the seminiferous epithelium in a stage-specific manner.

A distinct and well-defined population of germ cells interacts with Sertoli cells in a cyclic pattern. Sertoli cells secrete a large number of proteins and regulatory factors that are crucial for germ cell development [2], and apparently germ cells are needed to maintain that cycle of Sertoli cells [3,4]. Spermatogonies/spermatocytes in the basal compartment of seminiferous epithelium are in contact with the basal lamina and Sertoli cells, whereas pachytene spermatocytes and spermatids are located in the adluminal compartment. Mature spermatozoa are located apically in Sertoli cells. Both basal and adluminal compartments are formed by tight junctions between adjacent Sertoli cells known as the blood-testis barrier (BTB) [5]. As germ cells move into the adluminal compartment, tight junctions between adjacent Sertoli cells from behind the germ cells, creating the BTB. The BTB segregates meiotic and postmeiotic cells into immunologically privileged adluminal compartment [5]. This cellular arrangement indicates that chemical messages stimulating development and differentiation of germ cells in adluminal compartment should be processed by Sertoli cells, and that germ cells in the basal compartment receive chemical messages distinct from those in the adluminal compartment.

The Role of FSH in Spermatogenesis

It is well established that the FSH plays an important role in testicular development and spermatogenesis. In vivo FSH has been show as a survival factor [6,7]. In the adult rat and in humans, serum FSH levels remain relatively constant. However, the expression of FSH receptor (FSHR) on Sertoli cells varies more than three fold in a cyclical and stage-specific manner such that receptor levels are highest in stages XIII-II and minimal during stages VII-VIII [8,9]. Because FSH receptor are found exclusively in the Sertoli cells, but not in germ cells themselves, it is likely that the FSH stimulation of Sertoli cells results in the induction of an intratubular factor essential for the survival of the male germ cells, as early suggested by Billig *et al.*, [7].

FSH binding to its receptor is known to activate at least five signaling pathways in Sertoli cells: cAMP-PKA, MAP kinase, calcium, phophatidylinositol 3-kinase, and phospholipase A2 patways (reviewed in [10]). The testicular actions of FSH are mediated via a specific G protein-coupled cell surface receptor uniquely expressed in Sertoli cells [11]. Ligand-dependent activation of the FSHR initiates the production of intracellular cAMP [12] and Ca^2 [13] second messengers. The levels of FSH-induced cAMP production in Sertoli cells closely follows the levels of FSH receptor with highest levels of cAMP observed in stages XIV-VI [14]. FSHR in perinatal animals induces a molecular signaling cascade that induces phosphatidylinositol 3-kinase/AKT pathway, it promotes aromatase expression and estradiol production, which stimulate the proliferation of Sertoli cells [15]. One target for the increase in cAMP and PKA is a class of transcription factors that bind to cAMP response elements (CREs). Specifically, the CRE binding protein (CREB) transcription factor is rapidly activated after being phosphorylated on serine 133 by PKA in response to FSH stimulation [16].

CREB may induce transcription factors required to activate others genes supportive of spermatogenesis [17], hormones that regulate spermatogenesis [18], and apoptotic factors [19,20]. FSH and cAMP induce expression of the stem cell factor (SCF) in Sertoli cells, and there are three potential CREB-binding sites in the SCF promoter [21,22]. It is knowed that in absence of SCF stimulation, spermatogonia and spermatocytes undergo apoptosis [23,24].

The CREB transcripts in Sertoli cells fluctuates in a cyclical manner that depend on the specific cell association along the spermatogenic wave. In Sertoli cells of rats CREB mRNA is present in stages I-III of the seminiferous epithelium, highest in stages II-VI, and decreases to nearly undetectable levels during stages IX-XIV [25]. Sertoli cells do not appear to undergo apoptosis or necrosis when CREBm1 (mutant of CREB that cannot be phosphorylated on serine 133) is over expressed, implying that CREB is nor required for Sertoli cell survival [26]. In Sertoli cells, CREB is known to regulate directly or to be linked to the regulation of a number of genes required for differentiation of spermatozoa [17].

Because of the blood-testis barrier, the Sertoli cell must provide many factors required for the maintenance and development of germ cells [27]. Several genes that encode for proteins involved with microtubule and cytoskeleton rearrangement were identified as being regulated by FSH. These include actins, tubulins, lamin A, ankrin 3, and catenin alpha 2 [28]. The change in expression of few genes such as junctional adhesion molecule and scinderin whose products are involved in cell adhesion indicated a role for FSH in the adhesion of germ cells to Sertoli cells [28].

FSH has also been implicated in the expression of various isoforms of CREM (cAMP-responsive element modulator) that are expressed in spermatocytes and spermatids and are required for survival [29,30]. However, as suggested by prior researchers [10], this regulation of germ cell CREM by FSH must be indirectly mediated trough as yet unidentified Sertoli cell factors. CREM is undetectable in prepubertal animals but highly abundant in adults [31]. CREM accumulates in haploid spermatids and is required for transcriptional activation of several post-meiotic genes [31,32]. Most of these genes encode structural proteins required for differentiation of spermatozoa [17,28]. On transgenic male rats -/- CREM germ cell development is arrested in spermiogenesis and the cell are eliminated by apoptosis [33,30].

The CREB/CREM system can negatively autoregulate itself. Another CREM isoform, inducible cAMP early repressor (ICER), encodes a powerful repressor of cAMP-responsive genes [34]. In Sertoli cells, the only member significantly expressed is the ICER repressor, whereas, all the cAMP-dependent gene activation is carried out by CREB. In germ cells, however, the task of executing gene activation falls mainly on CREM. Seminiferous epithelium analysis of CREM-mutant male mice revealed post-meiotic arrest at the first step of spermiogenesis and a dramatic decline in post-meiotic gene expression. However, FSH and testosterone levels were not affected by the mutation [33].

FSH is generally considered a differentiation promoting agent for Sertoli cells [35]. This corresponds to ID proteins generally being considered as proliferation-promoting and differentiation-inhibiting factors. The ability of FSH to supress ID expression correlates with an increased differentiation and decreased proliferation of the cells [36]. A overexpression of ID and ID2 genes in a posmitotic, terminally differentiated cell type have the capacity to induce reentry into cell cycle [35].

The Role of Testosterone in Spermatogenesis

Testosterone and Androgen Receptor in the Seminiferous Epithelium

Hormonal regulation of the spermatogenic process occurs during three phases: initiation, maintenance and re-initiation [37]. Spermatogenesis in gonadotropin-deficient mice is qualitatively initiated by androgens [38]. Androgen is necessary for the completion of meiosis and differentiation of round spermatids into spermatozoa. If androgens are eliminated by hypophysectomy, there is a partial loss of round spermatids, with elongated spermatids observed [39], and if residual testosterone is eliminated by using a Leydig cell citotoxic such as dimethanesulphonate, or by a competitor of androgen such as flutamide, only a few round spermatids are observed and no elongated spermatids are present [40,41]. It is thus clear that androgen plays a very important role in the development and differentiation of germ cells.

The role of androgen in spermatogenesis is mediated by androgen receptor (AR). Circulating androgens dissociate from carrier proteins, specifically sex hormone-binding globulin, prior to diffusing into target cells and binding to intracellular AR. The unliganded receptor is an inactive oligomer complexed to heat shock proteins (eg. Hsp90, Hsp70) and located in the cytoplasm. The oligomeric complex dissociates when bound to hormone, and the receptor undergoes a conformational change while being transporting to the nucleus, and while binding as a homodimer to DNA hormone-responsive elements [42]. AR is located in Leydig, peritubular myoid, and Sertoli cells, although localization in germ cells is controversial [43,44,45]. In this sense, some studies indicate that the presence of AR in germ cells is not necessary for normal germ cell development [46,47] and fertility [46]. Paracrine communication between somatic and germ cells thus appears to be responsible for germ cell differentiation, although the mechanisms involved are poorly understood. AR protein expression in Sertoli cells varies as a function of the cycle of seminiferous epithelium For example, in rat Sertoli cells, AR is first detected at late stage IV or early stage V, and high expression is maintained until spermiation in stage VIII. Expression disappears completely by stage IX [48,49].

Sertoli, peritubular myoid, and Leydig cells have the ability to mediate the transcriptional effects of androgen, and receptor concentrations may be a limiting factor for responsiveness [50]. The cell-autonomous AR actions in Sertoli cells is an absolute requirement for androgen maintenance of complete spermatogenesis, and spermatocyte/spermatid development/survival depends critically on androgens. However, which factors are regulated by AR/androgen and how they stimulate spermatogenesis progression is poorly understand. Recently, the use of AR-knockout or ABP-overexpressing models and microarray technology have revealed which genes are AR-regulated and have shed light on the general role of AR/androgen interactions in the spermatogenesis process.

Lessons of Knockout Models in AR Function

Studies using total or cell-specific AR-knockout male mouse lines have demonstrated a differential effect of AR in spermatogenesis. A Sertoli cell-selective knockout of AR increased germ cell apoptosis and grossly reduced expression of genes specific for late spermatocyte or spermatid development (AR exon 2 locus, Cre, Pem, A-myb, Pro-Acrosin binding protein, PT1, PT2, Protamin 1 and 2, anti-Mullerian hormone, ABP, cyclin A1 and sperm-1) [51,52]. These findings indicate that AR expression and function in testicular cells other than germ cells, play a critical role in normal spermatogenesis. Differences in expression of cytoskeletal and extracellular matrix-related genes were found in Sertoli cell AR-knockout mice [53], and are related to the induction of a broad spectrum of functional defects, such as alteration in movements of developing germ cells, impaired functional integrity of the BTB, transport protein and paracrine factor production and/or secretion [54]. These defects impair nursery functions for developing cells and lead to the arrest of spermatogenesis before the second wave of meiosis [52,54,55,56]. In total or Sertoli cell AR-knockout mouse lines, AR is essential for the completion of meiosis. In the total AR-knockout line, spermatogenesis does not progress beyond the pachytene spermatocyte stage (i.e. before meiosis II initiation, stage X), whereas Sertoli cell AR-knockout germ cells develop until the diplotene spermatocyte (approximately coincident with meiosis initiation, stage XI) [46,52], and Leydig cell AR-knockouts reach the diplotene/meiosis II stage (round spermatid, stage XII). Interestingly, spermatogenesis is complete and the sperm produced are functional in germ cell AR-knockout mice [46]. Thus androgen/AR signaling in Sertoli cells plays the most important role in spermatogenesis [46]. Another hand, *in vitro* [57] and *in vivo* [46,52,55] non-functional AR leads to a diminished steroidogenic function and to spermatogenesis arrest, principally at the round spermatid stage.

Reduction of androgen levels in transgenic male mice overexpressing ABP increased the expression of a large number of genes, and only a small number were decreased [58], whereas in hypogonadal mice, testosterone treatment repressed rather than stimulated gene expression [59]. This phenomenon may represent a compensatory mechanism activated in testosterone-deprived Sertoli cells, or some of the activated genes may actually be negative regulators of Sertoli cell functions required for the support of spermatogenesis [58].

Chemical communication between androgen-producing Leydig and Sertoli cells is very important for the maintenance of spermatogenesis. Although androgen and AR have important roles in Sertoli cells, loss of functional AR in Leydig cells has a major influence on Leydig cell steroidogenic function (i.e. testosterone production decreased) leading to spermatogenesis arrest, predominately at the round spermatid stage [46].

AR and the Blood-Testis Barrier

Prior to puberty the Sertoli cells undergo active cell proliferation, and at the onset of puberty they become a terminally differentiated posmitotic cell population that support spermatogenesis. The molecular mechanisms involved in the posmitotic block of pubertal and adult Sertoli cells are unknown [35]. Sertoli cell fate is established in the embryonic gonad at the time of sex determination [60,61] and is followed by a phase of rapid proliferation and differentiation. During puberty, the final phase of Sertoli cell differentiation occurs, which is marked by cessation of proliferation, and irreversible changes in Sertoli cell morphology and physiology [62]. The changes associated with terminal differentiation of Sertoli cell at puberty include exit from the cell cycle and the formation of BTB [35]. This differentiated phenotype is needed for the proper microenvironment and cytoarchitectural support of the developing spermatogenic cells. Sertoli cell differentiation is accompanied by the expression of many gene products that are not present in immature cells [45,63,64,65,66,67].

During spermatogenesis, basally located spermatogonia differentiate into the preletotene/leptotene stages of primary spermatocytes, and initiate meiosis. These primary spermatocytes must migrate from the basal compartment to the BTB in late stage VII to early stage IX of the cycle of seminiferous epithelium in mouse testis [68].

Loss of functional AR in Sertoli cells causes a broad spectrum of functional defects that lead to the arrest of spermatogenesis before the second wave of meiosis [52,55,56]. A cause of this phenomenon is the disruption of the BTB [54]. In Sertoli cell AR-knockout mice the expression of tight junction-related genes such as cystatin-TE [69], ocludin [54], claudin 3 [5] and claudin II [54,69], is decreased, so it is clear that the absence of AR impairs functional tight junction formation and integrity.

Sertoli and Germ Cells as Regulators of Cycle of the Seminiferous Epithelium

It is as yet not clear whether Sertoli or germ cells play the dominant role in regulating the seminiferous epithelium cycle. Timmons et al., [70] showed in WT mice that the somatic epithelial cycle can be maintained in the absence of germ cells for at least the first week of neonatal life, but Meng et al., [5] suggest that androgen could regulate the microenvironment within the seminiferous epithelium, creating conditions that are permissive, rather than instructive, for germ cell differentiation. Notably, rat germ cells transplanted into a mouse testis differentiate with rat cell kinetics [71]. The fact that embryonic stem cells can differentiate into germ cells *in vitro* in the absence of seminiferous epithelium supports the latter hypothesis [5].

An early research [72] suggested that there is Sertoli cell cycle, whereas and as Franca et al., [71] observed also germ cell cycles. However, the communication mechanisms that would permit synchronization of the cycles, and which are the molecular players involved in the dynamic cell differentiation (Sertoli and germ cells) are not known. However, a series of studies support parallel cycles. The absence of a functional AR in Sertoli cells prevents the progression of germ cell development [46,52,55] is indicative of a tight communication/synchronization of cycles between somatic and germ cells, and every germ-cell type expresses a group of genes in a stage-specific manner during spermatogenesis development [73,74,75]. Some of these genes could be expressed in all germ cells that are regulated during the cycle of seminiferous epithelium [73], which supports the idea that differential communication could synchronize Sertoli and germ cell cycles.

From all above information, androgen could trigger a transcriptional response in Sertoli cell, leading to instructive signals to the germ cells. Microarray studies on androgen function (AR-knockout [53], ABP-overexpression [58] or hypogonadal model [59]) show both up- and down-regulation of a great number of genes. However, at present only two genes are known to be expressed in Sertoli cells that have androgen-responsive elements that respond directly to androgens: the placental and embryo oncofetal gene Pem [76] and the protooncogene c-Myc [77]. Alternative mechanisms of androgen action, for instance, via conversion into estrogens [78] or via non-genomic pathways [79,80,81] thus must be considered to explain the development and differentiation of germ cells. Cheng *et al.* [80] and Fix *et al.*, [81] demonstrated the activation of Src kinase and thus a MAPK cascade and CREB transcription factor by testosterone stimulus in Sertoli cells. Expression of three genes (LDH-A, CREB and Egr1) was observed in testosterone-stimulated Sertoli cells, none of which is regulated by androgen-responsive elements in their promoters, a finding that is consistent with the paradigm that androgen can activate Sertoli cell expression by means of MAP kinase pathway and CREB [81]. CREB and Egr1 are also regulated by a decrease in testosterone concentrations [58]. A scheme for testosterone signaling pathways in Sertoli cells has been proposed by Walker *et al.*, [10].

CONCLUSIONS

The need for testosterone stimulation in meiotic/postmeiotic differentiation in germ cells is clear. However, the testosterone-induced factors involved in the communication/ synchronization between Sertoli and germ cells are largely unknown.

Gene expression data from FSH or testosterone-regulation demonstrate the complexity of the regulation of differentiation in seminiferous epithelium. To date, the need of FSH and testosterone in stage-specific development is relatively known, but spatial regulation along a tubule has not been explored. As mentioned above, a cell association develops and remains in one region of the seminiferous tubule, but it remains to be demonstrated how the chemical communication between adjacent Sertoli cells occurs, and particularly how a different association along a wave of seminiferous epithelium is established. Another layer of complexity is added to this problem by the occurrence of modulations in a tubule, i.e. the linking of consecutive segments of a seminiferous tubule with no descending order.

Acknowledgments

This work was supported by a grant from the Universidad Autónoma del Estado de México (2332/2006), and by the Program of Institutional Consolidation of CONACyT. The authors are grateful to José Vazquez Armijo for preparation of the figure 2.

References

[1] Russell LD, Etllin RA, Sinha Hikim AP, Clegg ED. *Histological and histopathological evaluation of the testis*. Florida: Caché River Press; 1990.
[2] Griswold MD. The central role of Sertoli cells in spermatogenesis. *Semin Cell Dev Biol.* 1998 Aug;9(4):411–416.
[3] Jegou B, Pineau C. Current aspects of autocrine and paracrine regulation of spermatogenesis. *Adv Exp Med Biol.* 1995;377:67–86.
[4] Weinbauer GF, Wessels J. "Paracrine" control of spermatogenesis. *Andrologia.* 1999 Sep;31(5):249–262.
[5] Meng J, Holdcraft RW, Shima JE, Griswold MD, Braun RE. Androgens regulate the permeability of the blood-testis barrier. *Proc Natl Acad Sci USA.* 2005 Nov;102(46):16696–16700.
[6] Russell LD, Corbin TJ, Borg KE, De Franca LR, Grasso P, Bartke A. Recombinant human follicle-stimulating hormone is capable of exerting a biological effect in the adult hypophysectomized rat by reducing the numbers of degenerating germ cells. *Endocrinology.* 1993;133:2062–2070.
[7] Billig H, Furuta Y, Rivier C, Tapanaheim J, Parvinen M, Hsueh AJW. Apoptosis in testis germ cells: developmental changes in gonadotropin dependence and localization to selective tubule stages. *Endocrinology.* 1995;136:5–12.
[8] Heckert LL, Griswold MD. Expression of follicle-stimulating hormone receptor mRNA in rat testes and Sertoli cells. *Mol Endocrinol.* 1991 May;5(5):670–677.
[9] Linder CC, Heckert LL, Roberts KP, Kim KH, Griswold MD. Expression of receptors during the cycle of the seminiferous epithelium. *Ann N Y Acad Sci.* 1991;637:313–321.
[10] Walker WH, Cheng J. FSH and testosterone signaling in Sertoli cells. *Reproduction.* 2005 Jul;130(1):15–28.
[11] Simoni M, Gromoll J, Nieschlag E. The follicle-stimulating hormone receptor: biochemistry, molecular biology, physiology, and pathophysiology. *Endocr Rev.* 1997 Dec;18(6):739–773.
[12] Means AR, Dedman JR, Tash JS, Tindall DJ, van Sickle M, Welsh MJ. Regulation of the testis Sertoli cell by follicle stimulating hormone. *Annu Rev Physiol.* 1980;42:59–70.
[13] Gorczynska E, Handelsman DJ. The role of calcium in follicle-stimulating hormone signal transduction in Sertoli cells. *J Biol Chem.* 1991 Dec;266(35):23739–23744.

[14] Kangasniemi M, Kaipia A, Mali P, Toppari J, Huhtaniemi I, Parvinen M. Modulation of basal and FSH-dependent cyclic AMP production in rat seminiferous tubules staged by an improved transillumination technique. *Anat Rec.* 1990 May;227(1):62–76.

[15] McDonald CA, Millena AC, Reddy S, Finlay S, Vizcarra J, Khan SA, et al. Follicle-stimulating hormone-induced aromatase in immature rat Sertoli cells requires an active phosphatidylinositol 3-kinase pathway and is inhibited via the mitogen-activated protein kinase signaling pathway. *Mol Endocrinol.* 2006 Mar;20(3):608–618.

[16] Walker WH, Fucci L, Habener JF. Expression of the gene encoding transcription factor cyclic adenosine 3',5'-monophosphate (cAMP) response element-binding protein (CREB): regulation by follicle-stimulating hormone-induced cAMP signaling in primary rat Sertoli cells. *Endocrinology.* 1995 Aug;136(8):3534–3545.

[17] Sassone-Corsi P. Regulating the balance between differentiation and apoptosis: role of CREM in the male germ cells. *J Mol Med.* 1998;76(12):811–817.

[18] Najmabadi H, Rosenberg LA, Yuan QX, Bhatia N, Albiston AL, Burger H, et al. Transcriptional and posttranscriptional regulation of inhibin alpha-subunit gene expression in rat Sertoli cells by 8-bromo-3',5'-cyclic-adenosine monophosphate. *Mol Endocrinol.* 1993 Apr;7(4):469–476.

[19] Wilson BE, Mochon E, Boxer LM. Induction of bcl-2 expression by phosphorylated CREB proteins during B-cell activation and rescue from apoptosis. *Mol Cell Biol.* 1996 Oct;16(10):5546–5556.

[20] Pugazhenthi S, Nesterova A, Sable C, Heidenreich KA, Boxer LM, Heasley LE, et al. Akt/protein kinase B up-regulates Bcl-2 expression through cAMP-response element-binding protein. *J Biol Chem.* 2000 Apr;275(15):10761–10766.

[21] Taylor WE, Najmabadi H, Strathearn M, Jou NT, Liebling M, Rajavashisth T, et al. Human stem cell factor promoter deoxyribonucleic acid sequence and regulation by cyclic 3',5'-adenosine monophosphate in a Sertoli cell line. *Endocrinology.* 1996 Dec;137(12):5407–5414.

[22] Rossi P, Dolci S, Albanesi C, Grimaldi P, Ricca R, Geremia R. Follicle-stimulating hormone induction of steel factor (SLF) mRNA in mouse Sertoli cells and stimulation of DNA synthesis in spermatogonia by soluble SLF. *Dev Biol.* 1993 Jan;155(1):68–74.

[23] Packer AI, Besmer P, Bachvarova RF. Kit ligand mediates survival of type A spermatogonia and dividing spermatocytes in postnatal mouse testes. *Mol Reprod Dev.* 1995 Nov;42(3):303–310.

[24] Yan W, Suominen J, Toppari J. Stem cell factor protects germ cells from apoptosis in vitro. *J Cell Sci.* 2000 Jan;113 (Pt 1):161–168.

[25] Waeber G, Meyer TE, LeSieur M, Hermann HL, Gérard N, Habener JF. Developmental stage-specific expression of cyclic adenosine 3',5'-monophosphate response element-binding protein CREB during spermatogenesis involves alternative exon splicing. *Mol Endocrinol.* 1991 Oct;5(10):1418–1430.

[26] Scobey M, Bertera S, Somers J, Watkins S, Zeleznik A, Walker W. Delivery of a cyclic adenosine 3',5'-monophosphate response element-binding protein (creb) mutant to seminiferous tubules results in impaired spermatogenesis. *Endocrinology.* 2001 Feb;142(2):948–954.

[27] Jegou B, Sharpe RM. Paracrine mechanisms in testicular control. In: de Kretser D, editor. *Molecular biology of the male reproductive system*. Academic Press; 1993. p. 271–310.

[28] Sadate-Ngatchou PI, Pouchnik DJ, Griswold MD. Follicle-stimulating hormone induced changes in gene expression of murine testis. *Mol Endocrinol*. 2004 Nov;18(11):2805–2816.

[29] Foulkes NS, Schlotter F, Pévet P, Sassone-Corsi P. Pituitary hormone FSH directs the CREM functional switch during spermatogenesis. *Nature*. 1993 Mar;362(6417):264–267.

[30] Nantel F, Monaco L, Foulkes NS, Masquilier D, LeMeur M, Henriksen K, et al. Spermiogenesis deficiency and germ-cell apoptosis in CREM-mutant mice. *Nature*. 1996 Mar;380(6570):159–162.

[31] Foulkes NS, Mellstrum B, Benusiglio E, Sassone-Corsi P. Developmental switch of CREM function during spermatogenesis: from antagonist to activator. *Nature*. 1992 Jan;355(6355):80–84.

[32] Sassone-Corsi P. CREM: a master-switch governing male germ cells differentiation and apoptosis. *Semin Cell Dev Biol*. 1998 Aug;9(4):475–482.

[33] Blendy JA, Kaestner KH, Weinbauer GF, Nieschlag E, Schutz G. Severe impairment of spermatogenesis in mice lacking the CREM gene. *Nature*. 1996 Mar;380(6570):162–165.

[34] Molina CA, Foulkes NS, Lalli E, Sassone-Corsi P. Inducibility and negative autoregulation of CREM: an alternative promoter directs the expression of ICER, an early response repressor. *Cell*. 1993 Dec;75(5):875–886.

[35] Chaudhary J, Sadler-Riggleman I, Ague JM, Skinner MK. The helix-loop-helix inhibitor of differentiation (ID) proteins induce post-mitotic terminally differentiated Sertoli cells to re-enter the cell cycle and proliferate. *Biol Reprod*. 2005 May;72(5):1205–1217.

[36] Chaudhary J, Johnson J, Kim G, Skinner MK. Hormonal regulation and differential actions of the helix-loop-helix transcriptional inhibitors of differentiation (Id1, Id2, Id3, and Id4) in Sertoli cells. *Endocrinology*. 2001 May;142(5):1727–1736.

[37] Weinbauer GF, Niechslag F. Hormonal control of spermatogenesis. In: de Kretser, editor. *Molecular biology of the male reproductive system*. Academic Press. San Diego; 1993. p. 101–142.

[38] Singh J, O'Neill C, Handelsman DJ. Induction of spermatogenesis by androgens in gonadotropin-deficient (hpg) mice. *Endocrinology*. 1995 Dec;136(12):5311–5321.

[39] Ghosh S, Sinha-Hikim AP, Russell LD. Further observations of stage-specific effects seen after short-term hypophysectomy in the rat. *Tissue Cell*. 1991;23(5):613–630.

[40] Kerr JB, Maddocks S, Sharpe RM. Testosterone and FSH have independent, synergistic and stage-dependent effects upon spermatogenesis in the rat testis. *Cell Tissue Res*. 1992;268(179-189).

[41] Franca LR, Parreira GG, Gates RJ, Russell LD. Hormonal regulation of spermatogenesis in the hypophysectomized rat: quantitation of germ-cell population

and effect of elimination of residual testosterone after long-term hypophysectomy. *J Androl.* 1998;19(3):335–342.
[42] White R, Parker MG. Molecular mechanisms of steroid hormone action. *Endocr Relat Cancer.* 1998;5:1–14.
[43] Sar M, Lubahn DB, French FS, Wilson EM. Immunohistochemical localization of the androgen receptor in rat and human tissues. *Endocrinology.* 1990 Dec;127(6):3180–3186.
[44] Kimura N, Mizokami A, Oonuma T, Sasano H, Nagura H. Immunocytochemical localization of androgen receptor with polyclonal antibody in paraffin-embedded human tissues. *J Histochem Cytochem.* 1993 May;41(5):671–678.
[45] Bremner WJ, Millar MR, Sharpe RM, Saunders PT. Immunohistochemical localization of androgen receptors in the rat testis: evidence for stage-dependent expression and regulation by androgens. *Endocrinology.* 1994;135(3):1227–34.
[46] Tsai MY, Yeh SD, Wang RS, Yeh S, Zhang C, Lin HY, et al. Differential effects of spermatogenesis and fertility in mice lacking androgen receptor in individual testis cells. *Proc Natl Acad Sci U S A.* 2006 Dec;103(50):18975–18980.
[47] Johnston DS, Russell LD, Friel PJ, Griswold MD. Murine germ cells do not require functional androgen receptors to complete spermatogenesis following spermatogonial stem cell transplantation. *Endocrinology.* 2001;142(6):2405–8.
[48] Vornberger W, Prins G, Musto NA, Suarez-Quian CA. Androgen receptor distribution in rat testis: new implications for androgen regulation of spermatogenesis. *Endocrinology.* 1994 May;134(5):2307–2316.
[49] Zhu LJ, Hardy MP, Inigo IV, Huhtaniemi I, Bardin CW, Moo-Young AJ. Effects of androgen on androgen receptor expression in rat testicular and epididymal cells: a quantitative immunohistochemical study. *Biol Reprod.* 2000 Aug;63(2):368–376.
[50] Ku CY, Loose-Mitchell DS, Sanborn BM. Both Sertoli and peritubular cells respond to androgens with increased expression of an androgen response element reporter. *Biol Reprod.* 1994 Aug;51(2):319–326.
[51] Gendt KD, Swinnen JV, Saunders PTK, Schoonjans L, Dewerchin M, Devos A, et al. A Sertoli cell-selective knockout of the androgen receptor causes spermatogenic arrest in meiosis. *Proc Natl Acad Sci U S A.* 2004 Feb;101(5):1327–1332.
[52] Chang C, Chen YT, Yeh SD, Xu Q, Wang RS, Guillou F, et al. Infertility with defective spermatogenesis and hypotestosteronemia in male mice lacking the androgen receptor in Sertoli cells. *Proc Natl Acad Sci U S A.* 2004 May;101(18):6876–6881.
[53] Denolet E, Gendt KD, Allemeersch J, Engelen K, Marchal K, Hummelen PV, et al. The effect of a sertoli cell-selective knockout of the androgen receptor on testicular gene expression in prepubertal mice. *Mol Endocrinol.* 2006 Feb;20(2):321–334.
[54] Wang RS, Yeh S, M CL, Lin HY, Zhang C, Ni J, et al. Androgen receptor in Sertoli cells is essential for germ cell nursery and junctional complex formation in mouse testis. *Endocrinology.* 2006;147:5624–5633.
[55] De Gendt K, V SJ, Saunders PT, Schoonjans L, Dewerchin M, Devos A, et al. A Sertoli cell-selective knockout of the androgen receptor causes spermatogenesis arrest in meiosis. *Proc Natl Acad Sci USA.* 2004;101:1327–1332.

[56] Holdcraft RW, Braun RE. Androgen receptor function is required in Sertoli cells for the terminal differentiation of haploid spermatids. *Development.* 2004;131:459–467.
[57] Hales DB, Sha LL, Payne AH. Testosterone inhibits cAMP-induced de Novo synthesis of Leydig cell cytochrome P-450(17 alpha) by an androgen receptor-mediated mechanism. *J Biol Chem.* 1987 Aug;262(23):11200–11206.
[58] Petrusz P, Jeyaraj DA, Grossman G. Microarray analysis of androgen-regulated gene expression in testis: the use of the androgen-binding protein (ABP)-transgenic mouse as a model. *Reprod Biol Endocrinol.* 2005;3:70.
[59] Sadate-Ngatchou PI, Pouchnik DJ, Griswold MD. Identification of testosterone-regulated genes in testes of hypogonadal mice using oligonucleotide microarray. *Mol Endocrinol.* 2004 Feb;18(2):422–433.
[60] Hacker A, Capel B, Goodfellow P, Lovell-Badge R. Expression of Sry, the mouse sex determining gene. *Development.* 1995 Jun;121(6):1603–1614.
[61] Lovell-Badge R, Hacker A. The molecular genetics of Sry and its role in mammalian sex determination. *Philos Trans R Soc Lond B Biol Sci.* 1995 Nov;350(1333):205–214.
[62] Jegou B. The Sertoli cell in vivo and in vitro. *Cell Biol Toxicol.* 1992;8(3):49–54.
[63] Ketola I, Anttonen M, Vaskivuo T, Tapanainen JS, Toppari J, Heikinheimo M. Developmental expression and spermatogenic stage specificity of transcription factors GATA-1 and GATA-4 and their cofactors FOG-1 and FOG-2 in the mouse testis. *Eur J Endocrinol.* 2002 Sep;147(3):397–406.
[64] Holsberger DR, Jirawatnotai S, Kiyokawa H, Cooke PS. Thyroid hormone regulates the cell cycle inhibitor p27Kip1 in postnatal murine Sertoli cells. *Endocrinology.* 2003 Sep;144(9):3732–3738.
[65] Law GL, Griswold MD. Activity and form of sulfated glycoprotein 2 (clusterin) from cultured Sertoli cells, testis, and epididymis of the rat. *Biol Reprod.* 1994 Mar;50(3):669–679.
[66] Schlatt S, de Kretser DM, Loveland KL. Discriminative analysis of rat Sertoli and peritubular cells and their proliferation in vitro: evidence for follicle-stimulating hormone-mediated contact inhibition of Sertoli cell mitosis. *Biol Reprod.* 1996 Aug;55(2):227–235.
[67] Norton JN, Skinner MK. Regulation of Sertoli cell differentiation by the testicular paracrine factor PModS: potential role of immediate-early genes. *Mol Endocrinol.* 1992 Dec;6(12):2018–2026.
[68] Russell LD. Movement of spermatocytes from the basal to the adluminal compartment of the rat testis. *Am J Anat.* 1977;148:313–328.
[69] Tan KAL, Gendt KD, Atanassova N, Walker M, Sharpe RM, Saunders PTK, et al. The role of androgens in Sertoli cell proliferation and functional maturation: studies in mice with total or Sertoli cell-selective ablation of the androgen receptor. *Endocrinology.* 2005 Jun;146(6):2674–2683.
[70] Timmons PM, Rigby PWJ, Poirier F. The murine seminiferous epithelial cycle is prefigured in the Sertoli cells of the embryonic testis. *Development.* 2002 Feb;129(3):635–647.

[71] Franca LR, Ogawa T, Avarbock MR, Brinster RL, Russell LD. Germ cell genotype controls cell cycle during spermatogenesis in the rat. *Biol Reprod.* 1998 Dec;59(6):1371–1377.
[72] Elftman H. The Sertoli cell cycle in the mouse. *Anat Rec.* 1950;106:381–393.
[73] Anway MD, Li Y, Ravindranath N, Dym M, Griswold MD. Expression of testicular germ cell genes identified by differential display analysis. *J Androl.* 2003;24(2):173–184.
[74] Almstrup K, Nielsen JE, Hansen MA, Tanaka M, Skakkebaek NE, Leffers H. Analysis of cell-type-specific gene expression during mouse spermatogenesis. *Biol Reprod.* 2004 Jun;70(6):1751–1761.
[75] Clemente EJ, Furlong RA, Loveland KL, Affara NA. Gene expression study in the juvenile mouse testis: identification of stage-specific molecular pathways during spermatogenesis. *Mamm Genome.* 2006 Sep;17(9):956–975.
[76] Lindsey JS, Wilkinson MF. A testosterone- and LH-regulated homeobox gene expressed in mouse Sertoli cells and epididymis. *Dev Biol.* 1996;174:471–484.
[77] Lim K, Yoo JH, Kim kY, Kweon GR, Kwak ST, Hwang BD. Testosterone regulation of proto-oncogene c-myc expression in primary Sertoli cell cultures from prepubertal rats. *J Androl.* 1994;15:543–550.
[78] ODonnell L, Robertson KM, Jones ME, Simpson ER. Estrogen and spermatogenesis. *Endocrine Rev.* 2001;22(3):289–318.
[79] Silva FRMB, Leite LD, Wassermann GF. Rapid signal transduction in Sertoli cells. *Eur J Endocrinol.* 2002 Sep;147(3):425–433.
[80] Cheng J, Watkins SC, Walker WH. Testosterone activates mitogen-activated protein kinase via Src kinase and the epidermal growth factor receptor in Sertoli cells. *Endocrinology.* 2007 May;148(5):2066–2074.
[81] Fix C, Jordan C, Cano P, Walker WH. Testosterone activates mitogen-activated protein kinase and the cAMP response element binding protein transcription factor in Sertoli cells. *Proc Natl Acad Sci U S A.* 2004 Jul;101(30):10919–10924.

In: Cell Differentiation Research Developments
Editor: L. B. Ivanova, pp. 233-245
ISBN: 978-1-60021-939-9

Chapter XI

TRANSFORMATION – NORMALIZING-REDIFFERENTIATION – APOPTOSIS SEQUENCE AND THE ROLE OF THE MITOCHONDRIA IN THE RETINOID-PANCREATIC CANCER MODEL: IS IT OBLIGATORY?

Tarek H. EL-Metwally *

Departments of Medical Biochemistry, Assiut University, Faculty of Medicine, Assiut, Egypt.

ABSTRACT

Retinoic acid induced redifferentiation and apoptosis in pancreatic adenocarcinoma cell lines. Redifferentiation included early reversion into aerobic metabolism reflected by an increase of mitochondrial activity and -mass with normal membrane potential and terminal ductal cell differentiation. Cells in such state would attempt correcting their DNA abnormalities or otherwise commit suicide by apoptosis. In some cell systems such as the present case, the stem cell potential of ductal pancreatic cells gave them an alternative option, i.e., to transdifferentiate into functional endocrine normal cell type. Again due to impossibility of correcting highly corrupted genome, cells eventually succumbed apoptosis. Mitochondrial changes appear to be forcing factor for this process. Such a process, transformation - normalizing-redifferentiation - apoptosis sequence, has been shown by several studies utilizing various cell types, apoptotic inducers, biomarkers and time-frames. Although some studies have shown concomitant apoptosis and redifferentiation, others have reported apoptosis without prior redifferentiation. However, utilizing appropriate time-frame and markers of earlier mitochondrial changes, one would

* Correspondence concerning this article should be addressed to: Dr. Tarek Hassan EL-Metwally, PhD., Associated Professor, Medical Biochemistry. Department of Medical biochemistry, Faculty of Medicine, Assiut University, Assiut, Egypt. Phone: 0020882309566; Fax: 0020882332278; Email: thelmetwally@hotmail.com.

detect a scenario similar to the present retinoid model. This situation can be achieved by delaying apoptosis or reducing the inducer concentration in such systems. The final fate of normal differentiated cells is apoptosis. Therefore, it was suggested that a degree of normalizing redifferentiation of transformed cells might be expect prior to apoptosis, simulating the physiological fate.

Keywords: retinoids – mitochondrial – redifferentiation therapy – redifferentiation-apoptosis sequence.

INTRODUCTION

The progress in differentiation depends on an increase in the ratio between mitochondrial differentiation-promoting activity and nuclear differentiation preventing activity. This ratio is low in embryonic and stem cells, due to low mitochondrial content, but it increases by a rate of mitochondrial multiplication that is larger than a doubling of mitochondrial content per cell cycle. The rate of mitochondrial multiplication, therefore, determines the progress in differentiation and subsequent apoptosis as a physiological fate. This rate is modifiable by extracellular signals and cellular defects. Mutations and cytoskeletal changes are likely to decrease the rate to the extend that differentiation is arrested with ensuing neoplastic growth. Agents used in chemical therapy and ionizing radiation overcome this arrest by preferentially targeting the cell cycle. The mitochondria multiply during the transitory cell cycle inhibition, resulting in increased differentiation promoting activity. The finding of increase in mitochondrial mass and induction of differentiation prior to the release of cytochrome c and apoptosis points to integration of the initial molecular pathways of differentiation and apoptosis (von Wangenheim and Peterson, 1998). Differentiation is a short-term program of biogenesis responsible for the rapid changes in the bioenergetic phenotype of mitochondria. In contrast, proliferation is a long-term program responsible for the decrease in mitochondrial mass in the cell. Moreover, some tissues, such as fetal liver have many phenotypic manifestations in common with highly glycolytic tumor cells (Cuezva et al., 1997). The differentiation - dedifferentiation process during transformation is a gradual loss of tissue specific characteristics from benign, well-, moderately-, poorly- to undifferentiated state. Likewise, the process of dedifferentiation – re-/trans-differentiation should not be looked at as an on/off process but rather as a stepwise retrograde acquisition of the lost biochemical properties, the least of it is the reversion into aerobic metabolism featured as a normal mitochondrial membrane potential ($\Delta\Psi_m$) and increased mitochondrial mass and activity as observed in the retinoid-induced pancreatic cancer redifferentiation-apoptosis sequence (Ohashi et al., 1992; El-Metwally et al., 2005a and b, El-Metwally et al., 2006; Abelev and Lazarevich, 2006).

The importance of the mitochondria in cell life and death is clear from being a major target for the critical balance between pro- and antiapoptotic effectors particularly among the Bcl2 family. In principle, biochemically, cancer cell is cancer cell even in the submicroscopic stage. Tumor progression is usually associated with partial or complete loss of morphological and biochemical features of the original tissue by the process of dedifferentiation. To a certain extent, carcinomas preserve differentiation markers of normal tissue, and hemoblastoses precisely reflect the direction and differentiation level of their precursor cells. However, carcinomas and hemoblastoses retain the ability to differentiate (Abelev and Lazarevich, 2006). In cancer cells the activities, amounts, and pattern of key enzymes is stringently linked with transformation and progression. In parallel with the rise in tumor growth rates there is an increase in the activities of the key glycolytic enzymes in an aerobic manner (conversion of glucose to lactate); glucokinase/hexokinase, 6-phosphofructokine, pyruvate kinase (Weber, 2001). Accordingly, the glycolytic shift of cancer cell metabolism is an integral part of the transformation process rather than an adaptive measure (Walenta et al., 2004).

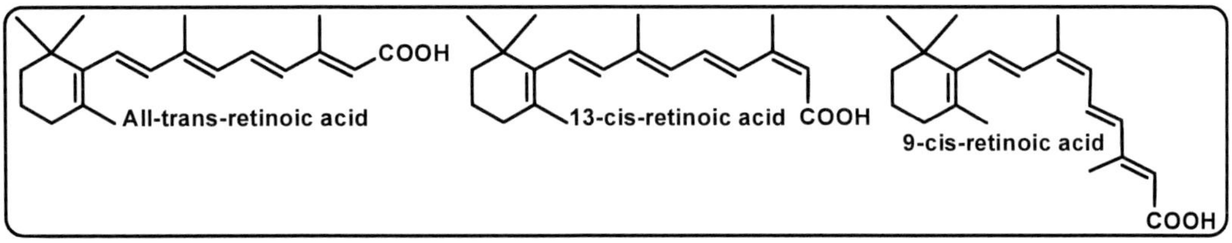

Figure 1. Chemical structures of the three utilized natural isomers of retinoic acid.

In our retinoic acid (RA, see figure 1) model using human pancreatic adenocarcinoma cells, it was noticed that the direct MTT (3-(4,5-dimethylthiazol-2-yl)-2,5-diphenyl-2H-tetrazolium bromide) cell proliferation assay greatly masked the antiproliferative effect of RA compared to cell count, total proteins or ^{3}H-thymidine incorporation (EL-Metwally and Adrian, 1999). Since, early transformation is associated with aerobic glycolysis and reduction of mitochondrial activity and mass, it was hypothesized that for cancer cell to succumb to apoptosis a minimum degree of redifferentiation, i.e., malignant transformation - redifferentiation - apoptosis sequence, contrasting, normal - transforming-dedifferentiation - immortalization sequence is required. Many would oppose this hypothesis, putting in mind a full-scale tissue-specific redifferentiation as a prerequisite. However, what seems to be of profound importance is the reversion into aerobic metabolism through mitochondria. This degree of aerobic redifferentiation could be the least required for induction of apoptosis that would accommodate any kind of apoptotic inducer or cell model. As a result, "immediate" features of redifferentiation that suite such scenario depends on time-frame of the apoptotic induction, type and concentration of the apoptotic inducer and cell type used. Therefore, the observed induction of reductive MTT activity when it was normalized to number of cells or total protein content was expectable because of the induction of the mitochondrial activity and mass correlating redifferentiation. RA-induced redifferentiation early in time was associated with several fold induction of MTT particularly using apoptotic concentrations of RA (Figure 2). This would reflect induction of mitochondrial dehydrogenase activities. Untreated pancreatic cancer cells growing in peripheral wells along the circumference of the tissue culture plates showed reduced glucose utilization, lactate production and proliferation compared to cells in internal wells (our unpublished observations). The former cells possibly received more O_2 from the circulating incubation gas than cells in the internal wells suggesting that the mere aerobiosis of cancer cells could be antiproliferative and redifferentiating. This contrasts the effect of hypoxia that correlates the adverse prognosis of cancer in vivo. Expectedly, increasing O_2 concentration in cancers at the tissue and cellular levels reduced the hypoxia-induced cancer cell aggression and expression of angiogenic genes, a good prognostic biomarker (Al-Hallaq et al., 1998 and Funasaka et al., 2005). In support, small size benign and well-differentiated cancers are well-vascularized and aerated. However, cells of these cancers and normally-appearing surrounding cells are already glycolytic. Despite that they do not have metastatic potential (Kolodin, 1977; Chevrollier et al., 2005; Mazzanti et al., 2006). Transformation of colonic epithelial cells is characterized by decreased mitochondrial activity, increased $\Delta\Psi_m$, and disrupted proliferation-apoptosis equilibrium. Decreased mitochondrial gene expression is an early marker of colonic cancer and tumorigenesis, thereby implicating alterations in mitochondrial function as an early event in the transformation of colonic epithelial cells, regardless of the etiology. Moreover, an intact $\Delta\Psi_m$ is essential for growth arrest and apoptosis induced by butyrate (Augenlicht et al., 1987; Faure Vigny et al, 1996; Heerdt et al., 2003). Similar to the observed induction of MTT reducing activity, RA promoted nitroblue tetrazolium-associated functional cell differentiation in prostate cancer PC-3 cells with re-activation of the tumor suppressor p75 neurotrophin receptor (Nalbandian et al., 2005). Additionally, the natural isoflavone genistein was antiproliferative on MCF-7, human breast tumor; Jurkat cells, human T-cell leukemia; L-929, mouse transformed fibroblasts cell lines *in vitro* with a G2/M cell cycle arrest and an increase in cell volume and in mitochondrial number and/or activity. Consequently, the significant influence of genistein on mitochondrial number and/or function resulted in a sequential increase in MTT reduction to formazan per cell (Pagliacci et al., 1993).

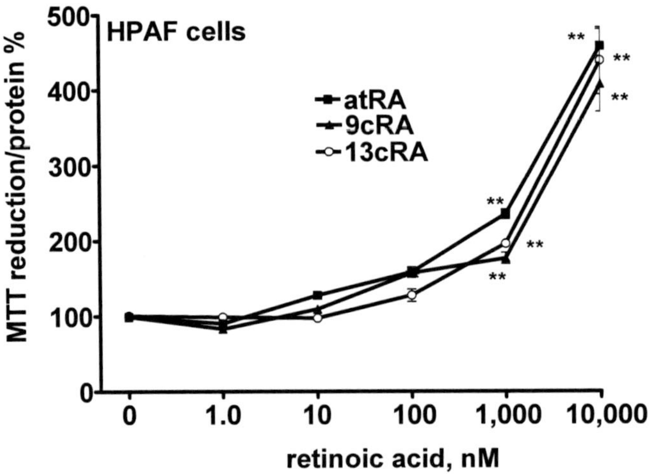

Figure 2. Induction of the MTT-reducing activity by retinoic acid. The extensive increase in the MTT optical density normalized to total protein content after 2 days of treatment with three natural isomers of retinoic acid (all-trans-, 9-cis- and 13-cis-) was measured and expressed as % of untreated controls in pancreatic HPAF cells (El-Metwally and Adrian 1999 and El-Metwally et al 2005a).

It has been suggested that the reduction in mitochondrial differentiation rather than hypoxia modulates prognosis of cancer cells (Walenta et al., 2004; Funasaka et al 2005; Mazzanti et al 2006). Moreover, it has been shown that solid tumors have massive differences in their metastatic potential although they all are hypoxic (von Wangenheim and Peterson 1998; Weber 2001). In all tumor types investigated, high molar concentrations of lactate were correlated with a high incidence of distant metastasis already in an early stage of the disease. This was due to the activation of hyaluronan synthesis by tumor-associated fibroblasts, upregulation of the vascular endothelial growth factor and of hypoxia-induced factor-1α, and direct enhancement of cellular motility which generates favorable conditions for metastatic spread (Walenta et al., 2004). Therefore, abnormalities in cytoskeleton and associated proteins could be the major modifier of mitochondrial biogenesis and activity. RA-induced apoptosis in T24 bladder cancer cells was associated with redistribution of Bax and Bcl-2 proteins in the subcellular compartment. This coincided with reorganization of the cytoplasmic intermediate filament network. Cleavage of cytokeratins by caspases in this model was associated with aggregation of mitochondria and lysosome (Chien et al., 2004). Moreover, chemical depolymerization of microtubules (MTs) invariably leads to the inhibition of mitochondrial volume and mass increases during interphase of cell cycle. The presence of stable MTs is most pronounced in various cellular processes which require a large amount of energy: neurite outgrowth; myotube differentiation; formation of monolayer of epithelial cells; pressure overloaded cardiac hypertrophy. Stabilization of MTs by prolonged exposure of the human osteosarcoma cell line 143B cells and rat liver-derived RL-34 cells to taxol induced an abnormal accumulation of mitochondria in cells arrested in G2/M phase of the cell cycle. Mitochondrial proliferation and degradation have been suggested to depend upon functional states of the organelles or energetic states of the cell. Accordingly, disorganization of these processes is often associated with abnormal accumulation of mitochondria in various models of cell death. On the contrary, depolymerization of microtubules by nocodazole and colchicine inhibited mitochondrial proliferation during G1 to G2 phase progression and arrested cells in G2/M phase. Co-treatment of cells with taxol and nocodazole or taxol and colchicines suppressed taxol-induced proliferation of mitochondria as was confirmed by increase in subunit VIII of human cytochrome c oxidase and by enhanced mtDNA replication. Two subpopulations of mitochondria were detected in taxol-treated cells: mitochondria with high $\Delta\Psi_m$ and those with low $\Delta\Psi_m$ reflecting apoptotic death. Treatment with herbimycin A or H_2O_2, known to induce the accumulation of mitochondria in mammalian cells, causes stabilization of MTs in 143B cells. Furthermore, the elevation of mitochondrial mass has been reported in HEP-2 cells cultured in the presence of cytotoxic necrotizing factor 1 (CNNF1), an activator of Rho GTPase, a sufficient factor required for the stabilization of MTs. 2-methoxyestradiol, which is structurally unrelated to nocodazole or colchicine, depolymerizes MTs and at the same time inhibits the proliferation of mitochondria (Karbowski et al., 2001). Leflunomide, an inhibitor of mitochondrial enzyme dihydroorotate dehydrogenase, causes an unrestrained proliferation of mitochondria both in human osteosarcoma cell line 143B cells and rat liver derived RL-34 cells with increased intracellular level of acetylated α-tubulin. In consequence, changes in the physicochemical properties of MTs may someway modulate the biogenesis of mitochondria (Spodnik et al., 2002).

We observed that antiproliferation and cell cycle arrest were integral parts of a redifferentiation - apoptosis sequence since low non-redifferentiation concentrations of RA failed to provoke apoptosis (El-Metwally et al., 2005a). The proposed aerobic redifferentiation was also confirmed by a few fold induction of mitochondrial mass compared to controls within 24 hrs. The $\Delta\Psi_m$ was normal at least for 3 days post-treatment. Significant but gradual loss of $\Delta\Psi_m$ was observed thereafter (Figure 3).

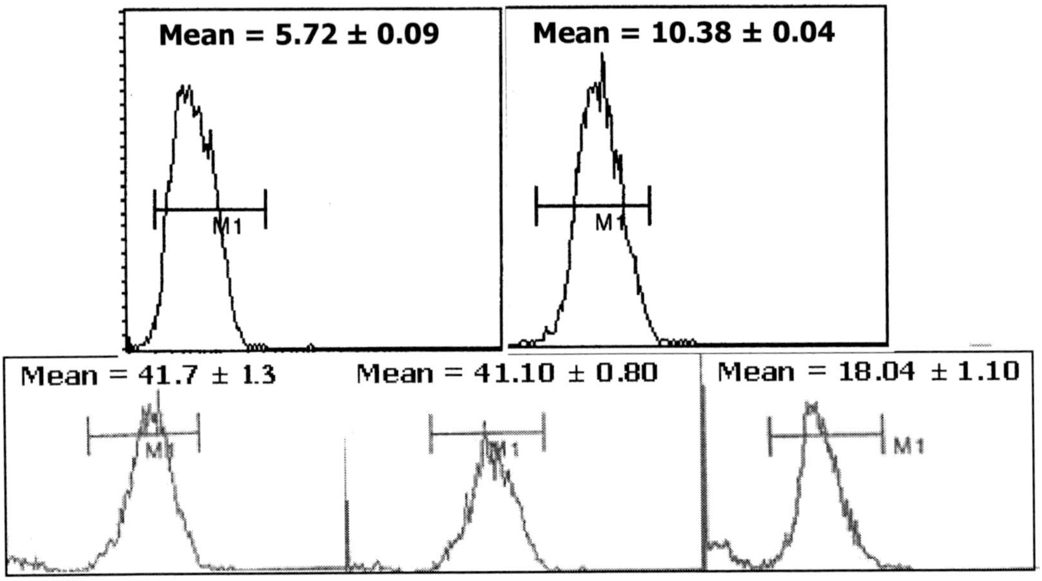

Figure 3. Increased mitochondrial mass and changes in mitochondrial membrane potential induced by all-trans-retinoic acid. Mitochondrial mass was doubled in HPAF cells treated with the highest apoptotic concentration used, 10 μM for 24 hrs, *right side*, compared to untreated control, *left side* (Upper panel, El-Metwally et al., 2005a). Mitochondrial membrane potential was normal after 24 hrs of treatment, *central part*, but was gradually reduced to reach significance at 3 days, *right part*, compared to control, *left part* (Lower panel, El-Metwally et al., 2005b).

Previous reports showed that cell cycle arrest and apoptotic cascade induced by butyrate is dependent upon the presence of a normal mitochondrial membrane potential (Bordonaro et al., 1999). And, $\Delta\Psi_m$ elevation, stabilization, or collapse results in delayed, decreased, or blocked apoptosis, respectively (Heerdt et al., 2000). The observed induction of mitochondrial activity and mass suggests an effector role for the mitochondria in the present RA-induced re-/trans-differentiation and subsequent apoptosis model. The mitochondrial redifferentiation process is a part of a more generalized reprogramming of gene expression towards normal cell behavior. Some of these normalized cells were able to functionally transdifferentiate into pancreatic endocrine cells despite the presence of apoptotic concentrations of RA (El-Metwally et al., 2006). This reflected the viability of the treated cells and suggested their stem-cell potential to the extent of escaping apoptosis by transdifferentiation into another cell type. Consequently, mitochondrial redifferentiation may provide energy and/or effector molecules for the whole process including the repair of DNA abnormalities re- and/or trans-differentiate. Apoptosis was induced a final fate because of the impossibility of such a DNA repair. Similarly, it was reported that retinoids, as a single agent, induce terminal differentiation phenotype well before cell death by apoptosis in a temporally defined order (Yin et al., 2005). The acyclic retinoid, all trans-3,7,11,15-tetramethyl-2,4,6,10,14-hexadecapentaenoic acid, prevented hepatocarcinogenesis in animal models and in a randomized clinical trial by eradicating premalignant and latent malignant clones of transformed cells from the liver. Production of albumin, a redifferentiation marker, was recovered while that of lectin-reactive isoform of alpha-fetoprotein, a dedifferentiation agent, was reduced after treatment of human hepatoma-derived cell lines for 2 days. The cells subsequently died due to apoptosis after 4 - 6 experimental days (Yasuda et al., 2002).

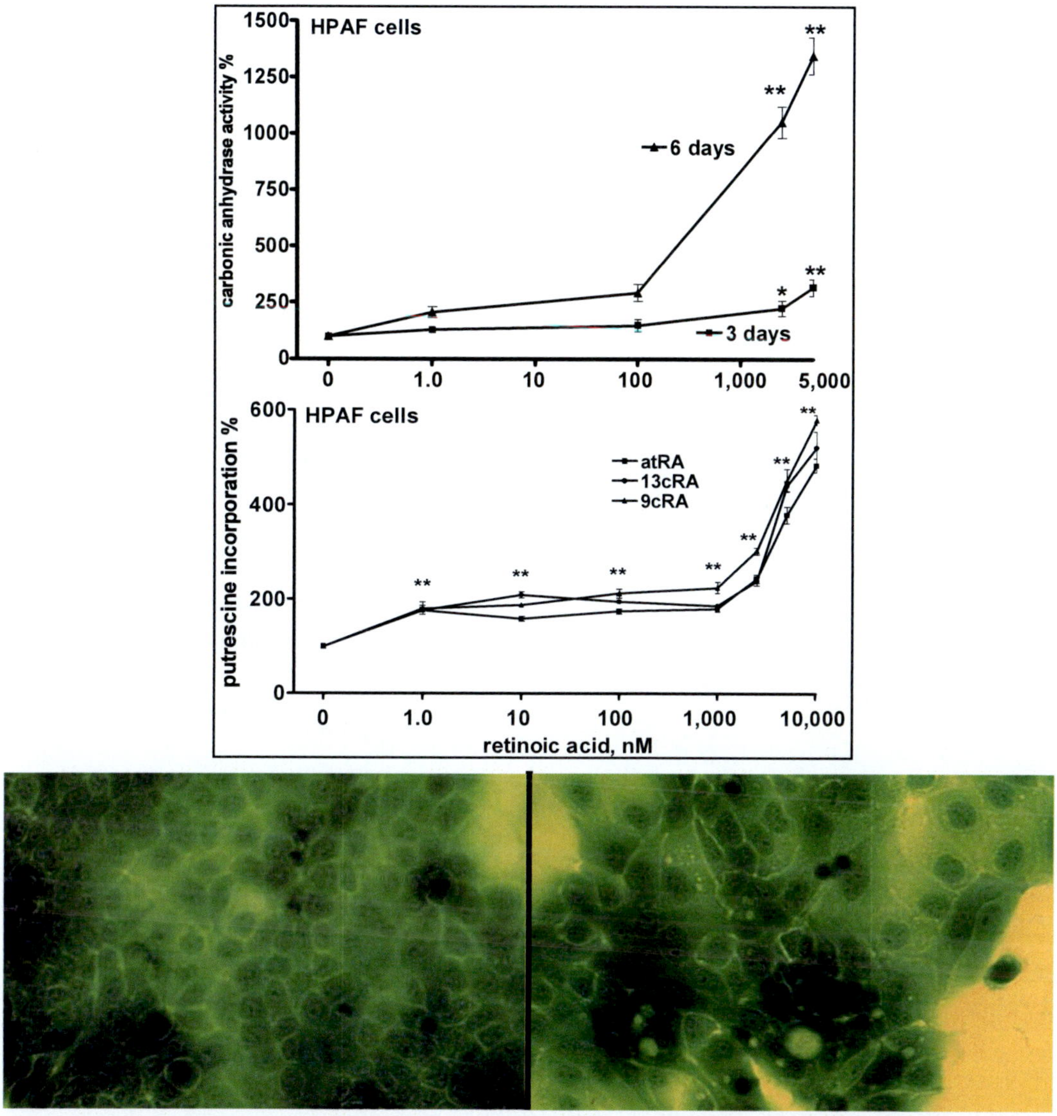

Figure 4. Induction of terminal biochemical and phenotypic differentiation by retinoic acid. Particularly treatment with the apoptotic concentrations of all-trans-retinoic acid induced extensive carbonic anhydrase II, upper part (El-Metwally et al., 2005a), and transglutaminase (for 4 days, measured as ^{3}H-putrescine incorporation, El-Metwally et al., 2005b), middle part, activities. Lower panel show morphology of control HPAF cells with polygonal unpolarized cells with massive nucleus-cytosol ratio and mitotic figures, left side, compared with 2.5 μM all-trans-retinoic acid treated cells for 3 days with typical polarized simple columnar epithelium with highly reduced nucleus-cytoplasm ration, secretory activity and a few apoptotic bodies, *right side* (El-Metwally et al., 2005a and b).

Normal spatial and temporal patterns of cell proliferation, differentiation, and apoptosis in the colonic mucosa are determined by developmentally programmed genetic and external signals generated by homo- and heterotypic cell interactions, humoral agents, and luminal contents. Mitochondrial function may play a pivotal role in integrating these signals and in determining probability of cells entering different maturation pathways (Augenlicht et al., 1999). Both doxorubicin and mitoxantrone were differentially potent inducers of apoptosis in H9C2 cardiomyocytes and MTLn3 breast cancer cells. In particular, the two drugs induced a progressive increase in mitochondrial mass in the cancer cells but not in the cardiac cells. The mitochondrial proliferation preceded the nuclear apoptosis. The proliferation of mitochondria could explain the higher toxicity of doxorubicin to cancer cells compared to cardiac cells (Kluza et al., 2004). Elevated mitochondrial gene expression is an early event in the switch from proliferation to differentiation of the human colon adenocarcinoma cell line, HT29, promoted by trehalose replacement of exogenous glucose (Lu and Seligy, 1992; Lu et al., 1992). The expression levels of COXIII, a mitochondrial gene encoding one of the 13 subunits of cytochrome c oxidase, are abnormally low in colonic tumors and colonic tissue at genetic risk for developing tumors. It was increased in HT29 human colonic adenocarcinoma cells treated with butyrate that may enhance the potential for cellular respiration (Heerdt and Augenlicht, 1991). Herbimycin-treated breast cancer cell line SKBr3 cells showed increased mitochondrial mass, with no corresponding increase in $\Delta\Psi_m$ before undergoing apoptosis (Mancini et al., 1998). Mitochondrial mass and mitochondrial DNA content were increased with differentiation of human embryonic stem cells, which was accompanied by the increase of the amount of ATP (4-fold) and its by-product reactive oxygen species (2.5-fold) supporting anaerobic - to - aerobic metabolic transition during differentiation (Cho et al., 2006). It was demonstrated that normal mitochondrial respiratory chain (MRC) plays an essential role in the interferon-β/retinoic acid-induced cancer cell death. They upregulate the expression of MRC complex I subunits (Huang et al., 2006). In the most successful application of redifferentiation as an anticancer therapies, all-*trans*-RA induces differentiation of acute promyelocytic leukemias carrying the t(1;19) translocation. Cells that have differentiated although remained viable for a few days, eventually apoptosed. Restoration of a normal differentiation program in cancer cells, in consequence, appears to also activate an apoptotic mechanism similar to the normal physiological process (Ohashi et al., 1992).

In our model, few cells started apoptosing with the beginning of treatment with apoptotic concentration of RA. If these apoptotic cells would have been collected separately and investigated, one could not detect the expected full-scale terminal differentiation. This emphasizes the value of utilizing earlier mitochondrial redifferentiation to indicate its necessity for the apoptosis (Augenlicht and Heerdt, 2001). Subsequently, remaining majority of cells exhibited phenotypic and biochemical (increased carbonic anhydrase II and transglutaminase activities) features of terminal redifferentiation (El-Metwally et al., 2005b, Figure 4). Similar anaerobic into aerobic reversion was also reported during induction of glucocorticoid-induced apoptosis with a biphasic course of lactate production. Prior to the onset of apoptosis, i.e., prior to the loss of $\Delta\Psi_m$, lactate production was reduced. A massive increase in lactate production was observed in the human acute lymphoblastic leukemia cell line CCRF-CEM, subsequent to loss of $\Delta\Psi_m$ (Tiefenthaler et al., 2001). Apoptosis in

Herbimycin A-induced differentiation of the poorly differentiated colorectal carcinoma, Colo-205 cells involves unrestrained mitochondrial proliferation and progressive $\Delta\Psi_m$ dysfunction preceding apoptosis (Mancini et al., 1997). Mitochondrial activity and glucose consumption were significantly stimulated after sodium butyrate-induced peritoneal carcinomatosis cells differentiation and prior to their apoptosis (Boisteau et al., 1996). Ultrastructural examination of the doxorubicin-treated human MCF-7 breast adenocarcinoma cells revealed morphological alterations consistent with the induction of differentiation (e.g., increased lipid content and mitochondrial density, appearance of tight junctions, and secretory ducts) with subsequent gradual loss of cells through apoptosis (Fornari et al., 1994). The differentiation of the inducible murine neuroblastoma C1300 clone, N1E-115, associated an important increase of the cellular content in mitochondria. This increase could be observed with differentiating N1E-115 cells maintained in suspension, i.e., under conditions where neurite outgrowth is prevented but other early stages of (biochemical) differentiation continue to occur (Vayssiere et al., 1992).

Nevertheless, several studies reporting cytotoxic treatment inducing apoptosis set a scenario formed of: a loss of mitochondrial membrane potential with or without cytochrome c release, oxidative stress, activation of caspases, increase in annexin V binding and DNA fragmentation without prior tissue specific differentiation. This sequence of events could be true considering the time-frame; redifferentiation indicator(s) investigated; cell type and apoptotic inducer used. Nevertheless, induction of confirmed apoptosis might had been preceded by redifferentiation in these models, if earlier redifferentiation markers at earlier time points and lower concentrations of the apoptotic inducer were used. For example, cytokinins - plant redifferentiation-inducing hormones - were very effective in inducing mature granulocytes in the human myeloid leukemia HL-60 cells. On the other hand, cytokinin ribosides were the most potent substances for growth inhibition and apoptosis by greatly reducing the intracellular ATP content and disturbing the $\Delta\Psi_m$ and the accumulation of reactive oxygen species, not observed with cytokinins. Coincubation with O_2^- scavenger, antioxidant or caspase inhibitor, apoptosis was significantly reduced and the obscured degree of differentiation with the ribosides along full-scale granulocytic phenotype was disclosed and enhanced when the acute apoptosis was slowed down (Ishii et al., 2002). Therefore, in such a model, the meaning of redifferentiation would depend largely on the stage at which the respective biomarker would be expressed and detected. If biologically basic and general biomarkers such as transient increase in oxidative mitochondrial activity or its effector subunits, mitochondrial mass and normal potential were investigated at a tight time-frames, it could be found that redifferentiation precede apoptosis. After differentiation of murine erythroleukemia there was initial overexpression of mitochondrial oxidative phosphorylation complexes II and IV mRNAs followed by a gradual decline after 36 hr incubation with DMSO and/or 2-(3-ethylureido)-6-methylpyridine (Vizirianakis et al., 2002). This is normal since apoptosis is the physiological fate of a normal differentiated cell or a normalized redifferentiated cell. Joshi et al. (1999) reported that a functioning mitochondrial respiratory chain was required for cellular sensitivity to BMD188, a novel prostate cancer chemotherapeutic drug. Resveratrol, a plant polyphenol triggers a p53-independent apoptotic pathway in colon carcinoma HCT116 cell line that may be linked to differentiation since, apoptosis was preceded by mitochondrial proliferation and signs of epithelial differentiation.

Physiological concentrations of n-butyrate induced apoptosis independently of p53 in HCT116 colon carcinoma cell line. The apoptosis was mediated through the mitochondria and was accompanied by mitochondrial proliferation and $\Delta\Psi_m$ changes (Mahyar-Roemer and Roemer, 2001). Reipert et al. (1995) also reported mitochondrial proliferation and increased mass at all stages of cell cycle in etoposide treated FDCP-mix, a pluripotent murine hematopoietic stem cell line. Subsequently, there was a decline in mitochondria mass and $\Delta\Psi_m$ in the later stages of apoptosis with G2 arrest. The G2/M phase arrest was associated with an increase in $\Delta\Psi_m$ whereas, treatment with a tenfold higher drug concentration trigger massive apoptosis and a collapse of $\Delta\Psi_m$ (Facompre et al., 2000). At the late stage of etoposide-induced apoptosis in HL-60 cells, mitochondria increased in numbers as an integral part of a cascade of apoptotic events. There was also a drastic increase in mitochondrial DNA content with decreased $\Delta\Psi_m$ and ATP content. The increase in mitochondrial DNA levels correlated with elevated expression of one of the regulators of mitochondrial DNA replication, mtSSB (Eliseev et al., 2003). Treatment of U937 cells with the flavonoid quercetin elicited three cell populations with different $\Delta\Psi_m$: (1) healthy cells, with normal $\Delta\Psi_m$, DNA content and physical parameters, high mitochondrial mass without apoptosis; (2) cells with intermediate $\Delta\Psi_m$ and normal DNA content, but with physical parameters typical of apoptotic cells and low mitochondrial mass; (3) cells with collapsed $\Delta\Psi_m$ that had low mitochondrial mass and were apoptotic. They represent different stages of preapoptosis and apoptosis, respectively (Lugli et al., 2005).

An exception of the proposed transformation - normalizing-redifferentiation - apoptosis sequence could be expected during induction of apoptosis in the hemoblastoses characterized by a specific stage of frozen differentiation rather than dedifferentiation as seen in solid tumor cells (Abelev, 2000). Therefore, their apoptosis may be induced without further de novo differentiation utilizing the basic degree of differentiation they already have. However, even in such model there is redifferentiation at mitochondrial and tissue-specific metabolic and morphological levels. Accordingly, continuous exposure to the antimetabolite 1-beta-D-arabino-furanosylcytosine inhibited proliferation and induced expression of the myelomonocytic differentiation marker CD11b in approximately 35% of human myelomonocytic leukemia U937 cells (Wang et al, 2000). Arsenic trioxide exerted dose-dependent dual effects on acute promyelocytic leukemia cells. Rapid apoptosis was evident when cells are treated with 0.5 - 2.0 μM of arsenic trioxide while partial differentiation is observed using low concentrations (0.1 - 0.5 μM) of the drug (Zhang et al., 2001). The histone deacetylase inhibitor, sodium butyrate, induced cell cycle arrest and differentiation following by apoptosis in U937 human monocytic leukemia cells (Rosato et al., 2001). In dexamethasone-induced apoptosis of the human acute lymphoblastic leukemia cell line CCRF-CEM, at least 24 hr prior to and up to the point of apoptosis detection (36 hr), concomitant with redifferentiation and prior to the loss of $\Delta\Psi_m$, lactate production proportional to viable cell number was reduced compared to untreated controls. However, subsequent to loss of $\Delta\Psi_m$ a massive increase in lactate production was observed. Similar pattern was observed in other cell lines (HL60, THP1 and OPM2) with various cytotoxic agents [doxorubicin, gemcitabine and vumon (VM26)] (Tiefenthaler et al., 2001). K562 erythroleukemia cells undergo apoptosis when induced to differentiate along the erythroid lineage with hemin (Diaz et al., 1999). Reflecting mitochondrial redifferentiation, lactate accumulation in the medium and glucose utilization decreased during the induction of in vitro differentiation of mouse erythroleukemia and human myeloid leukemia HL-60 cells (Wu et al., 1991).

In conclusion, the effect of the retinoid on pancreatic adenocarcinoma cells as a model for a transformation – normalizing-redifferentiation – apoptosis sequence suggest the important role of mitochondria and confirms the results from several redifferentiation-apoptosis models that utilized different types of cells and apoptotic inducers. The necessity of such sequence mandates further molecular investigations.

References

Abelev, G.I. and Lazarevich, N.L. (2006). *Advances in Cancer Research. 95*, 61-113.

Abelev, G.I. (2000). *Biochemistry (Moscow). 65(1)*, 107-16.

Al-Hallaq, H.A., River, J.N., Zamora, M., Oikawa, H., and Karczmar, G.S. (1998). *Int J Radiat Oncol Biol Phys. 41(1),*151-9.

Augenlicht, L.H., Bordonaro, M., Heerdt, B.G., Mariadason, J., and Velcich, A. (1999). *Ann N Y Acad Sci. 889,* 20-31.

Augenlicht, L. H., and Heerdt, B. G. (2001). *Nat. Genet. 28,* 104-105.

Augenlicht, L. H., Wahrman, M. Z., Halsey, H., Anderson, L., Taylor, J., and Lipkin, M. (1987). *Cancer Res. 47*, 6017-6021.

Boisteau, O., Lieubeau, B., Barbieux, I., Cordel, S., Meflah, K., and Gregoire, M. (1996). *Bull Cancer. 83(3)*,197-204.

Bordonaro, M., Mariadason, J.M., Aslam, F., Heerdt, B.G., and Augenlicht, L.H. (1999). *Cell Growth Differ. 10(10)*, 713-20.

Chevrollier, A., Loiseau, D., Chabi, B., Renier, G., Douay, O., Malthiery, Y., and Stepien, G. (2005). *J Bioenerg Biomembr. 37(5)*, 307-16.

Chien, C.L., Chen, T.W., Lin, Y.S., and Lu, K.S. (2004). *J Biomed Sci. 11(5)*, 631-40.

Cho, Y.M., Kwon, S., Pak, Y.K., Seol, H.W., Choi, Y.M., Park, J., Park, K.S., and Lee, H.K. (2006). *Biochem Biophys Res Commun. 348(4)*, 1472-8.

Cuezva, J.M., Ostronoff, L.K., Ricart, J., Lopez de Heredia, M., Di Liegro, C.M., and Izquierdo, J.M. (1997). *J Bioenerg Biomembr. 29(4)*, 365-77.

Diaz, C., Lee, A.T., McConkey, D.J., and Schroit, A.J. (1999). Cell Death Differ. *6(3)*, 218-26.

Eliseev, R.A., Gunter, K.K., and Gunter, T.E. (2003). *Exp Cell Res. 289(2)*, 275-81.

EL-Metwally, T.H, and Adrian, T.E. (1999). *Biochem. Biophys. Res. Commun. 257*, 596-603.

El-Metwally, T.H., Hussein, M.R., Abd-El-Ghaffar, SKh., Abo-El-Naga, M.M., Ulrich, A.B., and Pour, P.M. (2006). *J Clin Pathol. 59(6)*, 603-10.

El-Metwally, T.H., Hussein, M.R., Pour, P.M., Kuszynski, C.A., and Adrian, T.E. (2005a). *Cancer Biol. Ther. 4(5)*, 602-11.

El-Metwally, T.H., Hussein, M.R., Pour, P.M., Kuszynski, C.A., and Adrian, T.E. (2005b). *Cancer Biol. Ther. 4(4)*, 474-83.

Facompre, M., Wattez, N., Kluza, J., Lansiaux, A., and Bailly, C. (2000). *Mol Cell Biol Res Commun. 4(1)*, 37-42.

Faure Vigny, H., Heddi, A., Giraud, S., Chautard, D., and Stepien, G. (1996). *Mol. Carcinog. 16*, 165–172.

Fornari, F.A. Jr., Jarvis, W.D,, Grant, S., Orr, M.S., Randolph, J.K., White, F.K., Mumaw, V.R., Lovings, E.T., Freeman, R.H., and Gewirtz, D.A. (1994). *Cell Growth Differ. 5(7)*, 723-33.

Funasaka, T., Yanagawa, T., Hogan, V., and Raz, A. (2005). *FASEB J. 19*, 1422–1430.

Heerdt, B.G., and Augenlicht, L.H. (1991). *J Biol Chem. 266(28)*, 9120-6.

Heerdt, B.G., Houston, M,A,, Wilson, A.J., and Augenlicht. L.H. (2003). *Cancer Res. 63(19)*, 6311-9.

Heerdt, B. G., Houston, M. A., Mariadason, J. M., and Augenlicht, L. H. (2000). *Cancer Res. 60*, 6704–6713.

Huang, G., Chen, Y., Lu, H. and Cao, X. (2007). *Cell Death Differ. 14(2)*,327-37 [ahead of print].

Ishii, Y,, Hori, Y., Sakai, S., and Honma, Y. (2002). *Cell Growth Differ. 13(1)*, 19-26.

Joshi, B., Li, L., Taffe, B.G., Zhu, Z., Wahl, S., Tian, H., Ben-Josef, E., Taylor, J.D., Porter, A.T., and Tang, D.G. (1999). *Cancer Res. 59*, 4343-4355.

Karbowski, M., Spodnik, J.H., Teranishi, M., Wozniak, M., Nishizawa, Y., Usukura, J., and Wakabayashi, T. (2001*). J Cell Sci. 114(2)*, 281-91.

Kluza, J., Marchetti, P., Gallego, M.A., Lancel, S., Fournier, C., Loyens, A., Beauvillain, J.C., and Bailly, C. (2004). *Oncogene. 23(42)*, 7018-30.

Kolodin, V.I. (1977). *Vopr Onkol. 23(11)*, 79-88.

Lu, X., Walker, T., MacManus, J.P., and Seligy, V.L. (1992). *Cancer Res. 52(13)*, 3718-25.

Lugli, E., Troiano, L., Ferraresi, R., Roat, E., Prada, N., Nasi, M., Pinti, M., Cooper, E.L,, and Cossarizza, A. (2005). *Cytometry A. 68(1)*, 28-35.

Mahyar-Roemer, M., Katsen, A,, Mestres, P., and Roemer, K. (2001). *Int. J Cancer. 94(5)*, 615-22.

Mancini, M,, Anderson, B.O,, Caldwell, E., Sedghinasab, M., Paty, P.B., and Hockenbery, D.M. (1997). *J Cell Biol. 138(2)*, 449-69.

Mancini, M,, Sedghinasab, M., Knowlton, K., Tam, A., Hockenbery, D., and Anderson, B.O. (1998). *Ann Surg Oncol. 5(3)*, 287-95.

Mazzanti, R,, Solazzo, M., Fantappie, O., Elfering, S., Pantaleo, P., Bechi, P., Cianchi, F., Ettl, A., and Giulivi, C. (2006). *Am J Physiol Gastrointest Liver Physiol. 290(6)*, G1329-38.

Nalbandian, A., Pang, A.L., Rennert, O.M., Chan, W.Y., Ravindranath, N., and Djakiew, D. (2005). *Differentiation. 73(8),* 385-96.

Ohashi, H., Ichikawa A., Takagi H., Hotta T., Naoe T., Ohno R., and Saito H. (1992). *Leukemia (Baltimore). 6*:859–862.

Pagliacci, M.C., Spinozzi, F., Migliorati, G., Fumi, G., Smacchia, M., Grignani, F., Riccardi, C., and Nicoletti, I. (1993). *Eur J Cancer. 29A(11)*, 1573-7.

Reipert, S., Berry J., Hughes MF, Hickman JA, and Allen TD. (1995). *Exp. Cell Res. 221(2)*, 281–288.

Rosato, R.R., Wang, Z., Gopalkrishnan, R.V., Fisher, P.B., and Grant, S. (2001). *Int J Oncol. 19(1)*, 181-91.

Spodnik, J.H., Wozniak, M., Budzko, D., Teranishi, M.A., Karbowski, M., Nishizawa, Y., Usukura, J., and Wakabayashi, T. (2002). *Mitochondrion. 2(3)*, 163-79.

Tiefenthaler, M., Amberger, A., Bacher, N., Hartmann, B.L., Margreiter, R., Kofler, R., and Konwalinka, G. (2001). *Br J Haematol. 114(3)*, 574-80.

Vayssiere, J.L., Cordeau-Lossouarn, L., Larcher, J.C., Basseville, M., Gros, F., and Croizat, B. (1992). *In Vitro Cell Dev Biol. 28A(11-12)*, 763-72.

Vizirianakis, I.S., Pappas, I.S., and Tsiftsoglou, A.S. (2002). *Biochem Pharmacol. 63(5)*, 1009-17.

von Wangenheim, K.H., and Peterson, H.P. (1998). *J Theor Biol. 193(4)*, 663-78.

Walenta, S., Schroeder, T., and Mueller-Klieser, W. (2004). *Curr Med Chem. 11(16)*, 2195-204.

Wang, Z., Wang, S., Fisher, P.B., Dent, P., and Grant, S. (2000). *Differentiation. 66(1)*, 1-13.

Weber, G. (2001). *Biochemistry (Moscow). 66(10)*, 1164-1173.

Wu, H., Scher, B.M., Chu, C.L., Leonard, M., Olmedo, R., Scher, G.S., Stecker, S., Scher, W., and Waxman, S. (1991). *Differentiation. 48(1)*, 51-8.

Yasuda, I., Shiratori, Y., Adachi, S., Obora, A., Takemura, M., Okuno, M., Shidoji, Y., Seishima, M., Muto, Y., and Moriwaki, H. (2002). *J Hepatol. 36(5)*, 660-71.

Yin, W., Raffelsberger, W.., and Gronemeyer, H. (2005). *Int J Biochem Cell Biol. 37(8)*, 1696-708.

Zhang, T.D., Chen, G.Q., Wang, Z.G., Wang, Z.Y., Chen, S.J., and Chen, Z. (2001). *Oncogene. 20(49)*, 7146-53.

In: Cell Differentiation Research Developments
Editor: L. B. Ivanova, pp. 247-258
ISBN: 978-1-60021-939-9

Chapter XII

ANGIOTENSISN CONVERTING ENZYME INHIBITORS IMPROVE SPERMATOGENESIS: EVIDENCE FROM TISSUE LEVEL IN RATS

Basil N. Okeahialam[1,*], Kemakolam Amadi[2] and Alphonsus S. Ameh[3]
[1]Department of Medicine and Radiology, [2]Department of Human Physiology, [3]Animal House Unit, Faculty of Medical Sciences, University of Jos, Jos, Nigeria.

ABSTRACT

Background: Angiotensin converting enzyme inhibitors (ACEI) have pharmacological effects in all body tissues containing rennin – angiotensin system (RAS) and the related kallikrein – kinin system (KKS). In the male reproductive tract which also contains these paracrine systems, results of effects have been conflicting. Encouraged by a few successes in humans and confirmation of improved sperm count and motility in rats; we set out to study at tissue level, the effects of ACEI on spermatogenesis. These experiments have potential for human application.

Methods: Four groups of adult male Wistar rats received, 0.2 ml per oral of the following doses weekly: A – 5mg/ml, B – 2.5 mg/ml, C – 0.25 mg/ml, D – 0.05 mg/ml. A fifth group, serving as controls, did not receive any drug. After 2 weeks when one spermatogenetic cycle was complete, animals were sacrificed in each group weekly and testes excised for histology.

Results: The control group showed normal evidence of spermatogenesis. The group receiving 0.05 mg/ml showed a strange arrangement of seminiferous tubules; but evidence of normal spermatogenesis in the normal tubules. By 0.25 mg/ml, seminiferous tubules were of low population with incomplete spermatogenesis in many foci. At the 2.5 mg/ml dose, all stages of spermatogenesis were evident in good amounts; with abundance of glycogen deposits. The group receiving the highest dose of 5 mg/ml showed plenty sperm heads in the seminiferous tubules. Spermatogenetic response was

* Correspondence concerning this article should be addressed to: Basil N. Okeahialam E mail: basokeam@yahoo.com; Phone: 234 0803 7211 698.

robust. Effect at the end of the study was largely similar to the week two specimen except for the control group.

Discussion: The existence of tissue RAS and KKS in the male reproductive tract implies potential effect of ACEI drugs. Preliminary, albeit anecdotal reports in humans of improved semen quality, supported by a dose dependent improvement of spermatogenesis in rats has been borne out by this tissue based work. There appears to be a dose dependent entry of the ACEI drugs into these tissue paracrine systems resulting in stimulation of spermatogenetic process; by creating a favourable ambient for multiplication and viability of spermatozoa. These include improvement in Leydig and Sertoli cell function made possible by local accumulation of bradykinin and nitric oxide; as well as reduction in levels of Angiotensin II brought about by inhibition of the ACE system. The potential of these results for human application in treatment of male infertility should be explored in greater detail.

Keywords: Angiotensin converting enzyme inhibitor, spermatogenesis, stimulation, cellular level, rats.

INTRODUCTION

The rennin-angiotensin system [1] and its related kallikrein-kinin system [2] exist in the male reproductive tract. These paracrine systems play a role in sertoli cell function and sertoli cell – germ cell cross talk [3]. Components of the rennin – angiotensin system have been found on the testes and other parts of the male reproductive system [4]; and evidence is accumulating of their involvement in spermatogenesis, sperm maturation and fertilization [5]. This effect is likely to be independent of its production in the systemic circulation [6]. The beneficial effects of angiotensin converting enzyme inhibitors are said to derive from several mechanisms, the relative importance varying depending on the specific condition being addressed [7].

Following an anecdotal report of improvement in semen quality while treating hypertensive men with lisinopril [8], we decided to investigate the subject further. This was with a view to documenting consistent improvement in semen quality with angiotensin converting enzyme inhibitors; so as to extend such benefits to distraught couples where poor quality semen is responsible for infertility. Having earlier recorded a dose dependent response in sperm count in rats to angiotensin converting enzyme inhibitors [9], we decided to study this response at cellular level. This was an attempt to see if the aforementioned response could be supported by tissue changes. The ease of obtaining testicular tissue in lower animals makes this model more attractive, in contrast with human studies which are associated with ethical problems. Confirmation of such response may then provide a justification for recommending angiotensin converting enzyme inhibitors as treatment of poor semen quality in men seeking to raise families. The rennin – angiotensin system has proved to be more complex biochemically than anticipated; with diverse physiological and pathophysiological mechanisms of potential therapeutic benefit [10]. This may be the fulfillment of earlier speculations of a bright future for drugs in the angiotensin converting enzyme inhibitor group [7].

The part of the testis involved with spermatogenesis is the seminiferous tubule. Seminiferous tubules occur in clusters called lobules, into which fibrous septae divide the testis. The seminiferous tubules are enmeshed in a web of loose connective tissue which is richly endowed with blood vessels, lymphatic channels, nerves and interstitial cells of leydig. This last group of cells (interstitial cells of leydig) appear on sexual maturity, and are responsible for steroidogenesis. The seminiferous tubules themselves consist of a layer of fibrous connective tissue, a well defined basal lamina and a complex germinal epithelium [11]. The epithelium consists of two types of cells, namely the sertoli or supporting cells and spermatogonia. The latter constitutes the lineage for spermatogenesis. Spermatogenesis in mammals is a highly ordered process. It starts with primitive germ cells or spermatogonia which lie next to the basal lamina. At sexual maturity, they undergo a series of mitotic cycles. The newly formed cells follow either a path of undifferentiation (resting cells), and form the germ cell reservoir. These are type A spermatogonia. Alternatively, they may differentiate further to the cell line commited to spermatogenesis called type B spermatogonia. Type B spermatogonia undergo further differentiation to form primary spermatocytes. With meiosis setting in, these change to secondary spermatocytes. This is the spermatocytogenic stage. On further meiotic divisions, spermatocytes form spermatids with reduction by half in the number of chromosomes and deoxyribonucleic acid (DNA) per cell. In the final stage of spermiogenesis, spermatids undergo elaborate process of cyto-differentiation, ending up as spermatozoa. This process involves acrosome condensation and nuclear elongation, development of flagellum and loss of cytoplasm. After this comes spermiation. This stage involves the process of release of mature spermatozoa into the lumen of the seminiferous tubules.

This whole process takes place in close proximity to specialized cells called sertoli cells. These are elongated pyramidal cells which envelope differentiating germ cells incompletely along the line of spermatozoal production. At the spermatogonial level where there is unhindered access to substances circulating in blood, adjacent sertoli cells are bound by occluding junctions. Higher up, subsequent germ cells undergoing differentiation are protected from blood borne substances by a blood-testis barrier formed by occluding junctions between the sertoli cells. The sertoli cells therefore function in supporting, protecting and regulating the nutritional needs of developing spermatozoa. Since the differentiating cells are isolated from systemic circulation by the blood-testis barrier (formed by sertoli cells), their survival largely depends on sertoli cell intergrity.

Depending on the species of mammal, a cross section of the tubule may show only a single stage or up to six identifiable stages [12].

The lutenising hormone and follicle stimulating hormone initiate and control the process of spermatogenesis. These hormones, called pituitary gonadotropins, are produced in response to hypothalamic gonadotropin releasing hormone. This occurs in waves. Lutenising and follicle stimulating hormones act synergistically in this function. Follicle stimulating hormone plays critical roles in the progression of type A to type B spermatogonia, the germ cell line committed to further differentiation in the process of spermatogenesis. In synergy with testosterone, germ cell viability is regulated. Some workers actually think that spermatogonia are primarily regulated by follicle stimulating hormone [13]. It also controls proliferation of sertoli cells [14], permits adhesion of round spermatids to sertoli cells [15] and supports germ cell maturation in mid stages of spermatogenesis [16]. Follicle stimulating hormone secretion is regulated by negative feed back from some testicular hormones [17]. In conjunction with testosterone, follicle stimulating hormone controls spermiation [18]. Lutenising hormone on the other hand, stimulates the interstitial cells of leydig to produce an androgen called testosterone. The bulk of testosterone is produced at this site [19]. It can also be produced by sertoli cells under the influence of follicle stimulating hormone. Testosterone is required to initiate spermatogenesis and completion of meiosis [20]. After being produced by the leydig cells, testosterone is secreted into the plasma and transported to sertoli cells by androgen binding protein. There, it is reduced to dihydro-testosterone by 5 alpha reductase. The activity of this enzyme is stimulated by follicle stimulating hormone. Both are then delivered to target tissues or sites of spermatogenesis in the seminiferous tubules by a specific globulin, where protein synthesis and sperm development occur. This requires adequate levels of intra – testicular testosterone to manifest, otherwise there is a reduction in the level of production of spermatozoa [21]. Apart from lutenising and follicle stimulating hormones, many peptides and growth factors are secreted locally in the seminiferous tubules, which impact on steroidogenesis [22]. Testosterone in the presence of follicle stimulating hormone ensures that spermatogonia development goes on uninhibited; as suppression of intra – testicular testosterone has been shown to suppress different stages of spermatogonia at different rates. Germ cell number also progressively declines and spermatids fail to detach from sertoli cells in an ambient of low intra – testicular testosterone. This retention causes the sertoli cells to phagocytose them with failure of spermiation being the result [23]. Apoptosis of germ cells at various stages of differentiation also occurs. For these benefits of testosterone to occur, the rise has to be intra – testicular. This is the autocrine and paracrine effect. Exogenous administration of testosterone is known to suppress gonadotropins when supraphysiological serum levels are attained. Testosterone also regulates the adhesion between sertoli cells and spermatids. This is through the effects of cell adhesion molecules or intra – cellular junctional apparatus located on sertoli cells [24].

Materials and Methods

Twenty five adult male Wistar strain albino rats inbred at the University of Jos Animal House were randomly distributed into five groups of five animals each. Five different doses of lisinopril were administered to different groups as follows: Group A – 5mg/ml, Group B - 2.5mg/ml, Group C – 0.25mg/ml, Group D – 0.05mg/ml, Group E (control) – No drug. Every week, 0.2 ml of each dose was administered to the respective groups per oral. This was achieved by holding the upper incisor teeth of each animal with a strong thread tied to a water tap. The lower incisors were held down with another loop of thread tied to the thumb. With the mouth agape, 0.2 ml of each dose was delivered by means of a 1 ml syringe around the oesophagus. Additional 2 drops of water were given to wash in the drug. The administration was continued for six weeks during which rat chow was fed to the animal groups. The animals had access to tap water ad-libitum. They were also allowed equal periods of daylight and darkness. From the second week, one animal from each group was euthanised weekly by bleeding through direct cardiac puncture. The testes were removed, fixed in 10% formol saline and processed for histological study after the methods of Baker et al [25]. Micro-sections 5 micro-metres thick were stained with Haematoxylin and Eosin, dehydrated, cleared in Xylene and mounted on Canada balsam for microscopic study using the standard light microscope. The dose effect in relation to various times the drug was allowed to act was recorded. We report here the tissue response after two weeks of drug administration, and at the end of the study by six weeks.

RESULTS

Group A

There were abundant mitochondria and smooth endoplasmic reticulum. The heads (nuclei) of spermatozoa in seminiferous tubules were plentiful. Interstitial cells of Leydig were abundant. Sertoli cells (sustentacular cells) were many with glycogen deposits. Spermatogonia in mitosis were profuse. Lamina propria and basement membrane were well defined and robust. There was spermatogenesis at all stages. Arterioles and venules were in good supply (Figure 1). By six weeks, interstitial cells of leydig were still abundant. Sertoli cells were in good number. Density of spermatozoa was good in most tubules. Spermatogonia in mitosis were very evident. There was full spermatogenesis in most tubules. Lamina propria and basement membrane were well defined and robust.

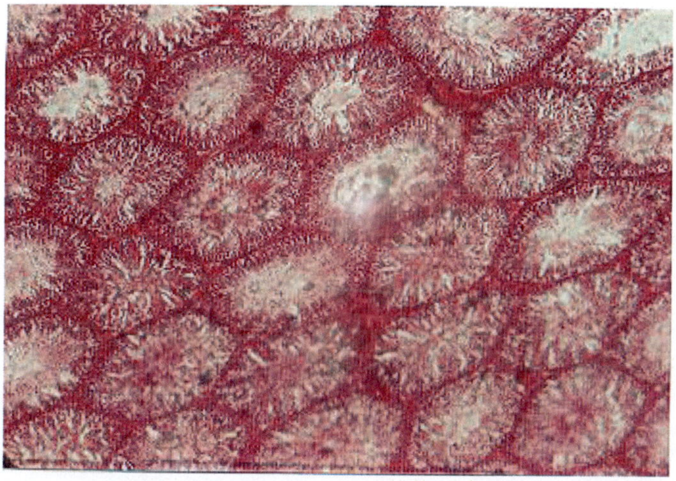

Figure 1. Seminiferous tubules showing plentiful spermatozoa filling the lumen. Leydig and Sertoli cells abound; with many dark brown glycogen deposits.

Group B

There was presence of all stages of spermatogenesis with glycogen deposits as a source of energy. There were good numbers of interstitial cells of Leydig and sertoli cells. The basement membrane was prominent. (Figure 2). By six weeks, there were numerous small sized tubules. Spermatogenesis was complete in most tubules. Spermatogonia in mitosis were evident. Sertoli cells were in good numbers. Interstitial cells of leydig were in clusters but prominent and abundant. Basal membrane and lamina propria were well defined.

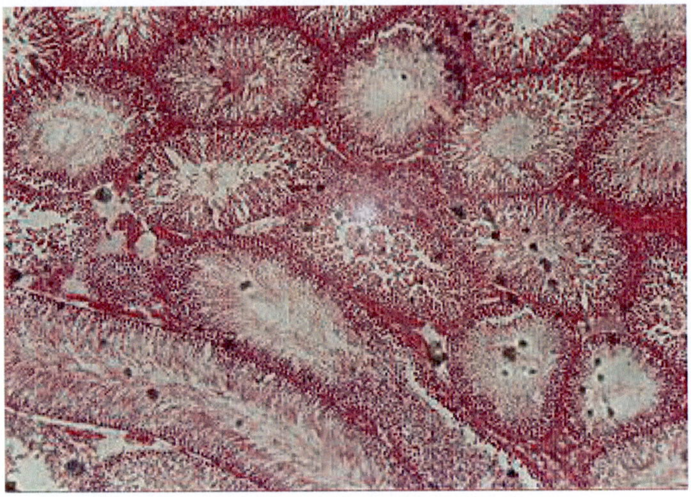

Figure 2. Seminiferous tubules showing all stages of spermatogenesis. Leydig and Sertoli cells are in good numbers. Dark brown glycogen deposits are evident.

Group C

Seminiferous tubules were relatively of low proportion. There was incomplete spermatogenesis in many foci. Fibroblasts were in abundance showing specks of metastatic calcification. Interstitial cells of Leydig were of good population. Sertoli cells were also plentiful (Figure 3). By six weeks, the calcification in the tubules masked the histological features. However complete spermatogenesis were seen in very few of the seminiferous tubules. Interstitial cells of Leydig and sertoli cells were few.

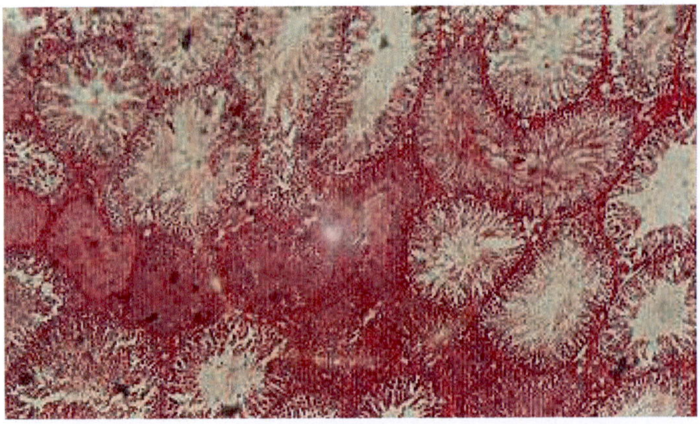

Figure 3. Seminiferous tubules are of low population. Spermatogenesis is incomplete in many foci. Leydig and Sertoli cells are evident. Fibrolasts showing specks of calcification are seen.

Group D

There was a strange grouping of seminiferous tubules. Elongated tubules were arranged inferiorly to the superiorly located ovoid and circular tubules; which showed dominance of sertoli cells and interstitial cells of Leydig when compared to the elongated tubules. Spermatogenesis was however complete in all tubules (Figure 4). By six weeks seminiferous tubules were robust with complete spermatogenesis in most of them. Many sertoli cells were evident.

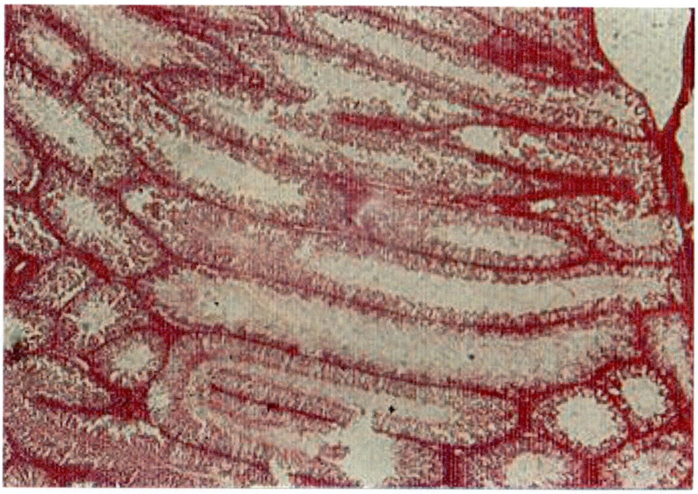

Figure 4. There are mixed shapes of seminiferous tubules. Whereas the normal ovoid/circular tubules show evidence of spermatogenesis, the rate was less in the elongated tubules.

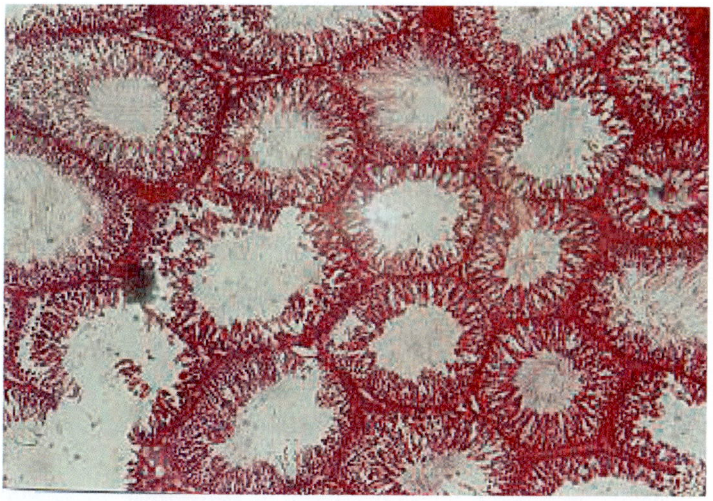

Figure 5. Normal seminiferous tubules with normal spermatogenesis; and evidence of dark brown deposits of glycogen.

Group E (Control)

There were mixed tubules of various sizes and shapes. Other features were normal but not as intense as the above doses. There were many glycogen deposits, proliferative spermatogonia and spermatocytes in many extra tubular foci. There was presence of interstitial cells of Leydig, sertoli cells arterioles and venules in good amounts (Figure 5). By six weeks, some tubules were either empty or show few clusters of mature spermatozoa. Other normal features were well represented.

Overall, the effects evident by two weeks remained at the end of the study, except in the control group where there appeared to be slowing down of spermatogenetic activities.

Discussion

We used the testicular tissue of adult Wistar rats after two weeks of administration because in this specie of mammals, 14 days was sufficient for one cycle of spermatogenesis [26]. Group E, the control group received no drug, and showed spermatogenesis without intervention. By six weeks, spermatogenesis seemed to be slowing down. This was not easily explainable, but may have to do with span of reproductive activity in this class of mammals. When administered drug from the 0.05 mg/ml Lisinopril group (D), there was elongation of most of the seminiferous tubules. Sertoli cells and interstitial cells of Leydig were present in the normal ovoid tubules with complete spermatogenesis. With increase to 0.25mg/ml (C), seminiferous tubules were relatively of low population with incomplete spermatogenesis in many foci. We noted specks of metastatic calcification which could occur in non pathologic cases [27]. Why it occurred only in the group receiving this dose is not clear to us. The increase to 2.5 mg/ml group (B), gave rise to robust presence of sertoli cells and interstitial cells of Leydig. Glycogen deposits that had started appearing became more prominent. By the highest dose of 5mg/ml, there was abundance of all organnels related to spermatogenesis. These changes persisted up till the end of the study, by six weeks. This bears out the dose related response of spermatogenesis reported earlier in lower animals [9]. In the human aspect of the study [8], Okeahialam and Ekwempu had also recorded variable response to the drug in sperm count and motility. In that study most of the subjects received a low dose because they were not hypertensive. The 2 in whose wives conception occurred received higher doses because they were hypertensive. Beyond a certain dose further improvements were not noted.

In the present study, increasing doses caused improvement in growth and function of the epididymis. This is the effect of improved Leydig and sertoli cell function.

Angiotensin converting enzyme inhibitors are drugs that act on substrates of the rennin angiotensin system. They are specific for the enzyme converting the inactive decapeptide angiotensin I to the potent pressor, angiotensin II [7]. Many tissues express messenger ribonucleic acids for rennin angiotensin and angiotensin converting enzyme [28]. This exists independent of the renal – hepatic based system, and have been detected in the male reproductive tract [29]; where they subserve autocrine and paracrine functions [1]. Angiotensin converting enzyme inhibitors in these locations as in others, block the conversion of angiotensin I to angiotensin II. At the same time they inhibit the break down of bradykinin leading to its accumulation [28]. This represents events in the kallikrein – kinin system which has been shown to exist locally in the male reproductive system; where it functions in the local regulation of spermatogenesis [30]. Accumulation of bradykinin is thought to be a local tissue and not a systemic effect [31]. The improved spermatogenesis could be variously explained. Angiotensin converting enzyme inhibitors block the formation of angiotensin II. In the presence of angiotensin II, Leydig cell function is suppressed. Lutenising hormone function becomes adversely affected, and testosterone production and function are suppressed [1].Testosterone promotes growth and function of the epididymis, one of which is spermatogenesis [19]. Therefore if angiotensin converting enzyme inhibitors are given and angiotensin II production is suppressed, leydig cell function would be enhanced. In addition, apart from blocking the conversion of angiotensin I to angiotensin II, angiotensin converting enzyme inhibitors also block the conversion of kinins (bradykinin) to inactive peptides; albeit partially [32]. The bradykinin accumulates alongside nitric oxide [33]; which has been reported to stimulate spermatogenesis [34]. As tissue hormones related to the rennin angiotensin system, bradykinins are known to stimulate spermatogenesis and sperm metabolism by activating sertoli cell function [5]. Sertoli cells are glycogen containing cells in the seminiferous tubules from which spermatozoa obtain nourishment. They play a key role in protecting and controlling the immediate environment around the germ cells [30]. Other sertoli cell functions under the influence of bradykinin include enhanced glucose uptake at cellular level; increased testicular blood flow, acceleration of glucose/fructose across sperm cell membranes and improvement of total sperm output [5]. The increased appearance of glycogen deposits from 0.25mg/ml dose depicts increase in sertoli cells; which are glycogen containing cells from which spermatozoa obtain nourishment. Other paracrine and autocrine functions of kinins include improvement of secretory activity of the sex glands. There is also a small influence on hormonal balance of the pituitary – gonadal axis, leading to increased intra – testicular testosterone levels. Up regulation of receptors for follicle stimulating hormone also occurs. These are said to favour epididymal maturation of spermatozoa; and being biologically active tissue hormones involved in cell proliferation of many tissues, are thought to be contributory to the observed positive spermatogenetic effects [5].

Going by this cellular evidence, there is no doubt that angiotensin converting enzyme inhibitors exemplified in these studies by Lisinopril can improve quality of semen. The appropriate dose would appear to be 5mg going by the effect both in humans and rats. Since the drug is an anti-hypertensive, it may suffice to give 2.5mg which has been shown to be well tolerated without untoward side effects in normotensive males [7]. Over time, the stimulation may build up to levels adequate for conception in their spouses; baring the control created by the blood testis barrier in the particular individual. The exception where 5 mg may be considered would be in individuals with low rennin status, in who as monotherapy, blood pressure lowering effect of angiotensin converting enzyme inhibitors is poor [7]. Further studies in the area of histochemistry should more clearly elucidate the mechanism of improvement in semen quality and the rate limiting effect of the blood testis barrier; which determines effective doses in individuals. Such studies are called for as full understanding of how production of specific angiotensin peptides in the various cell types expressing RAS components remain poor [6].

ACKNOWLEDGEMENTS

1. Dr. R. MacNeil of our Anatomy Department for loaning us her microscope, when the research microscope available to the group was stolen.

2. Mr. Goddy Njoku of our Medical Illustration Technology Unit for the photographing of study slides under the microscope

REFERENCES

[1] Leung PS, Sernia C. The rennin – angiotensin system and male reproduction: new functions for old hormones. *J. Mollec. Endocrin.* 2003; 30: 263 – 270.

[2] Monsees T, Miska W, Schill WB. Enzymatic digestion of bradykinin by rat sertoli cell cultures. *J. Androl.* 1996; 17: 375 – 381.

[3] Monsees TK, Schill WB, Miska W. Protease-protease inhibitor interaction in sertoli cell - germ cell cross talk. *Adv. Exp. Med. Biol.* 1997; 424: 111 – 123.

[4] Vinson GP, Saridogan C, Puddeforte JR, Diahanbakch O. Tissue renin – angiotensin systems and reproduction. *Hum. Reprod.* 1997; 12(4): 651 – 662.

[5] Schill WB. Medical treatment of male infertility. In V. Inster and B. Lunenfeld (Eds). *Infertility: male and female*. Edinburgh. London. Melbourne. New York. Churchhill Livingstone. 1986; p. 533 – 573.

[6] Diz DI. Approaches to establishing Angiotensin II as a neurotransmitter revisited. (Editorial Commentary). *Hypertens.* 2006; 47(3, Part 2): 334 – 336.

[7] Edward CRW, Padfield PL. Angiotensin converting enzyme inhibition: past, present and bright future. *The Lancet 1*. 1985; 8419: 30 – 34.

[8] Okeahialam BN, Ekwempu CC. Angiotensin converting enzyme inhibitor (ACE) inhibitor (Lisinopril) improves semen quality: a preliminary report. *J. Mollec. Cell Cardiol.* 2004; 37: 254. (Abstract)

[9] Okeahialam BN, Amadi K, Ameh AS. Effect of Lisinopril, an angiotensin converting enzyme (ACE) inhibitor on spermatogenesis in rats. *Arch Androl.* 2006; 52: 209 – 213.

[10] Ferrario CM. Angiotensin converting enzyme inhibitor 2 and angiotensin (1 – 7). An evolving story in cardiovascular regulation. *Hypertens.* (Supl.) 2006; 47 (3 part 2): 515 – 521.

[11] Janqueira LC, Carneiro J, Kelley RO. *Basic Histology*. 6th edition. Appleton and Lange Medical Publications. 1989. p. 422 – 438.

[12] McLachan RI, O'Donnel L, Meachem SJ, Stanton PG, de Krester DM, Pratis K, Robertson DM. Identification of specific sites of hormonal regulation in spermatogenesis in rats, monkeys and man. *Recent Progress in Hormone Research.* 2002; 57: 149 – 179.

[13] Meachem S, von Schonfeldt V, Schlatt S. Spermatogonia: stem cells with a great perspective. *Reproduction*. 2001; 121: 825 – 834.

[14] Orth JM. Cell biology of testicular development in the foetus and neonate. In: C. Desjardins, L.L. Ewing (Eds). *Cell and molecular biology of the testis*. New York. Oxford University Press. 1993. p. 3 – 43.

[15] Perryman KJ, Stanton PG, Loveland KL, McLachlan RI, Robertson DM. Hormonal dependency of neural cadherin in the binding of round spermatids to sertoli cell in vitro. *Endocrinology*. 1996; 137: 3877 – 3883.

[16] Meachem SJ, McLachlan RI, Stanto PG, Robertson DM, Wreford NG. Follicle stimulating hormone immunoneutralisation acutely impairs spermatogonial development in normal adult rats. *J. Androl.* 1999; 20: 765 – 762.

[17] Hayes FJ, De Cruz S, Seminara SB, Boepple PA, Crowley WF Jr. Differential regulation of gonadotropin secretion in the human male: absence of a negative feed back effect of testosterone on follicle stimulating stimulating hormone secretion. *J. Clin. Endocrinol. Metab.* 2001; 86: 53 – 58.

[18] Saito K, O'Donnell L, McLachlan RI, Robertson DM. Spermiation failure is a major contribution to early spermatogenic suppression caused by hormone withdrawal in adult rats. *Endocrinology*. 2001; 141: 2779 – 2785.

[19] Schlegel PN, Chang TSK. The testis, epididymis and ductus deferens. In: PC Walsh, AB Retik, TA Stamey, E Dracott-Vaughan (Eds). *Campbell's Urology*. Philadelphia, London, Toronto, Montreal, Sydney, Tokyo. W. B. Saunders Co.1992. p. 190 – 220.

[20] Steinberger E. Pathophysiology of the testis. In: V Insler, B Lunenfeld (Eds). *Infertility: Male and Female*. Edinburgh, London, Melbourne, New York. Churchhill Livingstone. 1986. p. 168 – 187.

[21] O'Donnell L, Pratis K, Stanton PG, Robertson DM, McLachlan RI. Testosterone dependent restoration of spermatogenesis in adult rats is impaired by 5 alpha reductase inhibitor. *J. Androl.* 1999; 20: 109 – 117.

[22] Swerdoff PS, Wang C. Physiology of male reproduction. Ibid. 19. p. 177 – 189.

[23] Russel LD. The perils of sperm release – " let my children go". *Intl. J. Androl.* 1991; 14: 307 – 311.

[24] O'Donnell L, McLachlan RI, Wreford NG, de Krester DM, Robertson DM. Testosterone withdrawal promotes stage specific detachment of round spermatids from the rat seminiferous epithelium. *Biol. Reprod.* 1996a; 55: 895 – 901.

[25] Baker FJ, Silverton RE, Luckock ED. *An Introduction to Medical Laboratory Technology*. 11th Ed. London. Butterworth Co. Publishers Ltd. 1994. London. P. 357.

[26] Stanbenfeldt GH. Reproductive cycles in animals. In: A.T. Phillipson, L.W. Hall, W.R. Pritchard (Eds). *Scientific foundations of Veterinary Medicine*. London. 1st edition. Heineman. 1980; P 14 – 25.

[27] Sodeman WA, Sodeman TM. *Pathologic Physiology*. 6th edition. London. Philadelphia. W.B. Saunders Co. 1984. p. 764.

[28] Jackson EK. Renin and angiotensin. In: JG Hardman, LE Limbird (Eds). Goodman and Gillman's The pharmacological basis of therapeutics. New York, Chicago, San Francisco,Lisbon, London, Madrid, Mexico City, Milan, New Delhi, San Juan, Seoul,

Singapore, Sydney, Toronto. *McGraw – Hill Medical Publishing Division.* 10th International edition. 2001. p. 819 – 841.

[29] Pauls K, Metzger R, Steger K, Klonisch T, Danilov S, Franke FE. Isoforms of angiotensin I converting enzyme in the development and differentiation of human testis and epididymis. *Andrologia.* 2003; 35(1): 32 – 43.

[30] Monsees TK, Blocher S, Heidorn F, Winkler A, Siems WE, Muller Esterl W, Hayatpour J, Miska W, Schill WB. Expression and location of the bradykinin B2 receptor in rat testis. *Biol. Reprod.* 2002; 67: 1832 – 1893.

[31] Johnson CI, Clappison BH, Anderson WP, Yasujima M. Effect of angiotensin converting enzyme inhibition on circulating and local kinin levels. *Am. J. Cardiol.* 1982; 49: 1401 – 1404.

[32] Monsees T, Miska W, Schill WB. Characterisation of kininases in testicular cell. *Immunopharmacol.* 1996; 32: 169 – 171.

[33] Bernstein KE. Views of the rennin – angiotensin system. Brilling, Mimsy and Slithy Tove. *Hypertens* (Suppl). 2006; 47 (3 part 2): 509 – 514.

[34] Tuteja N, Chandra M, Chandra R, Misra MK. Nitric oxide as a unique bioactive signaling messenger in physiology and pathophysiology. *J. Biomed. Biotech.* 2005; 4: 227 – 237.

INDEX

B

F

G

J

K

L

M

N

O

P

R

S